Universitext

Universitext

Editors: F. W. Gehring, P.R. Halmos, C.C. Moore

Chern: Complex Manifolds Without Potential Theory
Chorin/Marsden: A Mathematical Introduction to Fluid Mechanics
Cohn: A Classical Invitation to Algebraic Numbers and Class Fields
Curtis: Matrix Groups
van Dalen: Logic and Structure
Devlin: Fundamentals of Contemporary Set Theory
Edwards: A Formal Background to Mathematics I a/b
Edwards: A Formal Background to Higher Mathematics II a/b
Endler: Valuation Theory
Frauenthal: Mathematical Modeling in Epidemiology
Gardiner: A First Course in Group Theory
Godbillon: Dynamical Systems on Surfaces
Greub: Multilinear Algebra
Hermes: Introduction to Mathematical Logic
Kalbfleish: Probability and Statistical Inference I/II
Kelly/Matthews: The Non-Euclidian, The Hyperbolic Plane
Kostrikin: Introduction to Algebra
Lu: Singularity Theory and an Introduction to Catastrophe Theory
Marcus: Number Fields
Meyer: Essential Mathematics for Applied Fields
Moise: Introductory Problem Course in Analysis and Topology
Rees: Notes on Geometry
Reisel: Elementary Theory of Metric Spaces
Rey: Introduction to Robust and Quasi-Robust Statistical Methods (in prep)
Rickart: Natural Function Algebras
Schreiber: Differential Forms
Tolle: Optimization Methods

J. G. Kalbfleisch

Probability and Statistical Inference I

Springer-Verlag
New York Heidelberg Berlin

Dr. J. G. Kalbfleisch
Faculty of Mathematics
Department of Statistics
University of Waterloo
Ontario, Canada N2L 3G1

AMS Classifications (1980): 60-01, 62-01

With 35 illustrations

Library of Congress Cataloging in Publication Data

Kalbfleisch, J
Probability and statistical inference.

(Universitext)
Includes indexes.
1. Probabilities. 2. Mathematical Statistics.
I. Title.
QA273.K27 1979 519.2 79-22910

Printed in the United States of America.

9 8 7 6 5 4 3 (Third printing, 1984)

ISBN 0-387-90457-3 Springer-Verlag New York Heidelberg Berlin
ISBN 3-540-90457-3 Springer-Verlag Berlin Heidelberg New York

To Rebecca

PREFACE

This book is intended as a text for an introductory course in probability and statistics at the second or third year university level. It emphasizes applications and logical principles rather than mathematical theory. A good background in freshman calculus is sufficient for most of the material presented.

The book is in two parts. The first part (Chapters 1 - 8) deals with probability models and with mathematical methods for handling them. In the second part (Chapters 9 - 16), probability models are used as the basis for the analysis and interpretation of data. Several starred sections have been included as supplementary material. A large supply of practice problems is provided, and Appendix A contains answers to about one-third of these.

Computers and sophisticated pocket calculators are having a profound effect on statistical practice. Not only can the same arithmetic be done in a fraction of the time, but more importantly, statistical methods which were previously not feasible because of the amount of calculation required now present no difficulties. In particular, the likelihood function itself and procedures closely related to it now give simple and immediate solutions to problems which twenty years ago would have required complicated approximate solutions of doubtful accuracy. One reason for writing this book was to draw attention to these useful methods.

Most texts present statistical methods using the language and concepts of formal decision theory. One imagines that a statistical method is to be applied repeatedly under similar conditions, and attempts to minimize the frequencies of incorrect decisions over the long run. This repetitive decision model does not correspond to what is actually involved in most statistical applications. Usually one is interested in extracting information from a particular set of data. Although this information might subsequently be considered in arriving at one or more decisions, the main emphasis is on learning from the data at hand rather than on minimizing error frequencies in some hypothetical long-run of similar experiments. For this reason, I have not taken the usual approach to estimation and testing problems.

The earliest draft of this book was prepared during 1968-9 while I visited the University of Essex on a C.D. Howe Memorial Fellowship. It benefited greatly from many stimulating sessions with Professor George Barnard. The material has been revised and rewritten several times over the past ten years, and has been used as a text by several thousand mathematics undergraduates at the University of Waterloo. Their comments and those of their instructors have been most helpful. In particular, I wish to acknowledge the assistance and encouragement provided by Professor David Sprott and my brother Jack.

Very special thanks are due to Mrs. Annemarie Nittel for her expert typing of the manuscript.

I am grateful to the Biometrika Trustees for permission to reproduce material from Table 8 of *Biometrika Tables for Statisticians*, Vol. 1 (3rd edition, 1966); to John Wiley and Sons Inc. for permission to reproduce portions of Table II from *Statistical Tables and Formulas* by D. Hald (1952); and to the Literary Executor of the late Sir Ronald A. Fisher, F.R.S., to Dr. Frank Yates, F.R.S., and to Longman Group Ltd., London, for permission to reprint Tables I, III and V from their book *Statistical Tables for Biological, Agricultural and Medical Research* (6th edition, 1974).

CONTENTS—VOLUME I

CHAPTER 1. INTRODUCTION

1.1 Probability and Statistics

The purpose of this book is to introduce two important branches of modern applied mathematics: Probability Theory and Statistics. Both of these are relatively new subjects, the main developments having taken place within the last century, and both are rapidly expanding in theory and application. Indeed, it has now reached the point where some knowledge of these subjects is necessary if one is to read newspapers and magazines intelligently.

The study of probability was initially stimulated by the needs of gamblers, and games of chance are still used to provide interesting and instructive examples of probability methods. Today, probability theory finds applications in a large and growing list of areas. It forms the basis of the Mendelian theory of heredity, and hence has played a major part in the development of the science of Genetics. Modern theories in Physics concerning atomic particles make use of probability models. The spread of an infectious disease through a population is studied in the Theory of Epidemics, a branch of probability theory. Queueing theory uses probability models to investigate customer waiting times under the provision of various levels and types of service (e.g. numbers of checkout counters, telephone operators, computer terminals, etc.). Although it is impossible to deal thoroughly with such diverse and complex applications in an introductory book, it is possible to lay the groundwork and present some of the simpler applications. The first part of this book deals with methods of building probability models and handling them mathematically. This provides the foundation for the statistical methods described in later chapters, as well as for advanced study in Probability Theory itself.

Statistics was originally used to refer to the collection of data about the state or nation, such as size of the population, or the levels of trade and unemployment. Many statisticians are still involved in the important task of providing government with accurate statistical information on the basis of which the need for and effectiveness of their actions may be judged. However, the domain of application for statistical methods has increased rapidly during the twentieth century, and now includes virtually all areas of human endeavour where data are collected and analysed. The data may come from census results, questionnaires, surveys, or planned experiments in any field. There may be large quantities of data, as from a population census, in which case methods of accurately summarizing and simplifying the data are required. At the other extreme, many years of work and great expense may be necessary to obtain a few measurements in a scientific experiment. One may then wish to determine whether the data are in agreement with some general theory, or perhaps use the data to estimate physical constants. Because data are so difficult to obtain, it is important to extract the maximum possible amount of information from them.

This book deals primarily with problems of the latter type, where it is desired to draw general conclusions on the basis of a limited amount of data. Because they are based on limited data, such conclusions will be subject to uncertainty. The branch of Statistics which attempts to quantify this uncertainty using probability and related measures is called Statistical Inference. The last half of this book deals with two different problems in Statistical Inference: model testing and estimation. Having formulated a probability model, we will first wish to know whether it is in agreement with the data, and if not, which of the assumptions underlying the model require modification. Such questions may be investigated using Tests of Significance. Then, assuming that the model is satisfactory, one may wish to form estimates of unknown quantities, called parameters, whcih appear in the model. Such estimates will be subject to error. Determination of the likely magnitude of error in an estimate is an essential part of the estimation problem.

Statisticians are also concerned with the design of appropriate methods of data collection, so that upon analysis the data will yield the greatest possible amount of information of the type desired. Great care must be taken so that the data are free from unsuspected biases which might invalidate the analysis or cloud the interpretation. In many cases, asking a statistician to analyse improperly collected data is like calling the doctor after the patient has died: the most you can expect to learn is what the patient died of.

Statistical Decision Theory is a branch of Statistics which has received much attention since World War Two. It deals with the problem of selecting one among several possible courses of action in the face of uncertainties about the true state of nature. Both the costs of incorrect decisions and the information available from data and other sources are taken into account in arriving at a course of action which minimizes expected costs.

Many statistical problems involve both inferences and decisions. First we decide what data to collect. Having obtained the data, we try to learn as much as possible from it (Statistical Inference).

The information obtained might then be considered in deciding upon future courses of action. Nevertheless, it is important to distinguish carefully between the inferential and decision theoretic components. In decision problems, one is interested in learning from the data only if the information obtained can be used to reduce the anticipated cost of the particular action being considered. In Statistical Inference, one is interested in learning for its own sake, without reference to any particular decision problem in which the information obtained might subsequently be used.

In this book we shall consider inference problems rather than decision problems. Statistical Inference is a very controversial subject which is still very much in its formative stages. There are many fundamental differences of opinion concerning both the formulation of the problems to be considered, and the methods which should be employed in their solution. The approach taken in this book might be called Fisherian, because it is based primarily on the ideas of the British geneticist and statistician Sir Ronald Fisher (1890-1962). Most other writers have adopted either a decision-theoretic approach or a Bayesian approach. In the former, inferences are treated as if they were repetitive decision problems. In the latter, subjective prior opinions are combined with the data via Bayes's Theorem.

The increasing availability of high-speed electronic computers and associated graphic devices is having a profound effect on statistical theory and practice. It is now possible to perform analyses in only a few minutes which would previously have required months or years of laborious hand calculation. Thus one can now perform thorough analyses of much larger bodies of data than could formerly be handled, and one can analyse smaller bodies of data using exact methods where formerly it was necessary to rely on approximations, often with little guarantee of their accuracy. One aim of this book is to present methods suitable for use on the computer. Exact tests of significance, estimation from the likelihood function, and probability plotting are examples of techniques to be discussed whose usefulness has been greatly enhanced by the computer.

1.2 Observed Frequencies and Histograms

We shall be using the word *experiment* in a technical sense to mean some procedure giving rise to data. Some examples of "experiments" are: tossing a coin, rolling a die, dealing a poker hand of five cards from a well-shuffled deck, observing whether a certain telephone subscriber places a call during a one-minute interval, measuring the size of the crop in a field for a given variety of seed and amount of fertilizer, recording the number of hours of life of a television tube, asking a voter which candidate he prefers. The essential features of such experiments are that they have more than one possible outcome, and they may be considered repeatable. We may think of tossing the coin again, or observing the lifetime of a second tube which is identical to the first one. Of course, repeating the experiment will not necessarily result in repeating the outcome. The second toss of the coin may produce tails whereas the first toss resulted in heads. The lifetime of a second tube would not likely be exactly the same as that of the first.

When such an experiment is repeated n times, the result will be an ordered sequence of n outcomes, possibly not all different. For example, suppose that a cubical die with faces numbered $1, 2, \ldots, 6$ is rolled 25 times and the number on the up-face is recorded. The result of all 25 rolls will be a sequence of 25 numbers, such as

$$4\ 2\ 6\ 1\ 2\ 4\ 5\ 4\ 2\ 2\ 6\ 1\ 3\ 1\ 6\ 5\ 2\ 4\ 3\ 5\ 6\ 4\ 1\ 6\ 2\ .$$

When the order in which the various outcomes occurred is not of interest, it is often convenient to summarize such a data sequence in a _frequency table_. The frequency table records the number of times that each outcome occurs in the sequence. The observed frequencies may be found quickly using the tally method, as shown in Table 1.2.1. From the frequency table we can recover all the data except for the order in which the outcomes occurred.

Table 1.2.1

Frequency Table for 25 Rolls of a Die

Outcome	Tally	Observed Frequency	Relative Frequency
1	||||	4	0.16
2	~~||||~~		6	0.24
3	||	2	0.08
4	~~||||~~	5	0.20
5	|||	3	0.12
6	~~||||~~	5	0.20
Total		25	1.00

The last column of Table 1.2.1 gives the _relative frequency_ of each outcome. The relative frequency is the fraction of the time that the outcome occurred, and is obtained by dividing the observed frequency by the total frequency n.

If the number of possible outcomes is large in comparison with the number of repetitions n, most outcomes will occur once or not at all. Then a frequency table, as just described, may not give a very useful summary of the data. In such cases, it may be desirable to group together several of the possible outcomes in preparing the frequency table.

Example 1.2.1. The following are the 109 observed time intervals in days between 110 explosions in mines, involving more than 10 men killed, from 6 December 1875 to 29 May 1951. The first eight times are given in the first row, the next eight in the second row, etc. (The data are from an article by Maguire, Pearson, and Wynn in *Biometrika* [1952].)

378	36	15	31	215	11	137	4
15	72	96	124	50	120	203	176
55	93	59	315	59	61	1	13
189	345	20	81	286	114	108	188
233	28	22	61	78	99	326	275
54	217	113	32	23	151	361	312
354	58	275	78	17	1205	644	467
871	48	123	457	498	49	131	182
255	195	224	566	390	72	228	271
208	517	1613	54	326	1312	348	745
217	120	275	20	66	291	4	369
338	336	19	329	330	312	171	145
75	364	37	19	156	47	129	1630
29	217	7	18	1357			

A frequency table for these data as they stand would not give a useful summary of the data because most of the possible times either occur once or not at all. Consequently, we group the data into a reasonably small number of classes before preparing a frequency table. In order for the overall patterns in the data to be apparent, it is necessary that the number of classes be small enough so that most contain several of the observed times. On the other hand, the number of classes must not be too small, or else most of the information in the data will be lost.

A cursory inspection of the data shows that most times are less than 400, only a few exceed 700, and the largest is 1630. Consequently, we have taken classes of length 50 up to 400, with somewhat larger classes for times exceeding 400, as shown in Table 1.2.2. The observed frequencies are now easily obtained by the tally method. Note that the original observed times cannot be recovered from the grouped frequency table, so that there has been some loss of information. Also, the order in which the times were recorded has been lost. □

Frequency tables are often represented pictorially by means of frequency histograms. Suppose that we can associate disjoint in-

Table 1.2.2

Grouped Frequency Table for Accident Data

Class		Observed Frequency	Relative Frequency
[0,50)	𝍸 𝍸 𝍸 𝍸 𝍸	25	0.299
[50,100)	𝍸 𝍸 𝍸 𝍷𝍷𝍷𝍷	19	0.174
[100,150)	𝍸 𝍸 𝍷	11	0.101
[150,200)	𝍸 𝍷𝍷𝍷	8	0.073
[200,250)	𝍸 𝍷𝍷𝍷𝍷	9	0.083
[250,300)	𝍸 𝍷𝍷	7	0.064
[300,350)	𝍸 𝍸 𝍷	11	0.101
[350,400)	𝍸 𝍷	6	0.055
[400,600)	𝍸	5	0.046
[600,1000)	𝍷𝍷𝍷	3	0.028
[1000,2000)	𝍸	5	0.046
[2000,∞)		0	0.000
Total		109	1.000

tervals of real numbers with the classes in a frequency table in some natural way. To construct a frequency histogram, we mark out these class intervals on a horizontal axis, and above each interval we construct a rectangle whose area is equal to the observed frequency for that class. The total area of the resulting bar graph will be equal to the total frequency n.

Each of the classes in Table 1.2.2 already has associated with it an interval of real numbers, and the corresponding frequency histogram is given in Figure 1.2.1. The rectangle for class $[350,400)$ has base 50 and area 6, so that its height is $6/50 = 0.12$. The rectangle for class $[400,600)$ has base 200 and area 5, so that its height is $5/200 = .025$. Similarly, the height of the rectangle for class $[1000,2000)$ is $5/1000 = .005$. The histogram shows a long tail to the right which is characteristic of many types of waiting-time measurements.

The effect of combining two adjacent classes in the frequency table will be to replace two rectangles in the histogram by a single rectangle having the same total area. The height of this rectangle will be a weighted average of the heights of the two rectangles which it replaces. For example, if $[350,400)$ and $[400,600)$ were combined, we would obtain a single class $[350,600)$ with observed frequency $6 + 5 = 11$. The two rectangles of heights 0.12 and 0.025 would then be replaced by a single rectangle with area 11, base 250, and height $11/250 = 0.044$. Note that the units on the vertical axis will not be observed frequency, but rather observed frequency per unit of measurement on the horizontal axis.

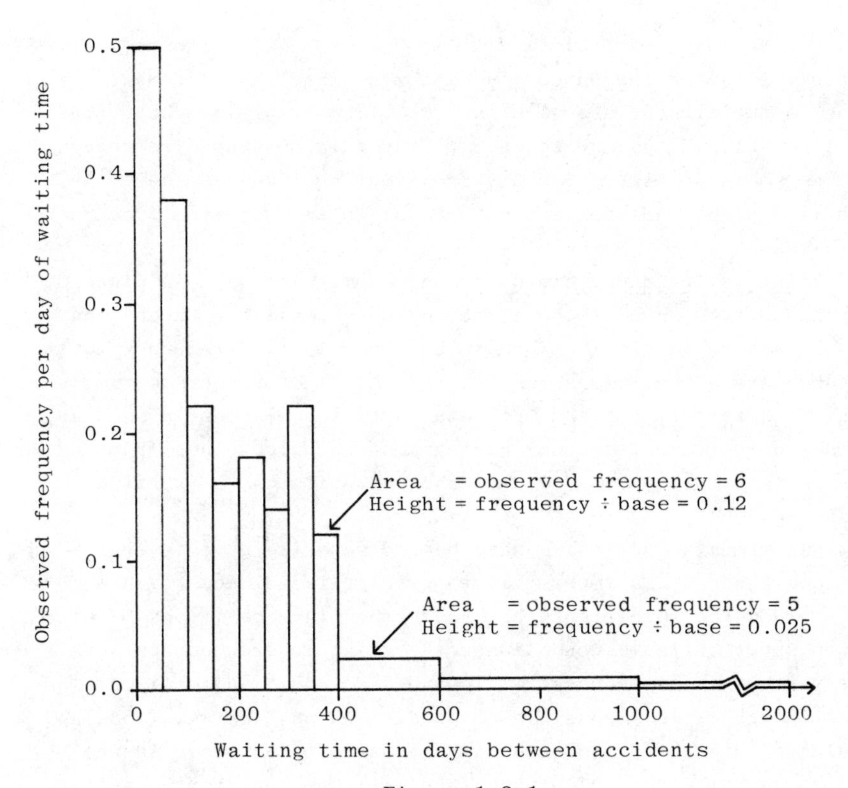

Figure 1.2.1

Frequency Histogram for Accident Data

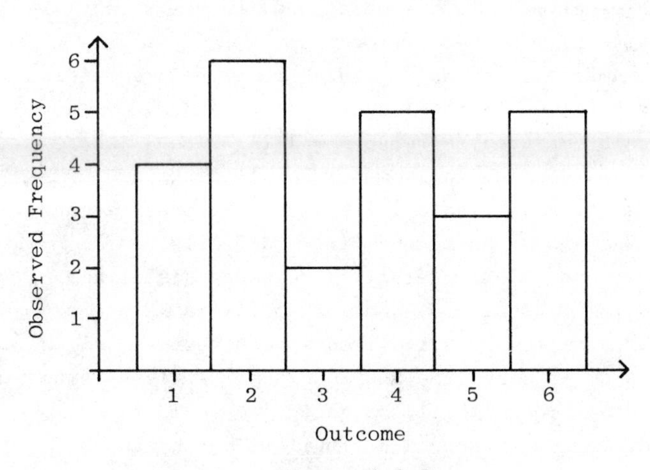

Figure 1.2.2

Frequency Histogram for 25 Rolls of a Die

For the data of Table 1.2.1, we may associate with each class a unit interval having the outcome as its midpoint. The interval corresponding to the first class will be $(0.5,1.5)$, that for the second class will be $(1.5,2.5)$, and so on. We may then obtain a frequency histogram as shown in Figure 1.2.2. Each rectangle now has base 1, and hence its height and area are both equal to the observed frequency for that class.

Relative frequency histograms are sometimes used in place of frequency histograms, and are constructed in the same way except that the area of each rectangle is taken to be the relative frequency for the corresponding class. As a result, the height of the histogram is reduced by a factor of n, and the total area becomes one. Frequency histograms and relative frequency histograms can be obtained from one another simply by multiplying or dividing the units on the vertical axis by n.

Experiments such as we have been discussing have several possible outcomes, and it is impossible to predict in advance which one will occur. However, experience has shown that in repetitions of such experiments under suitable conditions, the relative frequencies with which the various outcomes occur will stabilize and tend to fixed values. Although it is impossible to predict which face will come up when a balanced die is rolled once, we can say with some confidence that in a large number of rolls each face will come up about one-sixth of the time. No-one knows what will happen when a single coin is tossed once, but everyone knows that if we toss two tons of coins, roughly one ton of them will come up heads. Although the outcome of a single repetition of the experiment is unpredictable, there is regularity in the combined results of many repetitions.

To obtain an example of the way in which relative frequencies tend to stabilize at fixed values in the long run, I persuaded my daughter to spend part of a rainy Saturday afternoon rolling dice. She recorded the outcomes of 1000 rolls, and some of her findings are summarized in Tables 1.2.3 and 1.2.4. Table 1.2.3 gives the observed and relative frequencies of the six faces in the first n rolls, for $n = 25, 50, 100, 250, 500$, and 1000. Relative frequency histograms for these data are given in Figure 1.2.3. Note that initially there is considerable variability in the relative frequencies, but as n increases the frequencies seem to stabilize and tend towards fixed values near one-sixth. If the die were perfectly balanced, we would expect the relative frequency histogram to approach a uniform height as n became very large.

Table 1.2.3

Observed Frequencies in n Rolls of a Die

n	Observed Frequency 1	2	3	4	5	6	Relative Frequencies 1	2	3	4	5	6
25	3	6	5	2	6	3	0.120	0.240	0.200	0.080	0.240	0.120
50	7	9	10	7	11	6	0.140	0.180	0.200	0.140	0.220	0.120
100	16	15	14	20	22	13	0.160	0.150	0.140	0.200	0.220	0.130
250	36	26	52	48	49	39	0.144	0.104	0.208	0.192	0.196	0.156
500	88	69	86	81	95	81	0.176	0.138	0.172	0.162	0.190	0.162
1000	171	156	169	167	171	166	0.171	0.156	0.169	0.167	0.171	0.166

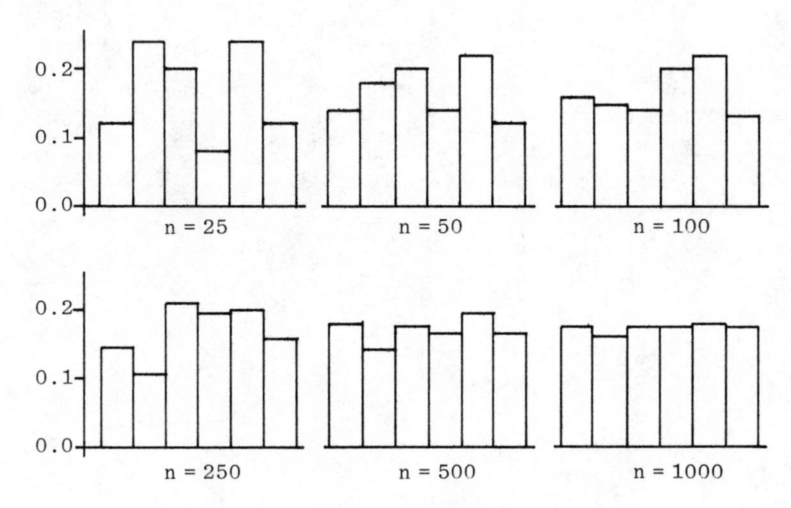

Figure 1.2.3

Relative Frequency Histograms for n Rolls of a Die

Table 1.2.4 gives the total observed frequency of outcomes 1,2 and 3 in the first n trials ($n = 25, 50, \ldots, 1000$), together with the corresponding relative frequencies. These relative frequencies are plotted against n in Figure 1.2.4. Initially they vary erratically from a high of 0.56 to a low of 0.433. However, as n increases the relative frequency changes more slowly, and appears to approach a fixed value near 0.5, which is the value expected for a balanced die.

In most applications, probabilities may be thought of as the limiting values approached by relative frequencies as n tends to infinity. The probability of an outcome is then the fraction of the time that the outcome would occur in infinitely many repetitions of the ex-

Table 1.2.4

Total Observed Frequency of Outcomes 1,2 and 3 in n Rolls of a Die

n	Observed Frequency	Relative Frequency	n	Observed Frequency	Relative Frequency
25	14	0.560	525	254	0.484
50	26	0.520	550	268	0.487
75	35	0.467	575	281	0.489
100	45	0.450	600	293	0.488
125	65	0.520	625	306	0.490
150	77	0.513	650	319	0.491
175	86	0.491	675	327	0.484
200	96	0.480	700	342	0.489
225	104	0.462	725	356	0.491
250	114	0.456	750	367	0.489
275	119	0.433	775	379	0.489
300	133	0.443	800	394	0.493
325	142	0.437	825	407	0.493
350	159	0.454	850	419	0.493
375	173	0.461	875	432	0.494
400	186	0.465	900	449	0.499
425	203	0.478	925	461	0.498
450	215	0.478	950	472	0.497
475	226	0.476	975	481	0.493
500	243	0.486	1000	496	0.496

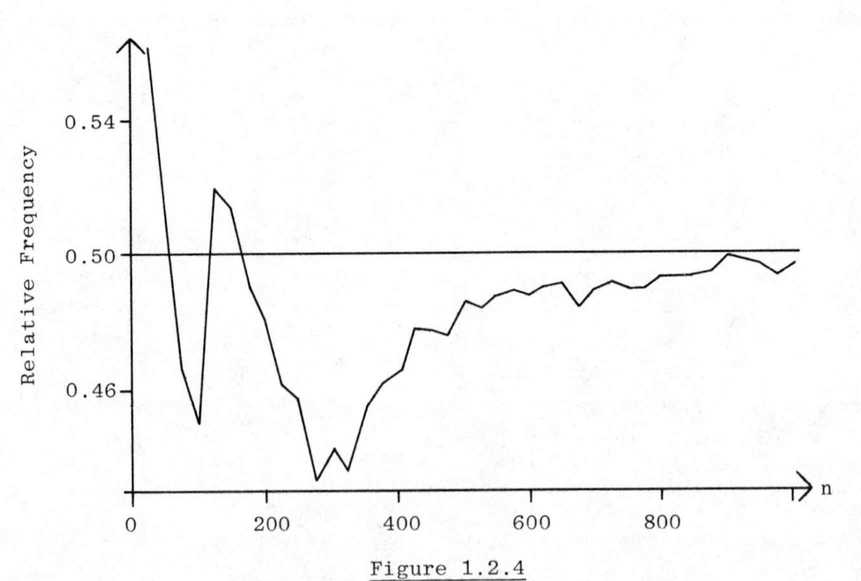

Figure 1.2.4

Relative Frequency of Outcomes 1,2,3 in n Rolls of a Die

periment. Probabilities are thus unknown quantities whose values are approximated, or estimated, by the corresponding relative frequencies. Although we can never determine the exact value of a probability, we can obtain as accurate an estimate as we wish by repeating the experiment sufficiently often. In this respect, probabilities are not unlike weights, lengths, temperatures, etc.; we can never know their exact values, but we can measure them with sufficient accuracy for many uses.

Problems for Section 1.2

1. As a class assignment, each student tossed a coin 90 times and kept track of the number of heads. The following are the numbers of heads recorded by the 83 members of the class:

44	46	45	40	44	39	52	50	42	45	43	53
44	49	45	49	44	45	51	57	38	42	46	51
45	50	42	46	43	41	43	48	44	45	46	40
46	43	44	42	46	48	51	51	44	43	47	43
41	42	52	45	52	50	50	50	58	48	44	46
46	43	51	41	53	39	56	43	42	45	45	52
53	43	47	50	36	43	48	44	48	47	43	

Prepare a frequency table without grouping the data, and draw the frequency histogram. On the same graph, show the frequency histogram that would result from using intervals of width 3, the central interval being (43.5,46.5).

2. The following are scores (out of 1000) achieved by 66 students in an English language examination:

345	395	563	505	402	472	691	624	523	461	490
530	516	444	604	406	475	523	582	575	439	523
556	479	629	490	730	611	468	574	420	596	585
354	494	439	446	505	585	468	578	603	417	585
593	417	486	604	515	523	545	505	527	384	431
574	494	560	464	549	541	468	629	607	490	549

Prepare a frequency table and histogram for these data using an interval width of 50.

1.3 Probability Models

The mathematical theory of probability is a deductive theory. As with classical mechanics or Euclidean geometry, the theorems may be deduced from a set of axioms without reference to any real-world applications. However, just as classical mechanics gains intuitive meaning if it is discussed in relation to the behaviour of real rigid bodies, the mathematical theory of probability gains from discussion in terms of real or conceptual experiments, such as those described in the preceding section. The essential features of such experiments are that they have several possible outcomes, and they may be considered repeatable.

The individual possible outcomes of an experiment are called simple events, or to use geometrical language, points. The first step in setting up a probability model for an experiment is to agree on the set of all possible outcomes, called (again using geometrical language) the sample space. As with any other mathematical model of a real situation, there is a certain degree of arbitrariness involved. The choice of the sample space depends upon one's intentions, and frequently entails some simplification. For example, when a coin is tossed the sample space will usually be taken to consist of two points corresponding to "heads" and "tails". The orientation of the coin and the possibility that it might land on its edge are deliberately ignored in order to simplify the mathematical model. When a poker hand is dealt, the sample space will be taken to consist of all possible combinations of five cards, the possibility of a misdeal in which four or six cards are dealt being ignored.

In Chapters 2,3,4 and 5 we shall deal only with discrete sample spaces; that is, we assume that there are only finitely many or countably many possible outcomes. The sample points may then be labelled with integers $1,2,3,\ldots$.

In Chapters 6 and 7 we shall consider some situations in which the theory allows uncountably many sample points. For example, in measuring quantities such as length, weight, and time, it seems natural to represent possible outcomes geometrically by points on the real axis. However, in practice, the resolving power of any measuring instrument will always be finite, and such measurements will be more properly regarded as given to a finite number of decimal places. Thus the sample space of any real experiment will be discrete, and will, in fact, contain only finitely many possible outcomes. Sample spaces with infinitely many points are allowed in the theory for reasons of mathematical convenience.

Suppose now that an appropriate discrete sample space S has been agreed upon, and that the points of S have been labelled $1,2,3,\ldots$. The next step is to assign to each point i of S a real number p_i which is called the probability of the outcome labelled i. The probabilities p_i must be non-negative and sum to one:

$$p_i \geq 0 \quad (i = 1,2,3,\ldots)$$

$$p_1 + p_2 + p_3 + \ldots = 1.$$

Any set of numbers $\{p_i\}$ satisfying these conditions is called a <u>probability distribution</u> on S because the total probability 1 has been distributed over the points of S.

The mathematical theory of probability assumes that the sample space S and probability distribution $\{p_i\}$ are given, and is concerned only with determining what probabilities should then be assigned to subsets of S. These subsets are called events. Thus an <u>event</u> A is, by definition, a subset of the sample space. Event A is said to occur if the outcome of the experiment belongs to A.

The <u>probability of event A</u>, or the probability that A occurs, is defined to be the sum of the probabilities p_i of all points i which belong to A, and will be denoted by $P(A)$. Symbolically,

$$P(A) = \sum_{i \in A} p_i.$$

Because the p_i's are non-negative with total 1, the sum of any subset of the p_i's will be a number between 0 and 1. Hence, for any event A, we have

$$0 \le P(A) \le 1.$$

The sample space S and empty set ϕ are considered to be events. Their probabilities are

$$P(S) = 1; \qquad P(\phi) = 0.$$

If A is any event defined on S, the <u>complementary event</u> $\bar{A}$ is the set of all sample points not belonging to A. Because every sample point belongs either to A or to its complement $\bar{A}$ but not to both, we have

$$\sum_{i \in A} p_i + \sum_{i \in \bar{A}} p_i = \sum_{i \in S} p_i.$$

It follows that $P(A) + P(\bar{A}) = P(S) = 1$, and hence

$$P(\bar{A}) = 1 - P(A).$$

The mathematical theory of probability does not depend upon any particular interpretation of the numbers p_i, nor is it concerned with how they should be measured or assigned. In applications, probabilities are usually interpreted as objective physical constants which may be measured, or estimated, by relative frequencies (see Section 1.2). Then the ratio of two probabilities p_i/p_j represents the fair odds for a bet on outcome i as against outcome j. For instance, if

$p_i/p_j = 10$, outcome i will occur 10 times as often as outcome j in the long run. If you win $1 whenever i occurs but lose $10 whenever j occurs, your average net gain per bet in a long sequence of bets will be zero. The ratio $p_i/(1-p_i)$ represents the fair odds in favour of outcome i; that is, the loss one should be willing to incur if i does not occur in return for a gain of $1 if i does occur.

Probabilities are sometimes used to measure personal or subjective belief in a particular event or proposition. Each individual determines his own personal probabilities by a process of introspection. Advocates of this approach recommend that you imagine yourself to be in the position of having to bet on a proposition. You then determine the odds at which you would be equally willing to bet either for or against the proposition, and from this your personal probability may be found. Different individuals will, of course, have different personal probabilities for a proposition, even when both are presented with the same data. Personal probabilities can be useful in the context of personal and business decisions, but a more objective and empirical approach seems better suited to the types of application which we shall be considering.

Example 1.3.1. Suppose that an experiment involves rolling a cubical die with faces marked 1,2,...,6. The sample space for the experiment will be $S = \{1,2,3,4,5,6\}$. If the die is carefully made of homogeneous material, one would expect each face to turn up approximately equally often in a large number of rolls, and the fair odds for one face versus another would be even (equal to one). Thus one would be led to assign equal probabilities $p_i = 1/6$ to the six points of the sample space. Of course, such a perfectly balanced and symmetrical die could not be constructed. Any real die will have some bias which may be detected by rolling it sufficiently often. Whether or not the assumed distribution $\{p_i\}$ is completely satisfactory can only be determined empirically - by actually performing the experiment and comparing the results observed with those predicted. Mathematical arguments, such as the one above based on symmetry, are useful only for suggesting theoretical models which must subsequently be evaluated empirically.

Any subset of the sample space is an event. For example, $A = \{1,2,3\}$ is the event corresponding to rolling a face numbered 3 or less. The probability of event A is

$$P(A) = p_1 + p_2 + p_3 = \frac{1}{6} + \frac{1}{6} + \frac{1}{6} = \frac{1}{2} .$$

The complement of A is $\overline{A} = \{4,5,6\}$, with probability

$$P(\overline{A}) = 1 - P(A) = \frac{1}{2}.$$

Note that the empirical results of Section 1.2 are in good agreement with the assumption of equally probable outcomes.

Example 1.3.2. Suppose that dice are made by cutting lengths from one centimeter square stock. If the distance between successive cuts is 1 cm., ordinary cubical dice are produced. If the distance is more or less than one centimeter, the dice will be "brick-shaped" or "tile-shaped" as illustrated in Figure 1.3.1. Such a die has four symmetrical uncut faces which we may number 1,2,3,4 and two symmetrical cut faces 5,6. The sample space for a single roll of the die

Brick-shaped die

Tile-shaped die

Figure 1.3.1

Brick-shaped and Tile-shaped Dice
(Example 1.3.2)

will be $S = \{1,2,3,4,5,6\}$ as in Example 1.3.1. However one should no longer assume that the six faces are equally probable. Instead, a reasonable probability distribution to assume in this case would be

$$p_1 = p_2 = p_3 = p_4 = p; \quad p_5 = p_6 = q.$$

Since the sum of all six probabilities must be one, we have $4p + 2q = 1$, and hence $q = 0.5 - 2p$. Since $q \geq 0$, we must have $0 \leq p \leq 0.25$.

If the distance between cuts is close to 1 cm., p and q will be close to 1/6. This suggests that we write

$$p = \frac{1}{6} + \theta$$

where θ is close to zero. It then follows that

$$q = \frac{1}{2} - 2p = \frac{1}{6} - 2\theta.$$

Since $0 \le p \le 0.25$, we must have $-1/6 \le \theta \le 1/12$. For each value of θ in this range we obtain a probability distribution:

$$p_1 = p_2 = p_3 = p_4 = \frac{1}{6} + \theta; \quad p_5 = p_6 = \frac{1}{6} - 2\theta. \qquad (1.3.1)$$

The size of θ is clearly related to the distance between cuts, with $\theta = 0$ for an ordinary cubical die. A theoretical value of θ might be obtained from the laws of mechanics. However, a more common procedure would be to regard θ as an unknown parameter of the distribution, whose value would be estimated on the basis of several rolls of the die. Such estimation problems will be discussed in Chapter 9.

Example 1.3.3. Consider an experiment in which a lightbulb is observed until it fails, and the number of completed hours of life is reported. Such an experiment cannot be continued indefinitely. If it is terminated at n hours, the possible outcomes are "0 hours", "1 hour",..., "n-1 hours", "at least n hours". However, it may not be possible to specify in advance the time n at which testing would cease if all lightbulbs had not yet failed. To avoid this difficulty and to gain mathematical simplicity, we consider an idealized experiment in which observation could conceivably continue forever. The sample space for the idealized experiment will be the set of all non-negative integers, $S = \{0,1,2,...\}$.

Some types of electronic components do not appear to age; that is, if one considers a large number of such components, the proportion α which survive a one-hour period does not depend upon the age of the components. If there are initially N components, the number failing in the first hour will be approximately $(1-\alpha)N$, and the number still operating at the beginning of the second hour will be αN. Of these, a proportion $1-\alpha$ fail during the second hour. Thus the number failing in the second hour will be approximately $(1-\alpha)\alpha N$, and the number still operating at the beginning of the third hour will be $\alpha^2 N$. Continuing in this fashion, we see that approximately $(1-\alpha)\alpha^{i-1}N$ components will fail in the ith hour, and will be reported as having completed $i-1$ hours of life. We are thus led to the following model:

$$p_i = (1-\alpha)\alpha^i \quad \text{for} \quad i = 0,1,2,\dots\ . \qquad (1.3.2)$$

Because probabilities are non-negative, we must have $0 \le \alpha < 1$. For each value of α in this range, the total probability is

$$p_0 + p_1 + p_2 + \dots = (1-\alpha)(1+\alpha+\alpha^2+\dots) = 1$$

and hence (1.3.2) defines a proper probability distribution. Because the probabilities form a geometric series, this is called a <u>geometric probability distribution</u>. The parameter α would usually be unknown and would be estimated from the available data.

The event $A = \{n, n+1, n+2, \ldots\}$ corresponds to an observed lifetime of at least n completed hours. Its probability is

$$P(A) = p_n + p_{n+1} + p_{n+2} + \cdots = (1-\alpha)\alpha^n(1 + \alpha + \alpha^2 + \ldots) = \alpha^n.$$

In the original experiment where observation stops after n hours, the sample space is $S = \{0, 1, \ldots, n-1, n^+\}$, where sample point "$n^+$" corresponds to a lifetime of at least n hours. The above considerations suggest the following distribution:

$$p_i = (1-\alpha)\alpha^i \quad \text{for} \quad i = 0, 1, \ldots, n-1; \quad p_{n+} = \alpha^n$$

where $0 \le \alpha < 1$. The total probability equals 1. We shall later take up methods which permit the suitability of this model to be investigated by comparing observed lifetimes with predictions from the model.

<u>Example 1.3.4</u>. When quantities such as time, weight, height, etc. are being considered, it is common to define probabilities as integrals. For instance, if T represents the waiting time between successive accidents in Example 1.2.1, the probability that T lies between a and b $(0 < a < b)$ can be defined as

$$P(a \le T < b) = \int_a^b f(t)dt$$

where f is a suitably chosen non-negative function, called a probability density function (see Chapter 6). If accidents were thought to be occurring randomly at a constant rate, then one would take

$$f(t) = \frac{1}{\theta} e^{-t/\theta}, \quad 0 < t < \infty$$

(see Sections 6.2 and 6.5). For this choice of f we have

$$P(a \le T < b) = \int_a^b \frac{1}{\theta} e^{-t/\theta} dt = \exp(-\tfrac{a}{\theta}) - \exp(-\tfrac{b}{\theta}).$$

The positive constant (parameter) θ represents the average waiting time between accidents over the long run. The data of Example 1.2.1 show a total of 109 accidents over a total period of 26,263 days. Hence, as a reasonable estimate, we take $\theta = 26263 \div 109 = 241$ days. Then

$$P(a \le T < b) = \exp(-a/241) - \exp(-b/241).$$

According to this model, the probabilities for the first two classes in Table 1.2.1 are

$$P(0 \le T < 50) = \exp(0) - \exp(-50/241) = 1 - 0.813 = 0.187$$
$$P(50 \le T < 100) = \exp(-50/241) - \exp(-100/241)$$
$$= 0.813 - 0.660 = 0.153.$$

The probabilities for the other classes may be found in a similar fashion. The total probability will be one because

$$\int_0^\infty f(t)dt = 1.$$

We shall discuss this example further in the next section.

Problems for Section 1.3

1. Four letters addressed to individuals W,X,Y, and Z are randomly placed in four addressed envelopes, with one letter in each envelope.
   (a) Describe a 24-point sample space for this experiment.
   (b) List the sample points belonging to each of the following events.

   A - "W's letter goes into the proper envelope";
   B - "no letters go into the proper envelopes";
   C - "exactly two letters go into the proper envelopes";
   D - "exactly three letters go into the proper envelopes".

   (c) Assuming that the 24 sample points are equally probable, compute the probabilities of the four events in (b), and determine the odds against event B.

†2. (a) Three balls are distributed at random into three boxes, there being no restriction on the number of balls per box. List the 27 possible outcomes of this experiment. Assuming these to be equally probable, find the probability of each of the following events:

   A - "first box is empty";
   B - "first two boxes are empty";
   C - "no box contains more than one ball".

   (b) Find the probabilities of events A,B and C when three balls are distributed at random into n boxes.
   (c) Find the probabilities of events A,B and C when r balls

† Answer given in Appendix A.

are distributed at random into n boxes.

3. If three identical balls are distributed at random into three boxes, there are only ten distinguishable arrangements. List these, and deduce their probabilities from the model in 2(a).

4. (a) Describe a 36-point sample space for the simultaneous rolling of a red die and a white die. Assuming that the dice are perfectly balanced, what probabilities should be assigned to the sample points?

   (b) Under the model set up in (a), find the probabilities of the following:

   A - "both dice show the same number";

   B - "red die shows a higher score than white die";

   C - "total score on the two dice is 6".

   (c) If two identical dice are rolled, what are the distinguishable outcomes, and what are their probabilities?

5. A balanced coin is tossed until "tails" comes up, or until "heads" occurs six times in a row. Set up a probability model for this experiment.

†6. Two teams play a best-of-seven series. Play stops as soon as one team has won four games. Describe a sample space for this experiment. If the teams are evenly matched, what probabilities should be assigned? What is the probability that the series will last the full seven games?

7. Three players A,B,C play a game in pairs as follows. At round one, A and B play while C observes. The winner plays C at round two while the loser observes. The game continues in this fashion, with the loser of one round sitting out the next round, until one player wins twice in succession, thus becoming the winner of the game.

   (a) Describe a sample space for this "experiment".

   (b) Suppose that each sample point which corresponds to a game consisting of k rounds is assigned probability 2^{-k} $(k = 2,3,4,\ldots)$. Show that the total probability is 1.

   (c) Show that A wins with probability $\frac{5}{14}$, B wins with probability $\frac{5}{14}$, and C wins with probability $\frac{2}{7}$.

1.4 Expected Frequencies

As was indicated in the last section, mathematical arguments are useful in devising probability models for experiments, but whether or not a model thus obtained is a good one must be determined empirically. This is done by comparing what has actually been observed in performing the experiment with what would be predicted or expected according to the model.

Suppose that an experiment is repeated n times, and that, according to the model, an event A has probability p of occurring at each repetition. Then the expected frequency, or theoretical frequency, of event A is defined to be np.

One common procedure for evaluating a model is to compare a table or histogram of observed frequencies with the corresponding table or histogram of expected frequencies. If the observed frequencies deviate too greatly from the expected frequencies, the model is judged to be unsatisfactory.

Exact agreement between the observed and expected frequencies is too much to hope for. Even if the model were correct, some discrepancy between the observed and expected frequencies would be anticipated owing to "chance variation" in the data. There are procedures for determining whether the observed deviations are too great to be attributed to chance. These are called tests of significance, and the discussion of them begins in Chapter 11.

Example 1.4.1. The probability model for a single roll of a balanced cubical die assigns probability $1/6$ to each face. Hence, in 100 rolls, the expected frequency of each face is $100(1/6) = 16.67$. The third row of Table 1.2.3 gives the observed frequencies in 100 actual rolls of a die. The observed and expected frequencies are compared in the following table:

Face							Total
Observed frequency	16	15	14	20	22	13	100
Expected frequency	16.67	16.67	16.67	16.67	16.67	16.67	100.02

In this case the expected frequencies are not integers, and exact agreement between observed and expected frequencies is impossible. A goodness of fit test (Chapter 11) shows that deviations as great as those observed could easily arise by chance, and hence the data show good agreement with the probability model.

Example 1.4.2. Consider the accident data of Example 1.2.1. According to the probability model considered in Example 1.3.4, the probability that T, the waiting time between successive accidents, lies between a and b is given by

$$P(a \le T < b) = \exp(-a/241) - \exp(-b/241).$$

Then, out of 109 waiting times, the number which would be expected to lie between a and b is $109 \cdot P(a \le T < b)$. For instance, the expected number in the interval $[0,50)$ is

$$109 \cdot P(0 \le T < 50) = 109[\exp(0) - \exp(-50/241)] = 20.42.$$

The expected frequencies for the other classes in Table 1.2.2 may be computed in a similar way, and are shown in Table 1.4.1.

Table 1.4.1

Comparison of Observed and Expected Frequencies for the Accident Data of Example 1.2.1

Class	Observed Frequency	Expected Frequency
[0,50)	25	20.42
[50,100)	19	16.60
[100,150)	11	13.49
[150,200)	8	10.96
[200,250)	9	8.91
[250,300)	7	7.24
[300,350)	11	5.88
[350,400)	6	4.78
[400,600)	5	11.69
[600,1000)	3	7.32
[1000,200)	5	1.69
[2000,∞)	0	0.03
Total	109	109.01

In the first six classes $(T < 300)$, there is close agreement between the frequencies observed and those expected under the model. However, the agreement is poor in four of the remaining six classes; the model does not give an accurate prediction of the frequencies with which the larger waiting times were observed.

At least part of the difficulty here is due to the fact that the average number of days between accidents seems to be increasing with time: the average of the first 24 waiting times is only 97.5, while the average of the last 24 waiting times is 280.4. Hence the accident rate seems to be decreasing whereas the model considered in Example 1.3.4 assumes a constant rate. A more complicated model may be required, in which θ, the average waiting time between accidents, is permitted to change with time.

Problems for Section 1.4

1. (a) Set up a probability model for the simultaneous tossing of a nickel, a dime, and a quarter.

(b) Find the probability of obtaining x heads when three balanced coins are tossed $(x = 0,1,2,3)$.

(c) The following are the numbers of heads observed in thirty tosses of three coins:

1 0 1 3 2 2 3 1 1 2 1 1 2 1 2
2 1 2 1 2 1 2 2 1 2 1 0 1 2 2

Prepare observed and expected frequency histograms.

(d) Toss three coins thirty (or more) times, and prepare histograms as in (c). Are your experimental results in reasonable agreement with the model?

CHAPTER 2. EQUI-PROBABLE OUTCOMES

This chapter discusses situations in which the sample space S contains N points ($N < \infty$), each of which is assigned the same probability $1/N$. By definition, the probability of an event A is equal to the sum of the probabilities of the points belonging to A. When all points are equally probable, this simplifies to

$$P(A) = \sum_{i \in A} p_i = \sum_{i \in A} \frac{1}{N} = \frac{\text{number of points in } A}{N} .$$

Hence, in this case, the probability of an event may be determined merely by counting the number of sample points which it contains. Section 1 records some useful definitions and identities from combinatorial analysis, which is the branch of mathematics dealing with counting methods. The discussion of probability theory is then resumed in Section 2.

2.1 Combinatorial Symbols

Let r be a non-negative integer and n a real number. The symbol $n^{(r)}$, read "<u>n to r factors</u>", is defined as follows:

$$\begin{aligned} n^{(r)} &= n(n-1)(n-2)\ldots(n-r+1), \quad r > 0 \\ n^{(0)} &= 1. \end{aligned} \tag{2.1.1}$$

Note that $n^{(r)}$ is the product of r factors. The first factor is n, and successive factors decrease by one, the rth factor being $n - (r-1) = n - r + 1$.

If n is a non-negative integer, $n^{(r)}$ is the number of arrangements, or permutations, of n different things taken r at a time, for which another common symbol is ${}_nP_r$. In particular, $n^{(n)} = {}_nP_n$ is the number of arrangements or permutations of n different things taken all at a time. The quantity $n^{(n)}$ is given a special symbol $n!$ and is called "<u>n factorial</u>":

$$n! = n^{(n)} = n(n-1)(n-2)\ldots(2)(1). \tag{2.1.2}$$

Note that $0! = 0^{(0)} = 1$ by definition.

<u>Examples</u>. $4^{(2)} = 4(3) = 12$

$2^{(4)} = 2(1)(0)(-1) = 0$

$(0.5)^{(4)} = (0.5)(-0.5)(-1.5)(-2.5) = -0.9375$

$5! = 5(4)(3)(2)(1) = 120.$

Let r be a non-negative integer and n a real number. The symbol $\binom{n}{r}$, read "<u>n choose r</u>", is defined as follows:

$$\binom{n}{r} = \frac{n^{(r)}}{r!} \quad \text{for} \quad r \geq 0. \tag{2.1.3}$$

Some simplification of summation formulae is possible if we also define this symbol for r a negative integer:

$$\binom{n}{r} = 0 \qquad \text{for} \quad r < 0. \tag{2.1.4}$$

It is easily verified from these definitions that the following identities hold for all integers r and real numbers n:

$$\binom{n+1}{r} = \binom{n}{r-1} + \binom{n}{r} \tag{2.1.5}$$

$$r\binom{n}{r} = n\binom{n-1}{r-1}. \tag{2.1.6}$$

More generally, if s is any non-negative integer, then

$$r^{(s)}\binom{n}{r} = n^{(s)}\binom{n-s}{r-s}. \tag{2.1.7}$$

Because of (2.1.9) below, $\binom{n}{r}$ is often called a <u>binomial coefficient</u>.

If r and n are non-negative integers, then $\binom{n}{r}$ is the number of ways to choose r items from n when the order of choice is unimportant; that is, $\binom{n}{r}$ is the number of different subsets of r elements which may be selected from a set containing n elements. This is sometimes referred to as the number of combinations of n things taken r at a time, and another frequently used symbol is ${}_nC_r$. Note that

$$\binom{n}{r} = \frac{n^{(r)}}{r!} = \frac{n(n-1)\ldots(n-r+1)}{r!}$$

$$= \frac{n(n-1)\ldots(n-r+1)(n-r)!}{r!(n-r)!} = \frac{n!}{r!(n-r)!}\ .$$

Hence if n and r are integers with $n \geq r \geq 0$, we have

$$\binom{n}{r} = \binom{n}{n-r} = \frac{n!}{r!(n-r)!} \tag{2.1.8}$$

Examples. $\binom{8}{6} = \binom{8}{2} = \frac{8!}{2!6!} = \frac{8(7)}{2} = 28$

$\binom{5}{-1} = 0$ by definition

$$\binom{0.5}{4} = \frac{(0.5)^{(4)}}{4!} = \frac{(0.5)(-0.5)(-1.5)(-2.5)}{24} = -\frac{5}{128}$$

$$\binom{-3}{4} = \frac{(-3)^{(4)}}{4!} = \frac{(-3)(-4)(-5)(-6)}{24} = 15$$

$$\binom{n}{0} = \frac{n^{(0)}}{0!} = 1 \quad \text{for all } n.$$

The Binomial Theorem. Under appropriate restrictions on n and t,

$$(1 + t)^n = \sum_{r=0}^{\infty} \binom{n}{r} t^r. \tag{2.1.9}$$

(i) If n is a positive integer, the binomial series is a finite sum, and thus converges for all real t.

(ii) If n is negative or fractional, the series converges for $|t| < 1$ and diverges for $|t| > 1$. It also converges for $t = +1$ provided that $n > -1$, and for $t = -1$ provided that $n > 0$.

For proofs of these results, see any advanced calculus textbook.

Example 2.1.1. If n is a positive integer, then

$$\binom{n}{0} + \binom{n}{1} + \binom{n}{2} + \ldots + \binom{n}{n} = 2^n.$$

Proof. The result follows by taking $t = 1$ in (2.1.9) and noting that all terms beyond the nth are zero.

Example 2.1.2. If n is a positive integer, then

$$\binom{n}{1} + 2\binom{n}{2} + 3\binom{n}{3} + \ldots + n\binom{n}{n} = n2^{n-1}.$$

Proof. The result may be obtained by differentiating (2.1.9) with respect to t and then setting $t = 1$. Alternatively,

$$\binom{n}{1} + 2\binom{n}{2} + 3\binom{n}{3} + \dots + n\binom{n}{n} = \sum r\binom{n}{r}$$

where the sum extends over all integers r. Now (2.1.6) gives

$$\sum r\binom{n}{r} = \sum n\binom{n-1}{r-1} = n\sum\binom{n-1}{r-1}$$

$$= n\{\binom{n-1}{0} + \binom{n-1}{1} + \dots + \binom{n-1}{n-1}\}$$

and the result follows from Example 2.1.1 with n replaced by $n-1$. □

The Hypergeometric Identity. Let a and b be real numbers, and let n be a positive integer. Then

$$\sum_{r=0}^{\infty} \binom{a}{r}\binom{b}{n-r} = \binom{a+b}{n}. \tag{2.1.10}$$

Proof. Let t be a real number with $|t| < 1$. Then by the binomial theorem (2.1.9),

$$(1 + t)^a = \sum\binom{a}{r}t^r; \quad (1 + t)^b = \sum\binom{b}{s}t^s$$

$$(1 + t)^{a+b} = \sum\binom{a+b}{n}t^n .$$

We also have

$$(1+t)^{a+b} = (1+t)^a(1+t)^b = \{\sum\binom{a}{r}t^r\}\{\sum\binom{b}{s}t^s\}$$

$$= \{\binom{a}{0} + \binom{a}{1}t + \binom{a}{2}t^2 + \dots\}\{\binom{b}{0} + \binom{b}{1}t + \binom{b}{2}t^2 + \dots\}$$

$$= \binom{a}{0}\binom{b}{0} + \{\binom{a}{0}\binom{b}{1} + \binom{a}{1}\binom{b}{0}\}t + \{\binom{a}{0}\binom{b}{2} + \binom{a}{1}\binom{b}{1} + \binom{a}{2}\binom{b}{0}\}t^2 + \dots$$

$$= \sum\{\sum\binom{a}{r}\binom{b}{n-r}\}t^n.$$

Now equating coefficients of t^n in the two power series expansion of $(1+t)^{a+b}$ gives (2.1.10).

Example 2.1.3 . If n is a positive integer, then

$$\binom{n}{0}^2 + \binom{n}{1}^2 + \dots + \binom{n}{n}^2 = \binom{2n}{n}.$$

Proof. From (2.1.8) we have

$$\sum\binom{n}{r}^2 = \sum\binom{n}{r}\binom{n}{r} = \sum\binom{n}{r}\binom{n}{n-r}.$$

The result now follows from (2.1.10) with $a = b = n$. □

Multinomial Coefficients.

Let $n, k, r_1, r_2, \ldots, r_k$ be non-negative integers with $r_1 + r_2 + \ldots + r_k = n$ and $k \geq 2$. The multinomial coefficient $\binom{n}{r_1 r_2 \ldots r_k}$ is defined as follows:

$$\binom{n}{r_1 r_2 \ldots r_k} = \frac{n!}{r_1! r_2! \ldots r_k!}. \tag{2.1.11}$$

It counts the number of arrangements or permutations of n things of k different kinds, there being r_1 of the first kind, r_2 of the second kind, ..., and r_k of the kth kind.

In the special case $k = 2$ we have $r_2 = n - r_1$, and hence

$$\binom{n}{r_1 r_2} = \frac{n!}{r_1! r_2!} = \frac{n!}{r_1!(n-r_1)!} = \binom{n}{r_1}.$$

Thus for $k = 2$, the multinomial coefficient reduces to a binomial coefficient.

Example 2.1.4. In how many different ways can the letters of the word MISSISSIPPI be arranged?

Solution. There are $n = 11$ letters of $k = 4$ different kinds, there being one M $(r_1 = 1)$, four I's $(r_2 = 4)$, four S's $(r_3 = 4)$, and two P's $(r_4 = 2)$. The number of different arrangements is therefore

$$\binom{11}{1\ 4\ 4\ 2} = \frac{11!}{1!4!4!2!} = 34{,}650.$$

The Multinomial Theorem.

If n is a positive integer and $t_1, t_2, \ldots, t_k$ are real numbers, then

$$(t_1 + t_2 + \ldots + t_k)^n = \sum\sum\ldots\sum\binom{n}{r_1 r_2 \ldots r_k} t_1^{r_1} t_2^{r_2} \ldots t_k^{r_k}, \tag{2.1.12}$$

where the summation extends over all non-negative integers

$r_1, r_2, \ldots, r_k$ with $r_1 + r_2 + \ldots + r_k = n$.

Example 2.1.5. Obtain the multinomial expansion of $(a+b+c)^3$.

Solution. The result may be obtained from (2.1.12) with $n = k = 3$.

r_1	r_2	r_3	$\binom{n}{r_1 r_2 r_3}$	r_1	r_2	r_3	$\binom{n}{r_1 r_2 r_3}$
3	0	0	1	1	2	0	3
0	3	0	1	0	2	1	3
0	0	3	1	1	0	2	3
2	1	0	3	0	1	2	3
2	0	1	3	1	1	1	6

$$(a+b+c)^3 = a^3 + b^3 + c^3 + 3(a^2b + a^2c + ab^2 + b^2c + ac^2 + bc^2) + 6abc.$$

Example 2.1.6. Find the coefficient of $x^5y^2z^3$ in the multinomial expansion of $(x+y+z)^{10}$.

Solution. The coefficient of $x^5y^2z^3$ will be

$$\binom{10}{5\ 2\ 3} = \frac{10!}{5!2!3!} = 2520.$$

Generalized Factorials: The Gamma Function.

The gamma function Γ is defined for real $n > 0$ as follows:

$$\Gamma(n) = \int_0^\infty x^{n-1} e^{-x}\, dx. \tag{2.1.13}$$

The improper integral converges for $n > 0$. This function has important applications not only in probability and statistics, but in other areas of applied mathematics as well, and it has been extensively studied and tabulated.

If $n > 1$ we may integrate by parts in (2.1.13) to obtain

$$\int_0^\infty x^{n-1} e^{-x}\, dx = \Big[-e^{-x} x^{n-1}\Big]_0^\infty + (n-1)\int_0^\infty x^{n-2} e^{-x}\, dx$$

from which it follows that, for $n > 1$,

$$\Gamma(n) = (n-1)\Gamma(n-1). \tag{2.1.14}$$

As a result, it is necessary to tabulate the gamma function only for values of n in some interval of length 1, such as $1 \le n \le 2$. One may then obtain $\Gamma(n)$ for any value of n by repeatedly applying (2.1.14).

If n is a positive integer, then repeated application of (2.1.14) gives

$$\Gamma(n+1) = n\Gamma(n) = n(n-1)\Gamma(n-1) = \ldots = n!\Gamma(1).$$

But note that

$$\Gamma(1) = \int_0^\infty e^{-x}\,dx = \left[-e^{-x}\right]_0^\infty = 1.$$

It follows that, for n a positive integer,

$$\Gamma(n+1) = n!. \qquad (2.1.15)$$

Because of this result, the gamma function is sometimes called the <u>generalized factorial</u>. Also, factorial notation is sometimes used when n is fractional; e.g. $(1.5)!$ could be used to denote $\Gamma(2.5)$.

<u>Stirling's Formula</u>.

If n is large, the following approximation is often useful:

$$n! \sim \sqrt{2\pi n}\; n^n\, e^{-n}. \qquad (2.1.16)$$

This may also be used to approximate $\Gamma(n+1)$ for large real n.

The symbol "$\sim$" in (2.1.16) implies that the ratio

$$r_n = \frac{\sqrt{2\pi n}\; n^n\, e^{-n}}{n!}$$

tends to one as $n \to \infty$. We find, for example, that

$$r_{10} = \frac{3,598,696}{3,628,800} = 0.9917$$

so that Stirling's formula is in error by less than 1% for n as small as 10. A similar calculation shows that the error is less than 0.5% for $n = 20$.

<u>Example 2.1.7</u>. The probability of obtaining n heads and n tails in $2n$ tosses of a balanced coin is given by

$$f(n) = \binom{2n}{n} 2^{-2n}.$$

For n large, (2.1.8) and (2.1.16) give

$$\binom{2n}{n}2^{-2n} = \frac{(2n)!}{n!n!}2^{-2n} \sim \frac{\sqrt{4\pi n}\,(2n)^{2n}\,e^{-2n}}{\{\sqrt{2\pi n}\,n^n\,e^{-n}\}^2}\,2^{-2n} = \frac{1}{\sqrt{\pi n}}\,.$$

Hence $f(n) \to 0$ as $n \to \infty$.

<u>Problems for Section 2.1</u>

†1. Evaluate the following:

$$0^{(5)}, 5^{(0)}, 6^{(3)}, (0.4)^{(3)}, (-2)^{(4)}, 7!, \binom{7}{3}, \binom{-7}{3}, \binom{7}{-3}.$$

†2. Evaluate the following:

$$\binom{10}{7\ 3}, \binom{12}{4\ 4\ 4}, \binom{14}{2\ 3\ 5\ 4}, \binom{-1.5}{4}.$$

3. Using tables of the gamma function, evaluate the following:

$$\Gamma(1.5),\ \Gamma(0.5),\ \Gamma(3.6),\ \Gamma(5.2).$$

4. Prove identities (2.1.5), (2.1.6), and (2.1.7).

5. Prove that, if n is a positive integer, then

(a) $\binom{n}{0} - \binom{n}{1} + \binom{n}{2} - \binom{n}{3} + - \ldots = 0;$

(b) $\binom{n}{1} - 2\binom{n}{2} + 3\binom{n}{3} - 4\binom{n}{4} + - \ldots = 0$ for $n \geq 2;$

(c) $\binom{2n}{n} = (-1)^n 2^{2n}\binom{-1/2}{n};$

(d) $\binom{-a}{n} = (-1)^n\binom{a+n-1}{n}$ for all real $a > 0$.

6. Prove that, for positive integers n and m,

$$\binom{n}{n} + \binom{n+1}{n} + \ldots + \binom{n+m}{n} = \binom{n+m+1}{n+1}.$$

7. Show that

$$\binom{0}{0} - \frac{1}{4}\binom{2}{1} + \frac{1}{16}\binom{4}{2} - \frac{1}{64}\binom{6}{3} + - \ldots = \frac{1}{\sqrt{2}}.$$

†8. How many terms will there be in the multinomial expansion of $(x+y+z+w)^5$? What will be the coefficient of xz^2w^2?

9. Let s be a positive integer. Show that

(a) $\sum r^{(s)}\binom{n}{r}t^r = n^{(s)}t^s(1+t)^{n-s}$;

(b) $\sum r^{(s)}\binom{a}{r}\binom{b}{n-r} = a^{(s)}\binom{a+b-s}{n-s}$.

10. Show that

(a) $1^2\binom{n}{1} + 2^2\binom{n}{2} + \ldots + n^2\binom{n}{n} = n(n+1)2^{n-2}$;

(b) $1^2\binom{n}{1}^2 + 2^2\binom{n}{2}^2 + \ldots + n^2\binom{n}{n}^2 = n^2\binom{2n-2}{n-1}$.

2.2 Random Sampling Without Replacement.

Some experiments involve selecting r items, called a *sample of size* r, from a set of n items called the *population*. The process of selecting the sample from the population is called *sampling*. For example, in trying to determine the unemployment rate in the country, it would be too expensive and time-consuming to interview each of the n members of the labour force. Instead, a sample of r workers would be examined, and estimates for the entire population would be based upon the sample results. The sampling design - the method by which the sample is selected - would need to be carefully thought out so that all regions, age groups and occupations were properly represented in the sample. If the sample is properly selected, one can use probability theory to calculate the probable accuracy of the estimates made. These calculations are useful in determining how many people to include in the sample to achieve the desired degree of accuracy. There is a substantial branch of Statistics which deals with the proper design and analysis of such sample surveys.

In this book we consider only *simple random sampling*, in which the method of sampling is such that all possible samples of size r have the same probability of being selected. This section and the next one consider the case where repetitions are not allowed in the sample. In Section 2.4 we consider sampling with replacement, in which the same member of the population can be selected repeatedly.

Sampling without replacement.

In sampling without replacement, the sample is chosen in such a way that no member of the population can be selected more than once;

that is, all r items in the sample must be different. If the order of selection is considered, the first item can be chosen in n ways, the second in $n-1$ ways (it must be different from the first), the third item in $n-2$ ways (it must be different from the first two items chosen), and so on. The rth item must be different from the first $r-1$, and can therefore be chosen in $n-(r-1)$ ways. The number of different ordered samples of size r is thus

$$n(n-1)(n-2)\dots(n-r+1) = n^{(r)}.$$

The sample is said to have been chosen <u>at random</u>, and is called a <u>random sample</u>, if each of the $n^{(r)}$ possible ordered samples has the same probability $1/n^{(r)}$ of being selected.

A set of r different items can be permuted in $r!$ ways to give $r!$ different ordered samples of size r. If each ordered sample has probability $1/n^{(r)}$, then each unordered sample of size r has probability

$$r!/n^{(r)} = 1/\binom{n}{r}.$$

Hence if a sample of size r is chosen at random without replacement from a set of n items, the $\binom{n}{r}$ unordered samples (subsets) of size r are equally probable. In problems involving random sampling without replacement, either ordered or unordered samples may be considered, whichever is more convenient.

In many card games, a hand consisting of r cards (the sample) is dealt from a deck of n cards (the population). A deck of n cards is called <u>well-shuffled</u> if the $n!$ possible arrangements of the cards are equally probable. Suppose that r cards are dealt from a well-shuffled deck. The number of arrangements of the deck in which these particular cards occupy r specified positions in the deck in the correct order is $(n-r)!$, the number of ways of permuting the remaining $n-r$ cards. Hence the probability of dealing a particular set of r cards in a particular order is

$$(n-r)!/n! = 1/n^{(r)}.$$

Each ordered set of r cards has probability $1/n^{(r)}$, and each unordered set will have probability $1/\binom{n}{r}$. Hence a hand of cards dealt from a well-shuffled deck will be a random sample. Of course, well-shuffled decks of cards are idealizations like perfect coins and dice. However, results derived on this assumption will be adequate for most practical purposes, provided that some care is taken over the shuffling.

Example 2.2.1. A bridge club has 12 members (6 married couples). Four members are randomly selected to form the club executive. Find the probability that

(a) the executive consists of two men and two women;

(b) the members of the executive are all of the same sex;

(c) the executive contains no married couple.

Solution 1. (using unordered samples)

The number of ways to select a sample of size 4 from a population of 12 is $\binom{12}{4} = 495$. Because the executive is selected at random, the 495 possible samples are equally probable. (The sample space contains 495 equally probable points.)

(a) There are 6 men from which two may be selected in $\binom{6}{2} = 15$ different ways. For each choice of the men, two women may be chosen from the 6 in $\binom{6}{2} = 15$ ways. The number of samples containing 2 men and 2 women is therefore $\binom{6}{2}\binom{6}{2} = 225$, and the probability that the executive consists of two men and two women is

$$\binom{6}{2}\binom{6}{2}/\binom{12}{4} = \frac{225}{495} = 0.455.$$

(b) An executive consisting entirely of men can be formed in $\binom{6}{4} = 15$ ways. Alternatively, an executive consisting entirely of women can be formed in $\binom{6}{4} = 15$ ways. The number of samples which consist either of four men or of four women is thus $15 + 15 = 30$, and the probability that the members of the executive are all of the same sex is

$$\left\{\binom{6}{4} + \binom{6}{4}\right\}/\binom{12}{4} = \frac{30}{495} = 0.061.$$

(c) An executive which contains no married couple must have one member from each of four different couples. The four couples to be represented may be chosen in $\binom{6}{4}$ ways. For each of the four couples selected there will be 2 choices - either the man or the woman can be selected. The number of samples containing no married couples is thus $\binom{6}{4}2^4 = 240$, and the probability that the executive contains no married couple is

$$\binom{6}{4}2^4/\binom{12}{4} = \frac{240}{495} = 0.485.$$

Solution 2. (using ordered samples)

The number of ordered samples of size 4 is $12^{(4)} = 11880$, and the sample space is taken to have 11880 equally probable points.

(a) The two positions on the executive which are to be occupied by men can be chosen from the four available in $\binom{4}{2} = 6$ ways. When this has been done, the first man may be selected in 6 ways, the second in 5 ways, the first woman in 6 ways, and the second woman in 5 ways. Hence the number of ordered samples containing 2 men and 2 women is $6 \times 6 \times 5 \times 6 \times 5 = 5400$, and the required probability is $5400/11880 = 0.455$ as before.

(b) An ordered sample of 4 men can be selected in $6^{(4)}$ ways. Alternatively, an ordered sample of 4 women can be selected in $6^{(4)}$ ways. The number of ordered samples with all members of the same sex is $6^{(4)} + 6^{(4)} = 720$, and the required probability is $720/11880 = 0.061$.

(c) The number of ordered samples containing no couple is $12 \times 10 \times 8 \times 6$. The first member may be selected in 12 ways. The second must come from one of the other five couples, and may be selected in 10 ways. The third must come from one of the four remaining couples and may be selected in 8 ways. Similarly, the fourth member may be chosen in 6 ways. The required probability is

$$\frac{12 \times 10 \times 8 \times 6}{11880} = 0.485.$$

Example 2.2.2. Poker Hand Probabilities.

A poker hand consists of 5 cards dealt from a standard deck of 52 cards. The following are the nine types of poker hands arranged in order of decreasing value:

(1) Straight flush - 5 cards in one suit, consecutive denominations
(2) Four of a kind - 4 cards of one denomination
(3) Full house - 3 cards of one denomination, 2 of a second denomination
(4) Flush - 5 cards in one suit
(5) Straight - 5 cards of consecutive denominations
(6) Three of a kind - 3 cards of one denomination
(7) Two pairs - 2 cards from each of two denominations
(8) One pair - 2 cards of one denomination
(9) None of the above

A hand is counted in the highest category to which it belongs. For example, the hand consisting of the 2,3,4,5, and 6 of hearts is counted as a straight flush (1), but not as a flush (4), nor as a straight (5). Aces may be played either at the bottom of a straight (A-2-3-4-5) or at the top (10-J-Q-K-A).

Poker players take into account the rarities of the various poker hands, as well as their knowledge of the betting habits of other players, in making their bets. We shall calculate the probabilities of the various poker hands on the assumption that the cards are dealt from a well-shuffled deck. Each of the $\binom{52}{5}$ unordered samples of 5 cards then has the same probability. The probability of a particular type of poker hand is obtained by counting the number of (unordered) hands of that type and dividing by $\binom{52}{5}$. The calculations are summarized in Table 2.2.1.

Table 2.2.1

Poker Hand Probabilities

Type	Number of Hands	Probability
(1)	$4(10) = 40$	0.0000154
(2)	$13(48) = 624$	0.000240
(3)	$13\binom{4}{3}12\binom{4}{2} = 3744$	0.00144
(4)	$4\binom{13}{5} - 40 = 5108$	0.00197
(5)	$10(4)^5 - 40 = 10,200$	0.00392
(6)	$13\binom{4}{3}\binom{12}{2}4^2 = 54,912$	0.02113
(7)	$\binom{13}{2}\binom{4}{2}^2 44 = 123,554$	0.04754
(8)	$13\binom{4}{2}\binom{12}{3}4^3 = 1,098,240$	0.42257
(9)	Difference = 1,302,540	0.50118
TOTAL	$\binom{52}{5} = 2,598,960$	1.00001

(1) There are 4 choices for the suit, and then ten choices (A,2,...,or 10) for the lowest denomination in the straight.

(2) There are 13 choices for the denomination of the four, and then 48 choices for the fifth card.

(3) There are 13 choices for the denomination of the three, and then $\binom{4}{3}$ choices for the three cards. Now the denomination of the pair may be chosen in 12 ways, and the two cards in $\binom{4}{2}$ ways.

(4) There are 4 choices for the suit, and then $\binom{13}{5}$ choices for the five cards in that suit. We must subtract 40 for the

straight flushes which are classified under (1).

(5) There are 10 choices for the smallest denomination of the straight, and 4 choices for the suit of each of the five cards. We must subtract the 40 straight flushes.

(6) There are 13 choices for the denomination of the three, and then $\binom{4}{3}$ choices for the three cards. The denominations of the remaining two cards are different, and can be selected in $\binom{12}{2}$ ways. There are now 4 choices for the suit of each of these two cards.

(7) There are $\binom{13}{2}$ choices for the denominations of the two pairs, and $\binom{4}{2}$ choices for the suits of the cards in each pair. The fifth card may be any one of the 44 in the other eleven denominations.

(8) The denomination of the pair may be selected in 13 ways, and then the cards of the pair in $\binom{4}{2}$ ways. The denominations of the other three cards are all different, and can be chosen in $\binom{12}{3}$ ways. Then there are $4 \times 4 \times 4$ choices for the suits.

(9) The number of hands of this type is obtained by subtracting the total number of hands of the above eight types from $\binom{52}{5}$.

Problems for Section 2.2

†1. The digits 1,2,...,7 are arranged randomly to form a 7-digit number. Find the probability that
   (a) the number is divisible by 2;
   (b) the number is divisible by 4;
   (c) digits 1,2,3 appear consecutively in the proper order;
   (d) digits 1,2,3 appear in the proper order but not consecutively.

2. If four cards are dealt from a well-shuffled deck of 52 cards, what is the probability that there will be one card from each suit?

†3. A box contains 100 lightbulbs of which 10 are defective. Two bulbs are selected at random without replacement. What is the probability that both are defective? that at least one of them is defective?

4. The letters of the word MISSISSIPPI are arranged at random in a row. What is the probability that they spell "MISSISSIPPI"?

5. Fifty-two people stand in line. Each (in order from the left) draws one card from a standard well-shuffled deck and keeps it. The ace of spades is the winning card. Is there any advantage to being first in line?

6. While dressing in the dark, I select two socks at random from a drawer containing five different pairs. What is the probability that my socks will match?

†7. While dressing in the dark, I select $2r$ socks at random from a drawer containing n different pairs. What is the probability that at least one pair will be chosen?

8. A club has 10 men and 10 women members. There are five married couples and ten singles. A committee of four people is formed at random. What is the probability that it contains
   (a) a married man, a single man, a married woman, and a single woman?
   (b) a married couple, a single man, and a single woman?

†9. The numbers $1,2,\ldots,10$ are written on ten cards and placed in a hat. Cards are then drawn one by one without replacement. Find the probabilities of the following events:

   A - "exactly three even numbers are obtained in the first five draws";

   B - "exactly five draws are required to get three even numbers" (i.e. there are two even numbers in the first four draws, followed by an even number on the fifth draw);

   C - "number 7 occurs on the 4th draw";

   D - "the largest number obtained in the first three draws is 5".

10. From a set of $2n + 1$ consecutively numbered tickets, three are drawn at random without replacement. What is the probability that their numbers are in arithmetic progression?

<u>Draw poker</u>: problems 11-13. After seeing his poker hand, a player may attempt to improve it by discarding 0,1,2, or 3 cards and then drawing replacements from the unused portion of the deck.

11. Initially, the poker hand contains two queens and three cards of different denominations. If the player keeps the queens but draws replacements for the other three cards, what is the probability that he will obtain
   (a) four of a kind; (b) a full house; (c) three of a kind;
   (d) two pairs; (e) no improvement?

12. Repeat problem 11 under the assumption that, in addition to the queens, the player keeps a "kicker" (a third card - usually a king or an ace).

†13. (a) Initially the hand contains cards of denominations 3,4,5,6,9 in various suits. A replacement is drawn for the 9 (drawing to an outside straight). What is the probability of completing the straight?

(b) Initially the hand contains cards of denominations 3,4,5,7,9 in various suits. A replacement is drawn for the 9 (drawing to an inside straight). What is the probability of completing the straight?

*14. In some types of poker, a four-card straight (four cards of consecutive denominations) beats a pair, a four-card flush (four cards in the same suit) beats a 4-card straight, and two pairs beats a four-card flush. Calculate the probabilities of a four-card flush, a four-card straight, one pair, and "nothing" in such a game. (Note that the ranking of hands does not match their probabilities.)

2.3 The Hypergeometric Distribution

Suppose that a sample of size n is drawn at random without replacement from an N-member population. Let us further suppose that the population contains just two different types of members - for example, men and women, or employed persons and unemployed persons, or good lightbulbs and defective lightbulbs. What is the probability that in the sample of size n we shall obtain x items of the first type and $n-x$ of the second type?

To make the discussion specific, let us think of the N members of the population as N balls in an urn (jar), there being a white balls and b black balls, where $a+b=N$. The balls are thoroughly mixed, and n balls are drawn. The selection is done without replacement, so that once a ball has been drawn it is kept out of the urn and cannot be drawn again. What is the probability that there will be x white balls and $n-x$ black balls in the sample?

Because we are interested only in the numbers of black and white balls obtained and not in the order in which they are obtained, we consider unordered samples. There are $\binom{a+b}{n}$ different samples, and they are equally probable because of the thorough mixing. Since x white balls may be chosen in $\binom{a}{x}$ ways and $n-x$ black balls in $\binom{b}{n-x}$ ways, there are $\binom{a}{x}\binom{b}{n-x}$ different samples containing x white and $n-x$ black balls. If we let $f(x)$ denote the probability of x white and $n-x$ black balls in the sample, then

$$f(x) = \binom{a}{x}\binom{b}{n-x}/\binom{a+b}{n}, \quad x = 0,1,2,\ldots \quad . \tag{2.3.1}$$

Note that, by the hypergeometric identity (2.1.10),

$$\sum_{x=0}^{\infty} f(x) = \sum_{x=0}^{\infty} \binom{a}{x}\binom{b}{n-x}/\binom{a+b}{n} = \binom{a+b}{n}/\binom{a+b}{n} = 1,$$

so that (2.3.1) distributes the total probability 1 over the non-negative integers. It is called the hypergeometric distribution. Note that $f(x) = 0$ for $x > a$ or $x > n$, so that there are only finitely many values with non-zero probability.

It is sometimes necessary to calculate $f(x)$ for several consecutive values of x. A convenient method of doing this is to compute $f(x)$ directly from (1) for the smallest x required, and then use the recursive formula

$$f(x) = r(x)f(x-1) \tag{2.3.2}$$

where $r(x)$ is the ratio of successive terms in (2.3.1):

$$r(x) = \frac{f(x)}{f(x-1)} = \frac{(a-x+1)(n-x+1)}{x(b-n+x)}\ . \tag{2.3.3}$$

This procedure is illustrated in Example 2.3.1 below. Approximations to $f(x)$ for $a+b$ large will be considered in Sections 2.5 and 6.8.

Example 2.3.1. In contract bridge, four hands of 13 cards each are dealt from a standard deck to four players sitting around a table. The players facing East and West form one partnership, and those facing North and South form another partnership.

(a) What is the probability that a bridge hand contains exactly x hearts?

(b) If my hand contains exactly 5 hearts, what is the probability that my partner's hand contains at least two hearts?

Solution. (a) This is an application of the hypergeometric distribution with $a = 13$, $b = 39$, $N = a+b = 52$, and $n = 13$. The 13 hearts in the deck are the white balls, and the other 39 cards are the black balls. Dealing a bridge hand from a well-shuffled deck corresponds to randomly drawing $n = 13$ balls without replacement. The probability that a bridge hand contains x hearts and $13-x$ other cards is thus

$$f(x) = \binom{13}{x}\binom{39}{13-x}\Big/\binom{52}{13}, \quad x = 0,1,2,\ldots,13.$$

These probabilities may be evaluated using the recursive method described above. First we evaluate

$$f(0) = \binom{39}{13}\Big/\binom{52}{13} = 39^{(13)}/52^{(13)} = 0.01279.$$

In this case the ratio of successive terms is

$$r(x) = \frac{(14-x)^2}{x(26+x)}$$

from (2.3.3). We now obtain

$$f(1) = r(1)f(0) = 13^2(0.01279)/27 = 0.08006;$$

$$f(2) = r(2)f(1) = 12^2(0.08006)/56 = 0.20587;$$

$$f(3) = r(3)f(2) = 11^2(0.20587)/87 = 0.28633;$$

and so on. The results are summarized in Table 2.3.1.

Table 2.3.1

Probability of x hearts in a bridge hand

x	f(x)	x	f(x)	x	f(x)
0	0.01279	5	0.12469	9	9.26×10^{-5}
1	0.08006	6	0.04156	10	4.12×10^{-6}
2	0.20587	7	0.00882	11	9.10×10^{-8}
3	0.28633	8	0.00117	12	7.98×10^{-10}
4	0.23861			13	1.57×10^{-12}

(b) Given that my hand contains 5 hearts and 8 other cards, the remainder of the deck contains 8 hearts and 31 other cards from which my partner's hand is selected at random without replacement. We therefore have an application of the hypergeometric distribution with $a = 8$, $b = 31$, $N = a + b = 39$, and $n = 13$. The probability that my partner's hand contains y hearts and $13 - y$ non-hearts is then

$$f(y) = \binom{8}{y}\binom{31}{13-y}/\binom{39}{13}, \quad y = 0,1,2,\ldots,8.$$

We find that

$$f(0) = \binom{31}{13}/\binom{39}{13} = 31^{(13)}/39^{(13)} = 0.0254;$$

$$f(1) = r(1)f(0) = \frac{8(13)}{19}(0.0254) = 0.1390.$$

The probability that my partner's hand contains at most one heart is therefore $f(0) + f(1) = 0.1644$, and the probability that he has at least two hearts is $1 - 0.1644 = 0.8356$. Hence if I have five hearts in my hand, the odds are 0.8356 to 0.1644, or better than 5 to 1, that my partner and I between us have the majority of the hearts.

Example 2.3.2. Capture-recapture methods.

The following procedure is sometimes used to estimate the size of an animal population - for example, the number of fish in a lake. First some fish are caught, marked or tagged to permit future identification, and returned to the lake. The lake then contains a tagged fish and b untagged fish, where a is known and b is unknown. A second sample of n fish is then taken. Assuming that this is a random sample, the probability that it contains x tagged and $n-x$ untagged fish will be given by the hypergeometric distribution (2.3.1). A reasonable estimate of the total number of fish in the lake is then na/x. It is possible to determine the probable accuracy of this estimate using the hypergeometric distribution.

The assumption that the second sample of fish is random may not be a very good one in practical situations. The fact that a fish has been caught and tagged may make it more (or less) likely to be caught again, and if this is the case the $\binom{a+b}{n}$ possible samples will not be equally probable. The hypergeometric distribution will then be inappropriate. Further difficulties may arise through movement of the fish into or out of the lake between the times that the two samples are taken.

Inspection Sampling in Industry

No manufacturing process is perfect; defective times are bound to occur. Many companies inspect the items they produce to ensure that proper quality is maintained, and that not too many defectives are passed on to their customers. It may not be possible to test every item produced - testing may be too expensive, or it may involve destruction of the product. In such cases a sampling inspection procedure can be used.

Suppose that items are either defective or good, and that they arrive for inspection in batches of size N. From each batch, a sample of size n is randomly selected for inspection. Suppose that a sample is found to contain x defectives. If x is large, it is likely that the batch from which it came contains a large number of defectives, and hence the batch should be rejected or subjected to further testing. If x is small, it is likely that the batch contains only a small number of defectives, and it should be accepted. This suggests a rule of the form "accept the batch if $x \le c$; otherwise reject it (or do additional testing)".

In using such a rule, there are two different types of error which can occur:

I: a batch containing a satisfactorily small number of defectives may be rejected unnecessarily;

II: a batch containing too many defectives may be accepted.

If the batch contains a defectives and b good items where $N = a + b$, then the probability that the sample contains x defectives and $n - x$ good items will be given by the hypergeometric distribution (2.3.1). From this, the probabilities of the two types of error may be computed as functions of a, n, and c. Each type of error will have an associated cost, as will the inspection procedure itself. These costs and probabilities can be combined to give the total expected cost of the inspection scheme, which we would then try to minimize by choice of n and c. (This can be done using Bayes's Theorem provided that we also know the relative frequencies with which the various values of a will occur in the sequence of inspections to be made.)

Example 2.3.3. Suppose that the batch size is $N = 50$, and that random samples of size $n = 10$ are inspected. The batch is accepted if the sample contains at most one defective; otherwise the batch is rejected. Determine the probability of accepting the batch as a function of a, the number of defectives it contains.

Solution. The probability that a random sample of size 10 contains exactly x defectives is

$$f(x) = \binom{a}{x}\binom{50-a}{10-x}/\binom{50}{10}, \qquad x = 0,1,2,\ldots .$$

The batch is accepted if $x = 0$ or $x = 1$. Hence the probability of accepting the batch is

$$f(0) + f(1) = \{\binom{50-a}{10} + \binom{a}{1}\binom{50-a}{9}\}/\binom{50}{10} = (41 + 9a)(50 - a)^{(9)}/50^{(10)}.$$

The probability of accepting the batch is 1 for $a = 0$ or 1, and may be determined recursively for larger values of a. Table 2.3.2 gives the probability of acceptance for several values of a.

Table 2.3.2

Acceptance probability as a function of a

a	4	8	12	16	20	24
P (accept batch)	0.826	0.491	0.236	0.094	0.031	0.008

Note that there is a fairly large probability (0.236) of accepting a batch with as many as 12 defective items. If a type II error probability of this magnitude were unacceptable, it would be necessary either to increase n or else to decrease c from 1 to 0. Increasing n would increase the cost of inspection, while decreasing c would increase the probability of a type I error. In order to determine the optimal inspection procedure it would be necessary to know the costs involved as well as the frequencies with which the various values of a were liable to arise.

Problems for Section 2.3

1. A batch of 20 transformers contains three defectives and seventeen non-defectives. If seven transformers are drawn at random without replacement, what is the probability of obtaining exactly two defectives? at least two defectives?
2. A manufacturer purchases transformers in batches of 20. Five transformers are selected at random from each batch and are tested. If a defective is found, the batch is returned to the factory, and otherwise it is accepted. Compute the probability of acceptance as a function of d, the number of defectives in the batch.

†3. Transformers are purchased in batches of 20, and a random sample of size 5 is inspected. The batch is accepted if there are no defectives, and sent back if there are two or more. If there is one defective in the sample, five more transformers are chosen and inspected. The batch is then accepted if the second sample contains no defectives, and is sent back otherwise. Compute the probability of acceptance as a function of d, the number of defectives in the batch.

4. A box contains d defective items and $25-d$ good items, where d is not known. Ten items are selected at random without replacement and are examined. None of them is defective. Calculate the probability of observing no defectives in ten for $d=0,1,2,3$, and 4. Do you think it likely that the box contains as many as 4 defectives?
5. During the past year, a doctor has successfully treated 20 patients for a certain skin condition. Of these, 9 received one drug (A), and 11 received a second drug (B). Unpleasant side effects were reported in a total of 8 cases.
   (a) If the drugs are equally likely to produce side effects, what is the probability that x of those reporting side effects

had received drug A? Calculate the probabilities for $x = 0, 1, \ldots, 8$.

(b) Only two of the patients reporting side effects had received drug A while six had received drug B. Should this be considered as proof of the first drug's superiority?

†6. A bridge hand of 13 cards is dealt from a well-shuffled standard deck. The bidding of the hand depends on the high cards held and the distribution of the cards over the four suits. Determine the probabilities of the following:

(a) 6 spades, 4 hearts, 2 diamonds, and 1 club;

(b) 6 of one suit, 4 of another, 2 of another, and 1 of the remaining suit (6-4-2-1 distribution);

(c) 6-3-2-2 distribution;

(d) 7-2-2-2 distribution.

7. Show that the hypergeometric probability (2.3.1) is greatest when x equals the integer part of $(a+1)(n+1) \div (a+b+2)$.
Hint: determine the condition under which $r(x) \geq 1$.

†8. A box contains 2 tulip bulbs, 3 crocus bulbs, and 4 daffodil bulbs. Three bulbs are picked at random and planted. What is the probability that there will be at least 1 tulip and at least 1 crocus?

9. There are six bad eggs in a batch of 120. The eggs are sent to ten different customers in cartons of a dozen. Find the probability that (a) one particular customer receives two or more bad eggs;

(b) no customer receives more than one bad egg;

(c) some customer receives two or more bad eggs;

(d) the first seven customers receive only good eggs.

†10. Four hands of bridge are dealt from a well-shuffled deck. What is the probability that North, South, East, and West get 6,4,2, and 1 spades, respectively? Compare with problem 6(a).

2.4 Random Sampling With Replacement

Suppose that a sample of size r is to be chosen from a population with n members. In sampling *without* replacement, no population member may be chosen more than once, so that the r items in the sample will all be different. In sampling *with* replacement, a member of the population may be chosen more than once, so that not all of the r items in the sample need be different. Indeed it is possible that the same item might be chosen every time, in which case the sample would consist of a single item repeated r times.

In problems involving sampling with replacement, it is usually more convenient to work with ordered sets (sequences) of outcomes rather than unordered sets. The first item in the sample can be selected in n ways. Because repetition is allowed, the second item can also be chosen in n ways, given n^2 choices for the first two. There will also be n choices for the 3rd item, giving n^3 choices for the first three. In general, there are n^r possible ordered samples of size r when the sampling is done with replacement. The sample is said to have been chosen <u>at random</u> and is called a <u>random sample</u> if the n^r possible ordered sequences are equally probable.

We may represent the population by an urn containing n balls numbered $1,2,\ldots,n$, and select a sample of size r by drawing r balls one at a time from the urn. If, after drawing a ball, we return it to the urn before making the next draw, we have sampling with replacement. If we do not return the balls drawn, we have sampling without replacement. To obtain a random sample, we thoroughly mix the balls in the urn before each draw.

In Section 2.2 we found that the number of ordered samples of size r such that all items of the sample are distinct is

$$n^{(r)} = n(n-1)(n-2)\ldots(n-r+1).$$

Hence if a sample of size r is drawn at random with replacement, the probability that all items of the sample are different is given by

$$q_r = \frac{n^{(r)}}{n^r} = \frac{n!}{(n-r)!n^r}. \qquad (2.4.1)$$

Note that $q_r = 0$ for $r > n$; in this case the sample must have a repeated member.

When n and $n-r$ are large, both of the factorials on the right hand side of (2.4.1) may be approximated by Stirling's formula (2.1.16) to give

$$q_r \sim \left(\frac{n}{n-r}\right)^{n-r+.5} e^{-r}. \qquad (2.4.2)$$

Keeping r fixed and letting $n \to \infty$, we see that

$$\left(\frac{n}{n-r}\right)^{n-r+.5} = \sqrt{\frac{n}{n-r}}\left(1 + \frac{r}{n-r}\right)^{n-r} \to 1\cdot e^r,$$

and hence $q_r \to 1$ as $n \to \infty$. Thus if r items are selected at random with replacement from a very large population, it is almost certain that they will all be different. This means that if the population size is much greater than the sample size, it makes very little difference whether the sampling is done with or without replacement. This finding will be used to justify an approximation to the hypergeometric distribution in the next section.

Example 2.4.1. The Birthday Problem

The birthdays of r people may be thought of as a sample drawn with replacement from the population of all the days in the year. As a reasonable first approximation, let us ignore leap years and consider a year to consist of 365 days. Furthermore, let us suppose that the 365^r possible sequences of r birthdays are equally probable. Then from (2.4.1), the probability that all r birthdays are different is given by

$$q_r = \frac{365^{(r)}}{365^r}.$$

This may be evaluated either directly or by using the approximation (2.4.2), with the following result:

r	10	15	20	25	30	35	40
q_r	0.883	0.747	0.589	0.431	0.294	0.186	0.109

For instance, at a party of 30 people, the probability that at least two people have the same birthday is $1 - 0.294 = 0.706$. In a party of forty people, the probability that at least two have the same birthday is $1 - 0.109 = 0.891$. These probabilities are somewhat higher than most people would intuitively expect them to be. The amount of clustering or grouping which one finds in random events is often surprisingly large. □

In random sampling with replacement, some members of the population may occur more than once in the sample. Suppose that the ith member of the population occurs r_i times in the sample $(i = 1,2,\ldots,n)$, where $\sum r_i = r$. From Section 2.1, the number of ordered sequences of r items of which r_1 are of the first kind, r_2 are of the second kind,..., and r_n are of the nth kind is

$$\binom{r}{r_1\, r_2 \ldots r_n} = \frac{r!}{r_1!r_2!\ldots r_n!}.$$

Since each ordered sequence has probability n^{-r}, the probability of such a sample is

$$P(r_1,r_2,\ldots,r_n) = \frac{r!}{r_1!r_2!\ldots r_n!} n^{-r}. \qquad (2.4.3)$$

The probability of a sample consisting of the first member of the population repeated r times is

$$P(r,0,\ldots,0) = \frac{r!}{r!0!\ldots 0!} n^{-r} = n^{-r}.$$

For $r < n$, the probability of a sample consisting of the first r members of the population in any order is

$$P(1,\ldots,1,0,\ldots,0) = \frac{r!}{(1!)^r(0!)^{n-r}} n^{-r} = r!n^{-r}.$$

In general, for random sampling with replacement, the probability of an unordered sample depends upon the frequencies with which repeated elements occur in it. In contrast, all unordered samples are equally probable under random sampling without replacement.

Occupancy Problems

Suppose that r identical balls are to be distributed at random into n boxes, there being no restriction on the number of balls in a box. This could be done by placing n slips of paper numbered $1,2,\ldots,n$ in a hat and then drawing r at random with replacement to give an ordered sequence of r box numbers for the r balls. There are n^r possible arrangements which we assume to be equally probable. The numbers $r_1,r_2,\ldots,r_n$ defined in the preceding paragraph now represent the numbers of balls placed in the 1st,2nd,...,nth boxes, and are called the <u>occupancy numbers</u>. Expression (2.4.3) above gives the probability of obtaining occupancy numbers $r_1,r_2,\ldots,r_n$. Many problems can be rephrased in terms of occupancy numbers. The birthday problem (Example 2.4.1) involved distributing r people into 365 boxes (possible birthdays), and we computed the probability q_r that none of the occupancy numbers exceeded one. Rolling r balanced dice is equivalent to randomly distributing r balls into 6 boxes, and r_i represents the number of times that the ith face comes up.

In some occupancy problems, the boxes as well as the balls are considered indistinguishable. For instance, in the birthday problem we are interested in multiple birthdays $(r_i > 1)$, but we do not care on which days of the year (i.e. in which boxes) these occur. In such cases, we need to find the probability that a set of occupancy numbers $\{r_i\}$ will occur in any order. To obtain this probability, we multiply (2.4.3) by the number of different arrangements of the n occupancy numbers $\{r_i\}$, taking due account of any repetitions among the r_i's. This procedure is illustrated in Examples 2.4.3 and 2.4.4 below.

Example 2.4.2. If twelve balanced dice are rolled, what is the probability that each face comes up twice?

Solution. Rolling twelve balanced dice is equivalent to an occupancy problem in which $r = 12$ balls are distributed in $n = 6$ boxes. By (2.4.3) the probability of obtaining occupancy numbers $r_1, r_2, \ldots, r_6$ is

$$P(r_1, r_2, \ldots, r_6) = \frac{12!}{r_1!r_2!\ldots r_6!} 6^{-12}$$

and the required probability is

$$P(2,2,\ldots,2) = \frac{12!}{2!2!\ldots 2!} 6^{-12} = 0.0034.$$

More generally, if $6m$ dice are rolled, the probability that each face occurs exactly m times is

$$P(m,m,\ldots,m) = \frac{(6m)!}{(m!)^6} 6^{-6m}.$$

Example 2.4.3. In the birthday problem (Example 2.4.1), what is the probability that exactly two out of the r people have the same birthday, and all the remaining birthdays are different?

Solution. This is an occupancy problem with $n = 365$. We want the probability that one occupancy number is two, $r - 2$ occupancy numbers are equal to one, and the remaining $365 - (r - 1)$ occupancy numbers are zero. By (2.4.3), the probability of obtaining these 365 occupancy numbers in a particular order is

$$P(2,1,\ldots,1,0,\ldots,0) = \frac{r!}{2!(1!)^{r-2}(0!)^{366-r}} 365^{-r} = \frac{r!}{2} 365^{-r}.$$

But there are 365 occupancy numbers of three kinds: one "two", $r - 2$ "ones" and $366-r$ "zeroes". These may be arranged in

$$\frac{365!}{1!(r-2)!(366-r)!}$$

different ways. Hence the required probability is

$$p_r = \frac{365!}{(r-2)!(366-r)!}\,\frac{r!}{2}\,365^{-r} = \frac{r(r-1)}{2(366-r)}\,q_r$$

where q_r is the probability of no multiple birthdays. Using the tabulated values of q_r from Example 2.4.1, we easily obtain the following:

r	10	15	20	25	30	35	40
p_r	0.112	0.223	0.323	0.379	0.380	0.334	0.260

It is easy to show that p_r exceeds q_r for $r \geq 27$. Initially p_r is small because there is a large probability of having no multiple birthdays. For large r there is likely to be more than one pair of identical birthdays, and p_r once again becomes small.

Example 2.4.4. List the possible outcomes and their probabilities when eight identical balls are distributed at random into six indistinguishable boxes.

Solution. There are twenty different sets of occupancy numbers $r_1, r_2, \ldots, r_6$ such that $\sum r_i = 8$, and these are listed in the first column of Table 2.4.1. We shall illustrate the computation of their probabilities for the set $\{2,2,2,1,1,0\}$. There are six occupancy numbers: three 2's, two 1's, and one 0. These may be arranged in $6!/3!2!1!$ ways to give ordered sets of occupancy numbers. By (2.4.3), the probability of a particular arrangement of these occupancy numbers is

$$\frac{r!}{r_1!r_2!\ldots r_n!}\,n^{-r} = \frac{8!}{2!2!2!1!1!0!}\,6^{-8} = \frac{8!}{2!2!2!}\,6^{-8}.$$

The probability of obtaining this set of occupancy numbers in any order is then

$$\frac{6!}{3!2!1!}\,\frac{8!}{2!2!2!}\,6^{-8} = 0.180041.$$

Computations are similar in the other nineteen cases, and the details are given in Table 2.4.1. (Terms $0!$ and $1!$ are omitted from Column 3.)

Table 2.4.1

Random Distribution of Eight Balls in Six Boxes

Occupancy numbers	Number of arrangements equals 6! divided by	Probability of one arrangement equals $8!6^{-8}$ divided by	Probability of occupancy numbers in any order
8 0 0 0 0 0	1!5!	8!	0.000004
7 1 0 0 0 0	1!1!4!	7!	0.000143
6 2 0 0 0 0	1!1!4!	6!2!	0.000500
6 1 1 0 0 0	1!2!3!	6!	0.002000
5 3 0 0 0 0	1!1!4!	5!3!	0.001000
5 2 1 0 0 0	1!1!1!3!	5!2!	0.012003
5 1 1 1 0 0	1!3!2!	5!	0.012003
4 4 0 0 0 0	2!4!	4!4!	0.000625
4 3 1 0 0 0	1!1!1!3!	4!3!	0.020005
4 2 2 0 0 0	1!2!3!	4!2!2!	0.015003
4 2 1 1 0 0	1!1!2!2!	4!2!	0.090021
4 1 1 1 1 0	1!4!1!	4!	0.030007
3 3 2 0 0 0	2!1!3!	3!3!2!	0.020005
3 3 1 1 0 0	2!2!2!	3!3!	0.060014
3 2 2 1 0 0	1!2!1!2!	3!2!2!	0.180041
3 2 1 1 1 0	1!1!3!1!	3!2!	0.240055
3 1 1 1 1 1	1!5!	3!	0.024005
2 2 2 2 0 0	4!2!	2!2!2!2!	0.022505
2 2 2 1 1 0	3!2!1!	2!2!2!	0.180041
2 2 1 1 1 1	2!4!	2!2!	0.090021
Total			1.000001

Problems for Section 2.4

1. A three digit number is formed by selecting its three digits at random with replacement from $0,1,\ldots,9$. Calculate the probabilities of the following events:

   A - "all three digits are equal";

   B - "all three digits are different";

   C - "the digits are all nonzero";

   D - "the digits all exceed 4";

   E - "the digits all have the same parity (all odd or all even)".

†2. Repeat problem 1 under the assumption that the first digit must be nonzero.

3. (a) If 22 balls are randomly distributed into 120 boxes, what is the probability that no box contains more than one ball?
   (b) In the 120-day period from November 1968 to February 1969 there were 22 hijackings of commercial airliners to Cuba. What is the probability that on some day there were two or more? The New York Times (Nov. 25, 1968) regarded the occurrence of two hijacks on the same day as a sensational and improbable coincidence. Is this claim justifiable?
4. What is the probability that, of 24 people, two have birthdays in each month? What is the probability that there are four months with one birthday, four with two birthdays, and four with three birthdays?
5. If a book of 300 pages contains 25 misprints, what is the probability that they are all on different pages? What is the probability that three pages contain two misprints each and all others contain zero or one misprint?

†6. Ten cars are parked at random in four large parking lots. What is the probability that
   (a) no cars are placed in the first lot?
   (b) two or more cars are placed in every lot?

2.5 The Binomial Distribution

Suppose that a sample of size n is drawn at random from a population which contains just two types of members. As in Section 2.3, we may represent the population by an urn containing a white balls and b black balls. We saw that when the sampling is done _without replacement_, the probability that the sample contains x white and $n-x$ black balls is given by the hypergeometric distribution (2.3.1). In this section we obtain an expression for this probability when the sampling is done _with replacement_.

Under random sampling with replacement, there are $(a+b)^n$ possible ordered samples of size n, and these are equally probable. Since there are a choices for each white ball and b choices for each black ball, a particular ordered sequence of x white and $n-x$ black balls such as WW...WBB...B may be chosen in $a^x b^{n-x}$ ways. Hence the probability of obtaining a particular sequence of x white and $n-x$ black balls is

$$\frac{a^x b^{n-x}}{(a+b)^n} = p^x(1-p)^{n-x}$$

where $p = \frac{a}{a+b}$ is the proportion of white balls in the urn. But a set of x W's and $n-x$ B's can be permuted in

$$\frac{n!}{x!(n-x)!} = \binom{n}{x}$$

ways to give different ordered sequences. Therefore, the probability of obtaining x white and $n-x$ black balls in any order is given by

$$f(x) = \binom{n}{x}p^x(1-p)^{n-x}, \quad x = 0,1,2,\ldots . \tag{2.5.1}$$

Note that $f(x) = 0$ for $x > n$ by the definition of $\binom{n}{x}$.

By the binomial theorem (2.1.9) we have

$$\sum_{x=0}^{\infty} f(x) = \sum_{x=0}^{\infty} \binom{n}{x}p^x(1-p)^{n-x} = (1-p)^n \sum_{x=0}^{\infty} \binom{n}{x}\left(\frac{p}{1-p}\right)^x$$

$$= (1-p)^n\left(1+\frac{p}{1-p}\right)^n = 1$$

so that (2.5.1) distributes the total probability 1 over the integers $0,1,2,\ldots,n$. It is called the <u>binomial distribution</u>. The binomial distribution plays the same role in sampling with replacement as does the hypergeometric distribution in sampling without replacement.

The binomial probabilities (2.4.1) may be computed recursively using $f(x) = r(x)f(x-1)$ where $r(x)$ is the ratio of successive terms,

$$r(x) = \frac{f(x)}{f(x-1)} = \frac{n-x+1}{x}\,\frac{p}{1-p} = \frac{n-x+1}{x}\,\frac{a}{b} .$$

Alternatively, for fairly small values of n, tables of the binomial distribution can be used. Approximations to the binomial distribution when the index n is large will be considered in Sections 4.3 and 6.8.

<u>Example 2.5.1</u>. An urn contains 20 balls of which 6 are white and 14 are black. Eight balls are drawn at random from the urn. Find the probability that the sample contains exactly x white balls if

(i) sampling is done without replacement;

(ii) sampling is done with replacement.

<u>Solution</u>. For sampling without replacement, the probability of obtaining x white balls is given by the hypergeometric distribution (2.3.1) with $a = 6$, $b = 14$, and $n = 8$:

$$P(x \text{ white balls}) = f_1(x) = \binom{6}{x}\binom{14}{8-x}\Big/\binom{20}{8}; \quad x = 0,1,\ldots .$$

For sampling with replacement, the probability of obtaining x white balls is given by the binomial distribution (2.5.1) with $n = 8$ and $p = 0.3$:

$$P(x \text{ white balls}) = f_2(x) = \binom{8}{x}(.3)^x(.7)^{8-x}; \quad x = 0,1,\ldots \quad .$$

These probabilities have been computed recursively using

$$f_1(x) = \frac{(7-x)(9-x)}{x(x+6)} f_1(x-1); \quad f_2(x) = \frac{6(9-x)}{14x} f_2(x-1),$$

and the results are shown in Table 2.5.1. In each case, the most probable number of white balls is 2, and the second most probable number is 3. However the distributions differ markedly in the tails (for large and small x). Under sampling without replacement, when several balls of one colour have been drawn from the urn the chance of drawing additional balls of that colour is reduced, and so the chance of obtaining a large number of balls of one colour is small. However, in sampling with replacement, the chance of drawing a white ball remains the same no matter how many white balls have previously been drawn and replaced, and hence the tail probabilities are somewhat larger. □

Table 2.5.1

Probability of x White Balls out of 8 for Sampling Without and With Replacement

x	$f_1(x)$	$f_2(x)$	x	$f_1(x)$	$f_2(x)$
0	0.0238	0.0576	5	0.0173	0.0467
1	0.1635	0.1977	6	0.0007	0.0100
2	0.3576	0.2965	7	0	0.0012
3	0.3179	0.2541	8	0	0.0001
4	0.1192	0.1361	Total	1.0000	1.0000

The Binomial Approximation to the Hypergeometric Distribution

In Section 2.4 we argued that when the population size is much greater than the sample size, it makes very little difference whether the sampling is done with or without replacement. It follows that, when $a + b$ is much larger than n, the hypergeometric distribution (2.3.1), which arises from sampling without replacement, will be well approximated by the binomial distribution (2.5.1), which is the corresponding result for sampling with replacement. We are thus led to the following:

Theorem 2.5.1. If n is much smaller than $a+b$, then

$$\binom{a}{x}\binom{b}{n-x}/\binom{a+b}{n} \approx \binom{n}{x}p^x(1-p)^{n-x}; \quad x=0,1,2,\ldots \tag{2.5.2}$$

where $p = \frac{a}{a+b}$.

An algebraic proof of this result is outlined in the exercises at the end of the section.

Example 2.5.2. A city is inhabited by 75000 adults of whom 500 are university professors. In a survey on higher education carried out by the local hotline radio show, 25 people are chosen at random without replacement for questioning. What is the probability that the sample contains at most one professor?

Solution. The probability that the sample contains x professors is given by the hypergeometric distribution (2.3.1) with $a=500$, $b=74500$, and $n=25$, so that

$$f(x) = \binom{500}{x}\binom{74500}{25-x}/\binom{75000}{25}; \quad x=0,1,2,\ldots .$$

The probability of at most one professor in the sample is thus

$$f(0)+f(1) = \left[\binom{74500}{25} + 500\binom{74500}{24}\right]/\binom{75000}{25}.$$

In this case n is much smaller than $a+b$, and the binomial distribution should give an accurate approximation. We have $n=25$ and $p=500/75000 = 1/150$, so that (2.5.2) gives

$$f(x) \approx \binom{25}{x}\left(\frac{1}{150}\right)^x\left(\frac{149}{150}\right)^{25-x}; \quad x=0,1,2,\ldots$$

$$f(0)+f(1) \approx \left(\frac{149}{150}\right)^{25} + \frac{25}{150}\left(\frac{149}{150}\right)^{24} = 0.98796.$$

A tedious computation shows the exact result to be 0.98798.

Example 2.5.3. A candidate obtains 52% of the N votes in an election, where N is very large. What is the probability that he leads in a poll of 100 votes?

Solution. There are $0.52N$ votes in favour of the candidate and $0.48N$ votes against him. Assuming that the 100 voters polled are a random sample without replacement from the population of voters, the probability that the candidate receives x votes will be given by the hypergeometric distribution (2.3.1) with $a = 0.52N$, $b = 0.48N$, and $n = 100$. For $N \gg 100$, we may use the binomial approximation (2.5.2) with $n = 100$ and $p = 0.52$ to obtain

$$P(x \text{ votes for candiate}) = f(x) \approx \binom{100}{x}(.52)^x(.48)^{100-x}; \quad x = 0,1,2,\ldots .$$

The probability that the candidate obtains 51 or more votes out of 100 is then

$$f(51) + f(52) + \ldots + f(100).$$

This sum of binomial probabilities can be evaluated on a computer. However this is unnecessary in the present case because the binomial distribution can itself be approximated by a normal distribution. See Section 6.8 for details.

Problems for Section 2.5

1. (a) Let r and s be integers with $0 \le s \le r$. Show that

$$(r - s)^s \le r^{(s)} \le r^s.$$

   (b) Show that the hypergeometric probability function (2.3.1) may be written as

$$f(x) = \binom{n}{x}a^{(x)}b^{(n-x)}/(a+b)^{(n)}.$$

   (c) Show that

$$\binom{n}{x}\left(p - \frac{x}{a+b}\right)^x\left(q - \frac{n-x}{a+b}\right)^{n-x} \le f(x) \le \binom{n}{x}p^x q^{n-x}\left(1 - \frac{n}{a+b}\right)^{-n}$$

   where $p = \frac{a}{a+b}$ and $q = 1 - p$. For what range of x values does this result hold?

   (d) Prove Theorem 2.5.1.

2. Of 1000 fish in a lake, 50 have previously been caught, tagged, and replaced. Give an expression for $f(x)$, the probability that a catch of 10 fish contains x tagged ones. Compute $f(0), f(1)$, and $f(2)$, and compare these with the values yielded by the binomial approximation.

†3. Of 8000 students at a university, 300 are under the age of 18. A sample of 40 students is selected at random without replacement.
   (a) Give an expression for the probability that
      (i) exactly four students sampled are under 18;
      (ii) fewer than 5% of the students sampled are under 18.
   (b) Use an approximation to evaluate the probabilities in (a).

4. Show that the binomial probability (2.5.1) is largest when x equals the integer part of $(n+1)p$, and that the maximum is unique unless $(n+1)p$ is an integer.

*2.6 The Theory of Runs

We consider random sequences of $r+s$ elements of two kinds, say r alphas and s betas. For instance, the following is a sequence with $r=6$ and $s=3$:

$$\alpha\beta\beta\alpha\alpha\alpha\beta\alpha\alpha$$

There are $r+s$ positions in the sequence, of which r may be chosen for the alphas in $\binom{r+s}{r}$ ways. The remaining positions must then be filled with betas. Each of these sequences has probability $1/\binom{r+s}{r}$.

A run is a subsequence of elements of the same kind which is both preceded and followed by either an element of the opposite kind or by no element. The above sequence contains three alpha runs of lengths 1, 3, and 2, and two beta runs of lengths 2 and 1:

$$\alpha|\beta\beta|\alpha\alpha\alpha|\beta|\alpha\alpha$$

The alpha runs and beta runs alternate, so that the numbers of alpha runs and of beta runs differ by at most one. We shall determine p_k, the probability of a total of k runs of both kinds.

We begin with k even, say $k = 2x$, and count the number of sequences with $2x$ runs. Since alpha and beta runs alternate, there must be x alpha runs and x beta runs. First we divide the r alphas into x runs. This is done by writing the alphas in sequence, $\alpha\alpha\alpha\ldots\alpha$, and choosing $x-1$ division points from the $r-1$ spaces between successive alphas in $\binom{r-1}{x-1}$ ways. Similarly, the s betas may be divided into x runs in $\binom{s-1}{x-1}$ ways. A given set of x alpha runs may be combined with a given set of beta runs in two ways: the first run may be either an alpha run or a beta run, but thenceforth they alternate. Thus the number of sequences containing $2x$ runs is $2\binom{r-1}{x-1}\binom{s-1}{x-1}$, and the probability of a sequence with $2x$ runs

*This section may be omitted on first reading.

is
$$p_{2x} = 2\binom{r-1}{x-1}\binom{s-1}{x-1}/\binom{r+s}{r} \qquad (2.6.1)$$

Now we count the number of sequences with $2x+1$ runs. Either there will be $x+1$ alpha runs and x beta runs, or there will be x alpha runs and $x+1$ beta runs. In the former case, the r alphas may be divided into $x+1$ runs in $\binom{r-1}{x}$ ways; the s betas may be divided into x runs in $\binom{s-1}{x-1}$ ways. A given set of $x+1$ alpha runs can be combined with a given set of x beta runs in only one way, because the sequence must begin with an alpha run. Hence there are $\binom{r-1}{x}\binom{s-1}{x-1}$ sequences in the first case. Similarly, there will be $\binom{r-1}{x-1}\binom{s-1}{x}$ sequences in the second case. The probability of a sequence with $2x+1$ runs is then

$$p_{2x+1} = \{\binom{r-1}{x}\binom{s-1}{x-1} + \binom{r-1}{x-1}\binom{s-1}{x}\}/\binom{r+s}{r}. \qquad (2.6.2)$$

Some applications of this theory will now be described.

(i) Weather. Suppose that the weather is observed over a period of $r+s$ consecutive days, of which r are sunny (S) and s are cloudy (C). The result is a sequence of r S's and s C's. If we assume that all $\binom{r+s}{r}$ possible sequences are equally probable, the probability of k runs of weather will be given by (2.6.1) and (2.6.2). If the observed number of runs were improbably small, there would be reason to doubt the initial assumption of equally probable sequences, and to claim evidence of persistence in the weather. For example, with $r=5$ and $s=9$, (2.6.1) and (2.6.2) give the probability p_k of k runs to be as follows:

k	2	3	4	5	6	7	8	9	10	11	Total
$2002p_k$	2	12	64	160	336	448	448	336	140	56	2002

Suppose that we observed the sequence

SS|CCCC|SSS|CCCCC

which has $k=4$ runs. The probability of 4 runs is $64/2002 = 0.032$, and the probability of 4 or fewer runs is $78/2002 = 0.039$. Because these probabilities are small, observing this sequence might cause one to question the assumption of a random distribution of weather (equally probable sequences).

(ii) Contagion. A row of $r+s$ plants contains r that are diseased and s that are normal. If all arrangements of the plants are assumed to be equally probable, the probability of k runs will be given by (2.6.1) and (2.6.2). If the number of runs actually observed were improbably small, one would have evidence of contagion.

(iii) Comparison of two treatments. Two drugs, A and B, are to be compared with respect to the increases in blood pressure which they cause. There are $r+s$ animals available for use in the experiment. Drug A is given to r of these selected at random, and drug B is given to the remaining s. Denote the observed blood pressure increases for drug A by $\alpha_1, \alpha_2, \ldots, \alpha_r$, and for drug B by $\beta_1, \beta_2, \ldots, \beta_s$. Now arrange all $r+s$ measurements in order of increasing magnitude. (We assume that there are no ties.) For instance, one might obtain

$$\alpha_1 < \alpha_2 < \alpha_3 < \beta_1 < \alpha_4 < \beta_2 < \ldots \quad .$$

If the two drugs are equally effective, all possible sequences will be equally probable, and the probability of k runs of like measurements will be given by (2.6.1) and (2.6.2). If the observed number of runs is improbably small, one has evidence that the drugs are not equally effective.

Problems for Section 2.6

*1. Consider a random sequence of r alphas and s betas. Show that the most probable number of runs is an integer k such that

$$\frac{2rs}{r+s} < k < \frac{2rs}{r+s} + 3.$$

Hint: consider the ratios p_{2x+2}/p_{2x} and p_{2x+1}/p_{2x-1}.

*2.7 Symmetric Random Walks

Two players, traditionally called Peter and Paul, repeatedly play a game. If W denotes a win for Peter and L a loss for Peter (a win for Paul), then the outcome of a series of x games may be represented by a sequence of W's and L's of length x. Let p and q denote the numbers of W's and L's in this sequence. Then $p \geq 0$, $q \geq 0$, and $p+q=x$. After x games, the difference between Peter's score and Paul's score will be $p-q=y$, say. Peter is in the lead after x games if $y>0$, Paul is in the lead if $y<0$, and there

*This section may be omitted at first reading; the material which it contains will not be used in the sequel.

is a tie if $y = 0$. Ties can occur only at even-numbered games.

Since p and q are non-negative integers satisfying

$$p = \frac{x+y}{2}, \quad q = \frac{x-y}{2}, \tag{2.7.1}$$

it follows that x and y are both even or both odd, and that $x \geq y$.

A sequence of x outcomes can be represented geometrically by a polygonal path in the plane beginning at point $(0,0)$ and terminating at point (x,y). Each time Peter wins a game, the path moves diagonally one unit upward and one unit to the right; each time Peter loses (Paul wins) it moves diagonally one unit downward and to the right. Figure 2.7.1 illustrates the path corresponding to the sequence WWLLLWWWWL, for which $p = 6$, $q = 4$, $x = 10$, and $y = 2$.

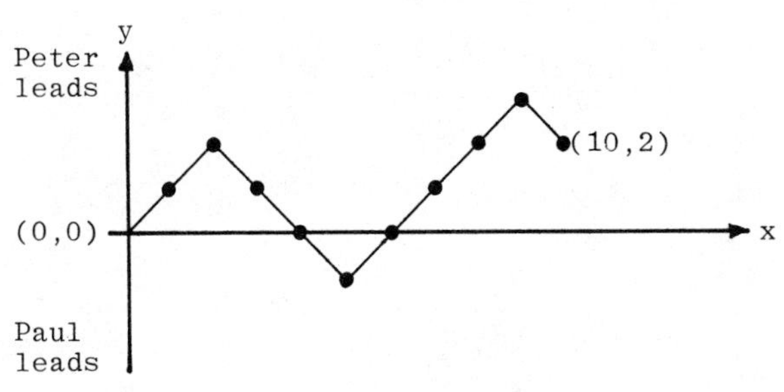

Figure 2.7.1

Path corresponding to the sequence WWLLLWWWWL

A tie occurs at game $2n$ if and only if the point $(2n,0)$ lies on the path. In Figure 2.7.1, ties occur at trials 4 and 6. Note that any path which crosses the axis necessarily contains a point on the axis; before Peter can lose the lead to Paul, there must first be a tie.

We can imagine a particle moving along the polygonal path as the games progress. The particle is said to be performing a random walk. There are 2^x possible paths (sequences) of length x. If neither player has an advantage and successive games are independent of one another, the 2^x possible paths are equally probable. In this case the particle is said to be performing a symmetric random walk. We shall investigate the frequency with which ties may be expected to occur in symmetric random walks. For a more general discussion of random walk problems, see Volume 1 of *An Introduction to Probability Theory and Its Applications* by William Feller.

We now prove a series of results which will eventually enable us to determine the probability that there are no ties in the first $2n$ games.

Lemma 2.7.1. Let $N(x,y)$ be the number of paths from $(0,0)$ to (x,y). Then

$$N(x,y) = \binom{x}{(x+y)/2}. \tag{2.7.2}$$

Proof. Each path from $(0,0)$ to (x,y) corresponds to a different sequence of p W's and q L's, where p and q are given by (2.7.1). From the $p+q$ places in the sequence, the p places to be occupied by W's may be chosen in $\binom{p+q}{p}$ ways, and substituting for p and q from (2.7.1) gives (2.7.2).

Lemma 2.7.2. The number of paths from (a,b) to (x,y) is $N(x-a,y-b)$.

Proof. There is a one-to-one correspondence between paths from (a,b) to (x,y) and paths from $(0,0)$ to $(x-a,y-b)$, as illustrated in Figure 2.7.2.

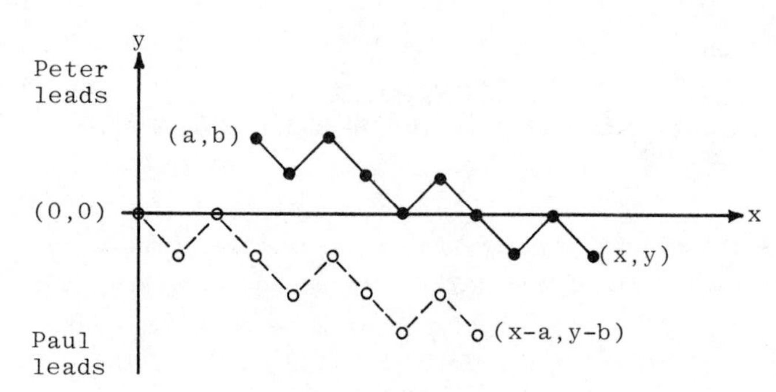

Figure 2.7.2
Correspondence between paths from (a,b) to (x,y) and paths from (0,0) to (x-a,y-b)

By Lemma 2.7.1, the number of paths from $(0,0)$ to $(x-a,y-b)$ is $N(x-a,y-b)$.

Lemma 2.7.3. Take $x > a \geq 0$, $y > 0$, and $b > 0$. Then the number of paths from (a,b) to (x,y) which touch or cross the x-axis is

$N(x-a, y+b)$.

Proof. Consider any path from (a,b) to (x,y) which touches or crosses the x-axis, as in Figure 2.7.3. This path must contain a point on the x-axis. Let the first such point be $(c,0)$, and reflect the portion of the path which lies to the left of this point in the x-axis. The result is a path from $(a,-b)$ to (x,y). Conversely, any

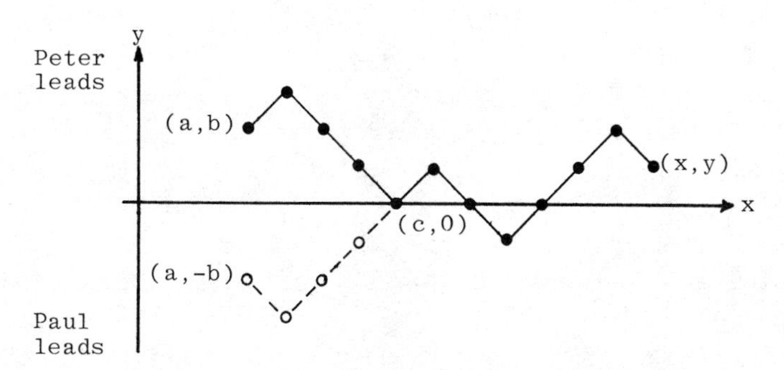

Figure 2.7.3

Correspondence between paths from (a,-b) to (x,y) and paths from (a,b) to (x,y) which meet the x-axis

path from $(a,-b)$ to (x,y) must cross the x-axis because b and y are positive. Let $(c,0)$ be the first point on the axis which the path contains, and reflect the portion of the path to the left of this point in the x-axis. The result is a path from (a,b) to (x,y) which touches or crosses the axis. Hence there is a one-to-one correspondence between the two types of path. By Lemma 2.7.2, the number of paths from $(a,-b)$ to (x,y) is $N(x-a, y+b)$, and the result follows.

Corollary. Under the conditions stated, the number of paths from (a,b) to (x,y) which do not touch or cross the axis is

$$N(x-a, y-b) - N(x-a, y+b).$$

Lemma 2.7.4. If x and y are positive, the number of paths from $(0,0)$ to (x,y) which do not touch or cross the x-axis is $\frac{y}{x} N(x,y)$.

Proof. Any such path lies entirely above the x-axis, and must pass through (1,1). By the preceding Corollary, the number of paths from (1,1) to (x,y) which do not touch or cross the axis is

$$N(x-1,y-1) - N(x-1,y+1).$$

Using (2.7.2) we may show that

$$\frac{N(x-1,y-1)}{N(x,y)} - \frac{N(x-1,y+1)}{N(x,y)} = \frac{x+y}{2x} - (1 - \frac{x+y}{2x}) = \frac{y}{x}$$

and the result follows.

Lemma 2.7.4 is called the Ballot Theorem. Suppose that x votes are cast in an election contest between Peter and Paul, and that Peter wins by a majority of y votes. The number of different orders in which the votes can be counted is equal to $N(x,y)$, the number of different paths from (0,0) to (x,y). The number of ways to count the votes so that Peter is always ahead is equal to $\frac{y}{x} N(x,y)$, the number of paths which do not touch or cross the axis. Thus, if the votes are counted in a random order, the probability that Peter always leads in the tally is y/x. For instance, if Peter gets 55% of the votes compared with 45% for Paul, there is a 10% chance that Peter leads at every stage in the count.

Lemma 2.7.5. The number of paths from (0,0) to (2n,0) which lie entirely above the x-axis is $\frac{1}{n}\binom{2n-2}{n-1}$.

Proof. Any such path must pass through (2n-1,1). By Lemma 2.7.4, the number of paths from (0,0) to (2n-1,1) which do not touch or corss the axis is

$$\frac{1}{2n-1} N(2n-1,1) = \frac{1}{2n-1}\binom{2n-1}{n} = \frac{(2n-2)!}{n!(n-1)!} = \frac{1}{n}\binom{2n-2}{n-1}.$$

Theorem 2.7.1. The probability that the first tie occurs at trial $2n$ is given by

$$f_n = \frac{2}{n}\binom{2n-2}{n-1}2^{-2n}, \qquad n = 1,2,3,\ldots \quad . \tag{2.7.3}$$

Proof. By symmetry, there are exactly as many paths from (0,0) to (2n,0) lying entirely below the axis as there are lying above the axis. Hence, by Lemma 2.7.5, the number of paths from (0,0) to (2n,0) which do not touch the axis at any intermediate point is $\frac{2}{n}\binom{2n-2}{n-1}$. Upon dividing this by 2^{2n}, the total number of paths of length $2n$, we obtain (2.7.3).

Lemma 2.7.6. Let $g_n = \binom{2n}{n}2^{-2n}$. Then $f_n = g_{n-1} - g_n$ for $n = 1,2,3,\ldots$.

Proof.

$$g_{n-1} - g_n = \binom{2n-2}{n-1}2^{-2n+2} - \binom{2n}{n}2^{-2n}$$

$$= 4\binom{2n-2}{n-1}2^{-2n} - \frac{(2n)(2n-1)}{n(n)}\binom{2n-2}{n-1}2^{-2n}$$

$$= \frac{2}{n}\binom{2n-2}{n-1}2^{-2n} = f_n.$$

Theorem 2.7.2. The probability that there is no tie up to and including game $2n$ is given by g_n.

Proof. The probability that the first tie occurs at game $2n$ or before is equal to $f_1 + f_2 + \ldots + f_n$. By Lemma 2.7.6,

$$f_1 + f_2 + \ldots + f_n = (g_0 - g_1) + (g_1 - g_2) + \ldots + (g_{n-1} - g_n)$$

$$= g_0 - g_n = 1 - g_n.$$

Hence the probability that there is no tie up to and including trial $2n$ is equal to $1 - (1 - g_n) = g_n$.

For large n we may apply Stirling's approximation (2.1.16) as in Example 2.1.7 to obtain

$$g_n = \binom{2n}{n}2^{-2n} \sim \frac{1}{\sqrt{\pi n}},$$

so that $g_n \to 0$ as $n \to \infty$. With probability one, a tie will eventually occur. However there is a surprisingly large probability that one will have to wait a very long time for the first tie, as the following table shows:

$2n$ = No. of games	10	20	50	200	800	3200	12800
g_n = Prob. of no ties	0.25	0.178	0.113	0.056	0.028	0.014	0.007

If Peter and Paul play 50 games, the probability that one of them stays in the lead for the entire 50 games is 0.113. If they play 800 games, there is 2.8% chance that one of them will stay in the lead for the entire time! Ties occur very infrequently, even in fair games where neither player has even a slight advantage.

Our finding that ties are rather unusual occurences runs counter to what many people expect. I have often heard players state that, although they have been losing, the "law of averages" is on their side and they will soon begin to catch up. In fact, this is quite unlikely. The law of averages says only that, in a fair game, the ratio of Peter's wins to Paul's should approach one in a large number of scores. However, with high probability the difference between their scores will be large.

CHAPTER 3. THE CALCULUS OF PROBABILITY

We recall from Section 1.3 that a probability model for an experiment has two ingredients: a sample space S and a probability distribution $\{p_i\}$. A subset of the sample space is called an event, and its probability is the sum of the probabilities of all the points it contains. In Chapter 2 we considered only cases in which all of the sample points were assumed to be equally probable. Now we return to the general case in which the p_i's need not be equal. In Sections 1 and 6 we develop formulae for the probability of a union of events. Sections 2 and 3 discuss the extremely important concepts of product models and independence while Sections 4 and 5 deal with conditional probability models.

3.1 Unions and Intersections of Events.

Suppose that A_1 and A_2 are two events defined on the same sample space S. These events may be combined via the set-theoretic operations of intersection and union to give new events. Their intersection A_1A_2 (or $A_1 \cap A_2$) is the set of all sample points belonging to both A_1 and A_2. Their union $A_1 \cup A_2$ is the set of all points belonging to A_1 or to A_2 or to both. These are represented by the shaded regions in the Venn diagrams of Figure 3.1.1.

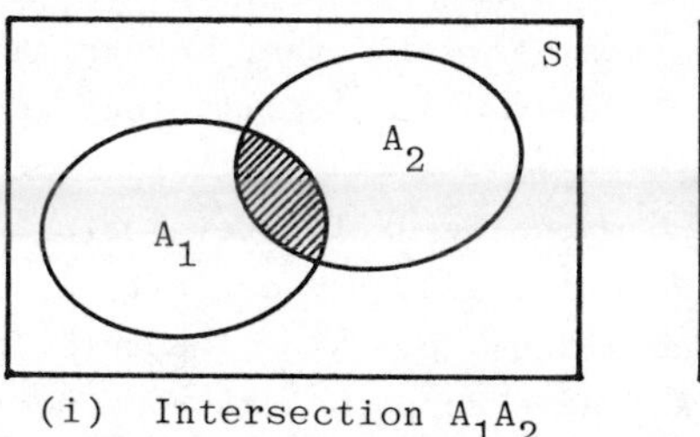

(i) Intersection A_1A_2

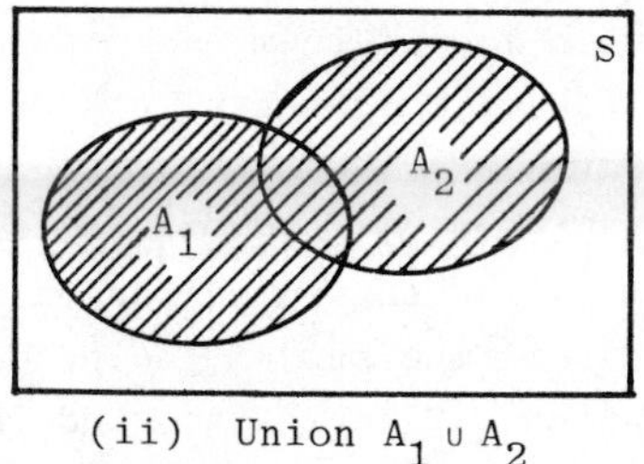

(ii) Union $A_1 \cup A_2$

Figure 3.1.1
Intersection and union of two events

More generally, given any collection of events $A_1, A_2, \ldots, A_n$ defined on the same sample space, we may define two new events: their intersection $A_1 A_2 \ldots A_n$, and their union $A_1 \cup A_2 \cup \ldots \cup A_n$. The intersection is the set of sample points belonging to all of the events $A_1, A_2, \ldots, A_n$. The union is the set of points belonging to one or more of $A_1, A_2, \ldots, A_n$. Venn diagrams for the intersection and union of three events are given in Figure 3.1.2.

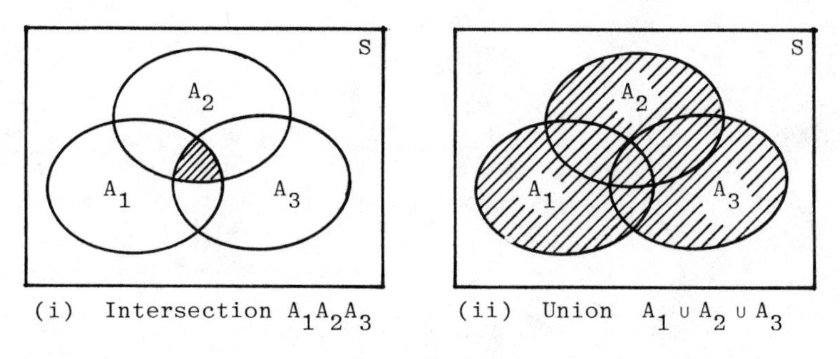

(i) Intersection $A_1 A_2 A_3$ (ii) Union $A_1 \cup A_2 \cup A_3$

Figure 3.1.2
Intersection and union of three events

Mutually exclusive events.

The n events $A_1, A_2, \ldots, A_n$ are called mutually exclusive, or disjoint, if $A_i A_j = \phi$ for all i and j with $i \neq j$. Then there is no sample point which belongs to more than one of $A_1, A_2, \ldots, A_n$, and hence no two of them can occur together. In other words, the occurence of one of the events A_i excludes the possibility that any of the others occurs.

The union $A_1 \cup A_2$ of two mutually exclusive events is represented by the shaded region in Figure 3.1.3. By definition, $P(A_1 \cup A_2)$ is the sum of the probabilities p_i of all points in the shaded region. Because A_1 and A_2 have no points in common, the sum may be broken into two parts:

$$P(A_1 \cup A_2) = \sum_{i \in A_1 \cup A_2} p_i = \sum_{i \in A_1} p_i + \sum_{i \in A_2} p_i = P(A_1) + P(A_2).$$

Hence the probability of a union of mutually exclusive events is equal to the sum of their probabilities. This result may be extended by induction to any finite number of events.

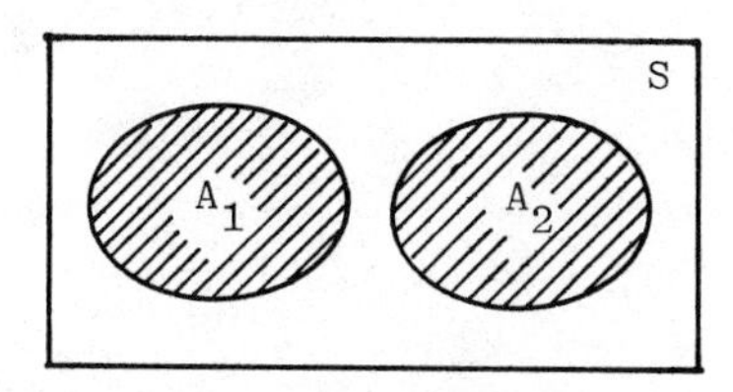

Figure 3.1.3
Union of Mutually Exclusive Events

Theorem 3.1.1. Let $A_1, A_2, A_3, \ldots$ be a finite sequence of mutually exclusive events on the same sample space. Then the probability of their union is equal to the sum of their probabilities:

$$P(A_1 \cup A_2 \cup A_3 \cup \ldots) = P(A_1) + P(A_2) + P(A_3) + \ldots \quad . \tag{3.1.1}$$

(In the general theory of probability, it is taken as axiomatic that (3.1.1) also holds for infinite sequences of mutually exclusive events.)

The probability of an event A is sometimes most easily obtained by expressing A as the union of mutually exclusive events $A_1, A_2, \ldots,$ and then using (3.1.1). This is illustrated by the following example.

Example 3.1.1. An executive of four members is chosen at random without replacement from a club consisting of 6 men and 6 women. Find the probability that there are at least two men on the executive.

Solution. Let A be the event that the executive contains at least two men. Then $A = A_2 \cup A_3 \cup A_4$, where A_x denotes the event that the executive contains exactly x men. If the order of selection is ignored, there are $\binom{12}{4}$ equally probable samples of size 4, of which $\binom{6}{x}\binom{6}{4-x}$ contain x men and $4-x$ women. Therefore

$$P(A_x) = \binom{6}{x}\binom{6}{4-x} / \binom{12}{4},$$

which gives $P(A_2) = \frac{225}{495}$, $P(A_3) = \frac{120}{495}$, and $P(A_4) = \frac{15}{495}$. Now since

A_2, A_3 and A_4 are mutually exclusive, we may apply (3.1.1) to obtain

$$P(A) = P(A_2) + P(A_3) + P(A_4) = \frac{360}{495} = 0.727.$$

Incorrect Solution. Counting arguments similar to the following are sometimes used by the unwary in such problems. We wish to determine the number of samples of size 4 which contain at least two men. First we select two men for the sample from the six available in $\binom{6}{2}$ ways. There are two members still to choose, and these may be either men or women. Thus any two of the ten remaining people may be selected, and the number of samples containing at least two men is $\binom{6}{2}\binom{10}{2} = 675$. The required probability is then $675/495 = 1.36$.

A probability greater than one is a sure indication that something has gone wrong! The difficulty with the above argument is that some possible samples are counted more than once; in fact, a sample is counted once for each pair of men it contains. There are $\binom{6}{2}\binom{6}{2} = 225$ samples with exactly 2 men, and each of them is counted once; there are $\binom{6}{3}\binom{6}{1} = 120$ samples with exactly 3 men, and each of them is counted $\binom{3}{2} = 3$ times; and there are $\binom{6}{4} = 15$ samples with 4 men, each of which is counted $\binom{4}{2} = 6$ times. The total count is thus

$$225(1) + 120(3) + 15(6) = 675.$$

Unfortunately, an incorrect counting argument will not always lead to a probability greater than one, and hence may be difficult to spot. As a general rule, when dealing with problems which demand the probability of "at most" or "at least" so many, it is wise to express the event as a union of mutually exclusive events and use (3.1.1). □

Union of two events.

The union of two events A_1, A_2 which are not mutually exclusive is represented by the shaded region in Figure 3.1.1(ii). By definition, the probability of $A_1 \cup A_2$ is equal to the sum of the probabilities p_i of all points i in the shaded region. First consider the sum

$$P(A_1) + P(A_2) = \sum_{i \in A_1} p_i + \sum_{i \in A_2} p_i.$$

If point i belongs to just one of the events A_1, A_2, then p_i

appears in only one of the sums on the right. However, if i belongs to A_1A_2 (that is, to both A_1 and A_2), then p_i occurs in both sums on the right. Hence the above expression exceeds $P(A_1 \cup A_2)$ by $\sum p_i$, where the sum extends over all i belonging to A_1A_2. We thus obtain the following result:

<u>Theorem 3.1.2</u>. Let A_1, A_2 be any two events defined on the same sample space. Then the probability that at least one of them occurs is

$$P(A_1 \cup A_2) = P(A_1) + P(A_2) - P(A_1A_2). \qquad (3.1.2)$$

<u>Union of three events</u>.

The union of three events A_1, A_2, A_3 is illustrated in Figure 3.1.2(ii), and by definition, $P(A_1 \cup A_2 \cup A_3)$ is the sum of the probabilities p_i of all points i in the shaded region. Consider the sum

$$P(A_1) + P(A_2) + P(A_3) = \sum_{i \in A_1} p_i + \sum_{i \in A_2} p_i + \sum_{i \in A_3} p_i.$$

If point i belongs to just one of the three events, p_i appears only once in this sum, as it should. However, if i belongs to both A_1 and A_2, p_i appears twice in this expression, and to adjust for this we must subtract off $P(A_1A_2)$. Similarly, we deduct $P(A_1A_3)$ and $P(A_2A_3)$ to obtain a new expression.

$$P(A_1) + P(A_2) + P(A_3) - P(A_1A_2) - P(A_1A_3) - P(A_2A_3).$$

Any point i belonging to one or two of the events contributes p_i to this expression. However, a point i which belongs to all three of the events contributes p_i to each of the six terms, for a net contribution of $3p_i - 3p_i = 0$. Hence the expression must be adjusted by adding $\sum p_i$ where the sum ranges over all points i in $A_1A_2A_3$. This leads to

<u>Theorem 3.1.3</u>. Let A_1, A_2, A_3 be any three events defined on the same sample space. Then the probability that at least one of them occurs is given by

$$P(A_1 \cup A_2 \cup A_3) = P(A_1) + P(A_2) + P(A_3) - P(A_1A_2) - P(A_1A_3) - P(A_2A_3) + P(A_1A_2A_3). \qquad (3.1.3)$$

Union of n events.

The extension of formulae (3.1.2) and (3.1.3) to the case of n events will be considered in Section 6. For the present we consider only two partial results.

Since $A_1 \cup A_2 \cup \ldots \cup A_n$ consists of all points belonging to at least one of the events $A_1, A_2, \ldots, A_n$, its complement consists of all points not belonging to any of the events $A_1, A_2, \ldots, A_n$; that is,

$$\overline{A_1 \cup A_2 \cup \ldots \cup A_n} = \bar{A}_1 \bar{A}_2 \ldots \bar{A}_n.$$

It follows that

$$P(A_1 \cup A_2 \cup \ldots \cup A_n) = 1 - P(\bar{A}_1 \bar{A}_2 \ldots \bar{A}_n). \tag{3.1.4}$$

This result is particularly useful when the events $A_1, A_2, \ldots, A_n$ are mutually independent (see Section 3).

A useful inequality can be obtained by noting that $P(A_1 A_2) \geq 0$ in (3.1.2), and hence

$$P(A_1 \cup A_2) \leq P(A_1) + P(A_2).$$

This result may now be extended by induction to give

$$P(A_1 \cup A_2 \cup \ldots \cup A_n) \leq P(A_1) + P(A_2) + \ldots + P(A_n). \tag{3.1.5}$$

Example 3.1.2. If three letters are assigned randomly, one to each of three envelopes, what is the probability that at least one letter is placed in the correct envelope?

Solution. There are $3! = 6$ equally probable arrangements of the letters. Define A to be the event that at least one letter is placed in the correct envelope. Then $A = A_1 \cup A_2 \cup A_3$, where A_i is the event that the ith letter is placed in the correct envelope. If the ith letter is placed correctly, the others may be arranged in $2!$ ways, so that $P(A_i) = 2/6$ $(i = 1,2,3)$. If both the ith and jth letters are placed correctly, the remaining letter must also be placed correctly, so that $P(A_i A_j) = 1/6 = P(A_1 A_2 A_3)$. Now (3.1.3) gives

$$\begin{aligned} P(A) &= P(A_1) + P(A_2) + P(A_3) - P(A_1A_2) - P(A_1A_3) - P(A_2A_3) + P(A_1A_2A_3) \\ &= \frac{1}{3} + \frac{1}{3} + \frac{1}{3} - \frac{1}{6} - \frac{1}{6} - \frac{1}{6} + \frac{1}{6} = \frac{2}{3}. \end{aligned}$$

This result may be verified by listing the six arrangements and verifying that in only two of them are all three letters out of position. The more general problem of distributing n letters randomly to n envelopes will be considered in Section 6.

Example 3.1.3. What is the probability that a bridge hand contains

(a) at least one 7-card suit?

(b) at least one 6-card suit?

Solution. (a) Let A be the event that a bridge hand contains at least one 7-card suit, and let A_i be the event that it contains 7 cards in the ith suit. Then $A = A_1 \cup A_2 \cup A_3 \cup A_4$. Since a bridge hand contains only 13 cards, it is impossible to have two 7-card suits. Hence A_1, A_2, A_3, A_4 are mutually exclusive events, and (3.1.1) applies. There are $\binom{52}{13}$ equally probable bridge hands, of which $\binom{13}{7}\binom{39}{6}$ contain 7 cards from a particular suit and 6 from the other three suits. Therefore,

$$P(A_i) = \binom{13}{7}\binom{39}{6}/\binom{52}{13} = 0.00882.$$

The required probability is then

$$P(A) = P(A_1) + P(A_2) + P(A_3) + P(A_4) = 4(0.00882) = 0.0353.$$

(b) Let B be the event that a bridge hand contains at least one 6-card suit, and let B_i be the event that it contains 6 cards in the ith suit. Then $B = B_1 \cup B_2 \cup B_3 \cup B_4$, but now B_1, B_2, B_3, B_4 are not mutually exclusive because it is possible for a bridge hand to contain two 6-card suits. Any two events intersect, but no three of them have a common point, as illustrated in Figure 3.1.4.

The probability that a bridge hand contains exactly 6 cards from a particular suit is

$$P(B_i) = \binom{13}{6}\binom{39}{7}/\binom{52}{13} = 0.04156.$$

Now (3.1.5) gives an upper bound for $P(B)$:

$$P(B) \le P(B_1) + P(B_2) + P(B_3) + P(B_4) = 4(.04156) = 0.1662.$$

In order to obtain the exact result, it is necessary to subtract off

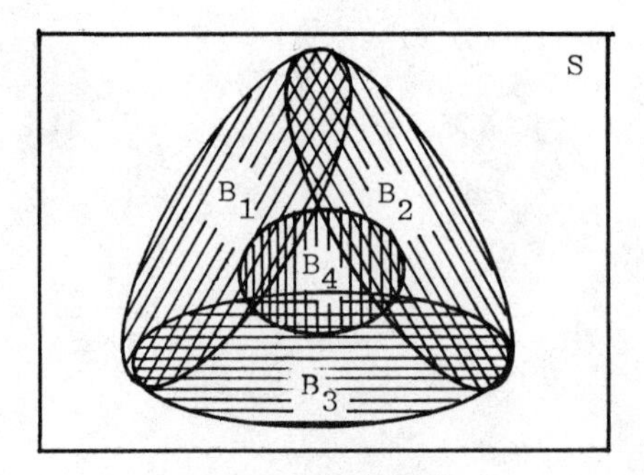

Figure 3.1.4
Venn diagram for Example 3(b)

the probabilities of the six intersections B_iB_j $(i < j)$. The number of bridge hands which contain exactly six hearts, six spades, and one club or diamond is $\binom{13}{6}\binom{13}{6}\binom{26}{1}$, and hence

$$P(B_iB_j) = \binom{13}{6}\binom{13}{6}\binom{26}{1}/\binom{52}{13} = 0.00012.$$

The required probability is then

$$P(B) = 0.1662 - 6(0.00012) = 0.1655.$$

The upper bound given by (3.1.5) is quite good in this case because the intersections all have small probabilities.

Problems for Section 3.1

1. Let A and B be mutually exclusive events with $P(A) = 0.25$ and $P(B) = 0.4$. Find the probabilities of the following events:

$$\bar{A},\ \bar{B},\ A \cup B,\ A \cap B,\ \bar{A} \cup \bar{B},\ \bar{A} \cap \bar{B},\ \overline{A \cap B}.$$

2. In a certain factory, 63% of the workers are male, 68% are married, and 45% are married men. What percentage of workers are married females? single females? male or married or both?

†3. Sixty percent of people drink, 40% smoke, and 50% swear. Only 25% have none of these habits, and 20% have all three. Furthermore, 25% smoke and swear, and 40% drink and swear. What is

is the probability that a randomly selected individual drinks and smokes? drinks and smokes but doesn't swear?

4. In Problem 2.4.1, evaluate the probabilities of the following events:

$$BE,\ B \cup D,\ B \cup D \cup E,\quad (A \cup B)D,\quad A \cup (BD).$$

Show the latter two events on Venn diagrams.

5. Balls numbered $1,2,\ldots,20$ are placed in an urn and one is drawn at random. Let A,B,C and D denote the events that the number on the ball drawn is divisible by 2, divisible by 5, prime, and odd, respectively. Show these four events on a Venn diagram, and compute the probabilities of the following:

$$A \cup B,\quad A\bar{B},\quad A \cup D,\quad AD,\quad (A \cup B)(C \cup D),\quad (AB) \cup (CD).$$

†6. A hand of six cards is dealt from a well-shuffled deck. What is the probability that it contains exactly two aces, exactly two kings, or exactly two queens?

3.2 Independent Experiments and Product Models

Consider two experiments:

Experiment 1 - sample space S, probability distribution $\{p_i\}$;
Experiment 2 - sample space T, probability distribution $\{q_j\}$.

A single composite experiment may be defined to consist of first performing experiment 1 and then experiment 2. For example, if the first experiment involves tossing a coin and the second involves rolling a die, the composite experiment would involve first tossing the coin and then rolling the die. Our aim is to set up an appropriate probability model (sample space and probability distribution) for the composite experiment.

We shall restrict attention to the situation in which there is no carry-over effect from the first experiment to the second, so that the outcome of the second experiment is not influenced in any way by the outcome of the first. The experiments are then called independent. For example, one would not expect the outcome on the die to depend upon whether the coin landed heads or tails. The coin will show "heads" one-half of the time in a large number of tosses. On one-sixth of these occasions, the die will show "1". Hence the relative frequency of obtaining "heads" on the coin and "1" on the die in a large number of repetitions of the composite experiment should be $\frac{1}{6}(\frac{1}{2}) = \frac{1}{12}$. More generally, let i be an outcome of the first experiment having probability p_i, and let j be an outcome of the second experiment having probability q_j. Then if there is no carry-over effect, the probability of outcome i followed by outcome j will be given by the product $p_i q_j$. One can recognize many real situations where the results of two experiments will be independent, and the product rule can then be verified by observation. Thus it has almost the status of an empirical or physical law.

Corresponding to this empirical law, we require a mathematical definition of independent experiments.

An outcome of the composite experiment may be represented by an ordered pair (i,j) where i is an outcome of the first experiment and j is an outcome of the second. Thus the sample space for the composite experiment will be the set of all pairs (i,j) where $i \in S$ and $j \in T$. This is called the <u>Cartesian product</u> (or combinatorial product) of sets S and T, and will be denoted by $S \times T$. If we take S to be a set of points on the horizontal axis and T a set of points on the vertical axis, then $S \times T$ corresponds to a rectangular lattice of points with Cartesian coordinates (i,j) where $i \in S$ and $j \in T$. This is illustrated in Figure 3.2.1.

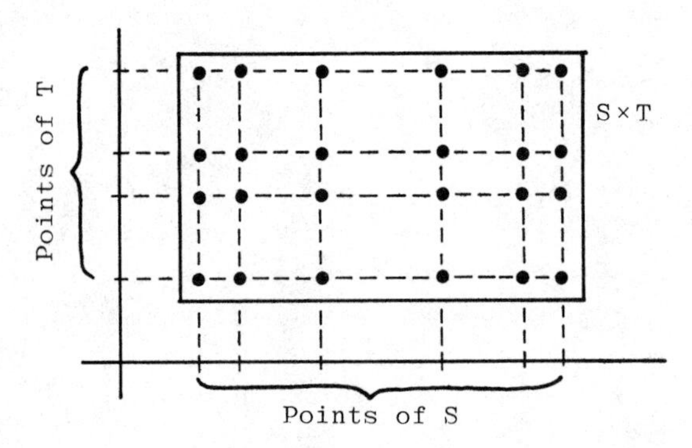

Figure 3.2.1
Cartesian Product of Sets S and T

In order to complete the definition of a probability model, we must define a probability distribution on the sample space $S \times T$. The two experiments are called independent if and only if the probabilities are assigned according to the product rule

$$P\{(i,j)\} = p_i q_j \tag{3.2.1}$$

for all points $i \in S$ and $j \in T$. The probability model which this defines is called the <u>product model</u> for the composite experiment.

Clearly, $P\{(i,j)\} \geq 0$ for all points $(i,j) \in S \times T$. Also, since

$$\sum_{i \in S} p_i = 1 = \sum_{j \in T} q_j$$

we find that

$$\sum_{(i,j) \in S\times T} P\{(i,j)\} = \sum_{i \in S,\, j \in T} \sum p_i q_j = \sum_{i \in S} p_i \sum_{j \in T} q_j = 1.$$

Therefore, (3.2.1) defines a proper probability distribution on the sample space $S \times T$.

Independent Repetitions

Consider now the important special case in which the second experiment is a repetition of the first one, so that $T = S$ and $\{q_j\} = \{p_i\}$. Taken together, the two repetitions form a composite experiment whose sample space is S^2, the Cartesian product of S with itself:

$$S^2 = S \times S = \{(i,j);\ \ i \in S,\ j \in S\}.$$

The repetitions are called <u>independent repetitions</u>, or <u>trials</u>, if and only if

$$P\{(i,j)\} = p_i p_j$$

for all points (i,j) in S^2. For n repetitions of the experiment, the sample space is S^n, the n-fold Cartesian product of S with itself:

$$S^n = S \times S \times \ldots \times S = \{(i,j,\ldots,k);\ \ i \in S,\ j \in S,\ldots,k \in S\}.$$

The n repetitions are called independent repetitions, or trials, if and only if

$$P\{(i,j,\ldots,k)\} = p_i p_j \ \ldots\ p_k$$

for all points $(i,j,\ldots,k)$ in S^n.

<u>Example 3.2.1</u>. Suppose that the composite experiment involves two

rolls of a balanced die. The sample space for a single roll is $S = \{1,2,\ldots,6\}$, each point having probability $\frac{1}{6}$. Because the die has no memory, the outcome of the second roll will not be influenced by the outcome of the first roll. Successive rolls of the die will thus be independent trials. The sample space for two rolls will be the Cartesian product

$$S^2 = \{(i,j);\ 1 \le i \le 6,\ 1 \le j \le 6\}.$$

Each of the 36 points in S^2 will be assigned probability $\frac{1}{6}(\frac{1}{6}) = \frac{1}{36}$. The probabilities of events in S^2 may now be calculated in the usual way, by adding up the probabilities of the points they contain. For instance, let A be the event that the total score in two rolls is 6. Then A contains five sample points

$$A = \{(1,5),(2,4),(3,3),(4,2),(5,1)\}$$

and $P(A) = 5/36$.

If the die were brick-shaped (see Example 1.3.2), the sample space for two rolls would be the same, but the probabilities obtained from (3.2.1) would now be as follows:

$$P\{(i,j)\} = \begin{cases} (\frac{1}{6} + \theta)^2 & \text{for } i \le 4 \text{ and } j \le 4; \\ (\frac{1}{6} - 2\theta)^2 & \text{for } i > 4 \text{ and } j > 4; \\ (\frac{1}{6} + \theta)(\frac{1}{6} - 2\theta) & \text{for } i \le 4 < j \text{ or } j \le 4 < i. \end{cases}$$

The probability of event A would then be

$$P(A) = 2(\frac{1}{6} + \theta)(\frac{1}{6} - 2\theta) + 3(\frac{1}{6} + \theta)^2 = (\frac{1}{6} + \theta)(\frac{5}{6} - \theta),$$

which also gives $P(A) = 5/36$ when $\theta = 0$ (a balanced die).

<u>Bernoulli Trials</u>

Independent repetitions of an experiment having just two possible outcomes S (success) and F (failure) are called Bernoulli trials. The sample space for a single trial contains two points S,F with probabilities p and $1-p$, where p is the same for all repetitions.

Bernoulli trials are used as a model in a great many situations. Successive tosses of a coin, with "heads" as success, will be Bernoulli trials. So will successive rolls of a die provided that we consider only one characteristic of the outcome - e.g. whether an even number (success) or odd number (failure) is obtained. Bernoulli trials might be used as a model for an inspection scheme, where items are tested as they come off an assembly line and classified as good (success) or defective (failure). The assumptions involved would be that successive items were independent, and that the proportion of good items produced remained constant over time. The Bernoulli trials model might, of course, prove unsatisfactory if these assumptions were too badly violated.

The sample space for two Bernoulli trials contains four points SS,SF,FS,FF with probabilities p^2, $p(1-p)$, $(1-p)p$, and $(1-p)^2$. (We omit unnecessary brackets and commas, and write (S,S) as SS.) The sample space for three Bernoulli trials contains 8 points as follows:

SSS	with probability p^3;
SSF,SFS,FSS	each with probability $p^2(1-p)$;
FFS,FSF,SFF	each with probability $p(1-p)^2$;
FFF	with probability $(1-p)^3$.

The sample space for n Bernoulli trials contains the 2^n possible sequences of n S's and F's. There are $\binom{n}{x}$ different sequences which contain x S's and $n-x$ F's, and each such sequence has probability $p^x(1-p)^{n-x}$. Therefore, <u>the probability of obtaining exactly x successes in n Bernoulli trials is given by the binomial distribution</u>,

$$f(x) = \binom{n}{x}p^x(1-p)^{n-x}; \quad x = 0,1,2,\ldots,n.$$

We previously encountered the binomial distribution in Section 2.5. There we were considering random sampling with replacement from an urn containing white and black balls. Under sampling with replacement, the proportion p of white balls in the urn remains constant from one trial to the next. Thorough mixing of the balls before each draw ensures the independence of successive draws. Hence, under sampling with replacement, successive draws are Bernoulli trials. This is not the case for sampling without replacement, for then successive draws are not independent. If the first ball drawn is white, the proportion of white balls in the urn decreases, and the probability that the second ball will be white is less than p. On the other hand, if the first ball drawn is black, the chance of a white ball on the second draw will be greater than p. The probability of obtaining a white ball on the second draw is therefore dependent upon the outcome of the first draw.

Example 3.2.2. Compute the probability of obtaining (a) exactly 4 heads and (b) at least 6 heads in eight tosses of a balanced coin.

Solution. Successive tosses of a coin are Bernoulli trials. The probability of obtaining exactly x heads (successes) will thus be given by a binomial distribution with $n = 8$ and $p = \frac{1}{2}$:

$$f(x) = P(x \text{ heads}) = \binom{8}{x}\left(\frac{1}{2}\right)^x\left(\frac{1}{2}\right)^{8-x} = \binom{8}{x}/256.$$

The probability of obtaining exactly 4 heads is then

$$f(4) = \binom{8}{4}/256 = 70/256;$$

the probability of obtaining at least 6 heads is

$$f(6) + f(7) + f(8) = (28+8+1)/256 = 37/256.$$

Problems for Section 3.2

1. Prepare a table showing the probability of obtaining x sixes in five rolls of a balanced die $(x = 0, 1, \ldots, 5)$.
2. What is the probability that three balanced dice show a total of more than 15?

†3. A balanced coin is tossed until the first head is obtained. What is the probability that exactly x tosses are needed? at most x?

4. Two teams play a best-of-seven series. The first two games are to be played on A's field, the next three on B's field, and the last two on A's field (if required). The probability that A wins a game is 0.7 for a home game and 0.5 on B's field. What is the probability that
   (a) A wins the series in 4 games? 5 games?
   (b) the series does not go to 6 games?
5. Three players A,B,C take turns rolling a balanced die in the order A,B,C,A,B,C,A,... . The player who first rolls a "six" is declared the winner, and the game ends. Find the probability of winning for each player.

†6. In order to win a game of tennis, a player requires four points and in addition must have at least two more points than his oppo-

nent. When he plays against B, A has probability p of winning any point, different points being independent. What is the probability that a game lasts longer than six points? What is the probability that A wins?

7. A slot machine pays off with probability p, and successive plays are independent. What is the probability that the first payoff will occur on the nth play?

8. John and Mary simultaneously play two slot machines. John's machine pays off with probability p_1, and Mary's with probability p_2. They decide to continue playing until one of them wins.
   (a) What is the probability that John wins before Mary?
   (b) What is the probability that they win simultaneously?

†9. A and B take turns shooting at a target, with A having the first shot. The probability that A misses the target on any single shot is p_1; the probability that B misses the target on any single shot is p_2. Shots are mutually independent, and the first to hit the target is declared the winner. Find the probability that A wins.

3.3 Independent Events

If two experiments are independent, probabilities of outcomes in the composite experiment are calculated from the product rule (3.2.1). It is natural to call two events independent if their probabilities satisfy a similar product rule.

Definition. Two events A and B defined on the same sample space are called independent events if and only if

$$P(AB) = P(A)P(B). \tag{3.3.1}$$

As applied to events, "independent" is no more than a convenient name to be applied if (3.3.1) is satisfied. It is not always obvious whether or not two events are independent until the three probabilities in (3.3.1) have been calculated. In general, whether or not two events A and B are independent will depend upon which probability distribution is assigned to the points of the sample space. The independence of two events does not necessarily imply any factorization of the sample space as a Cartesian product. On the other hand, when we speak of two independent experiments, we imply that the sample space may be represented as the Cartesian product of two other sample spaces S and T. The independence of the two experiments does not depend upon the choice of probability distributions $\{p_i\}$ on S and $\{q_j\}$ on T.

Example 3.3.1. A card is dealt from a standard deck of 52. Let A be the event "ace", and let B be the event "spade". If we assume that all 52 outcomes are equally probable, then

$$P(A) = \frac{4}{52}, \quad P(B) = \frac{13}{52}, \quad P(AB) = \frac{1}{52},$$

and hence (3.3.1) is satisfied. If the outcomes are equally probable, then A and B are independent events. However, if a different probability distribution is assumed, it may happen that A and B are no longer independent. For instance, suppose that the ace of spades has been marked, and that the dealer is able to make it turn up twice as often as any other card. The ace of spades then has probability 2/53, and every other card has probability 1/53. Now

$$P(A) = \frac{5}{53}, \quad P(B) = \frac{14}{53}, \quad P(AB) = \frac{2}{53}.$$

Since $P(AB) \neq P(A)P(B)$, events A and B are no longer independent.

Example 3.3.2. The brick-shaped die of Example 1.3.2 is rolled once. Let A be the event "outcome is even", and let B be the event "outcome is divisible by 3". Then

$$P(A) = 1/2, \quad P(B) = \frac{1}{3} - \theta, \quad P(AB) = \frac{1}{6} - 2\theta.$$

Note that

$$P(A)P(B) = \frac{1}{6} - \frac{\theta}{2}$$

which is different from $P(AB)$ unless $\theta = 0$. Thus events A and B are independent if $\theta = 0$ (a perfect die); otherwise they are not independent.

Events in Composite Experiments.

Now consider two independent experiments with sample spaces S and T and probability distributions $\{p_i\}$ and $\{q_j\}$. The sample space for the composite experiment is then the Cartesian product $S \times T$, and probabilities are assigned to points of $S \times T$ by the product rule,

$$P\{(i,j)\} = p_i q_j, \tag{3.3.2}$$

for all points (i,j) in $S \times T$.

Corresponding to any event A in S there is an event A' in $S \times T$ consisting of all points (i,j) with $i \in A$ and $j \in T$ (see Figure 3.3.1). Thus A' is the event that A occurs in the first experiment together with any outcome in the second experiment.

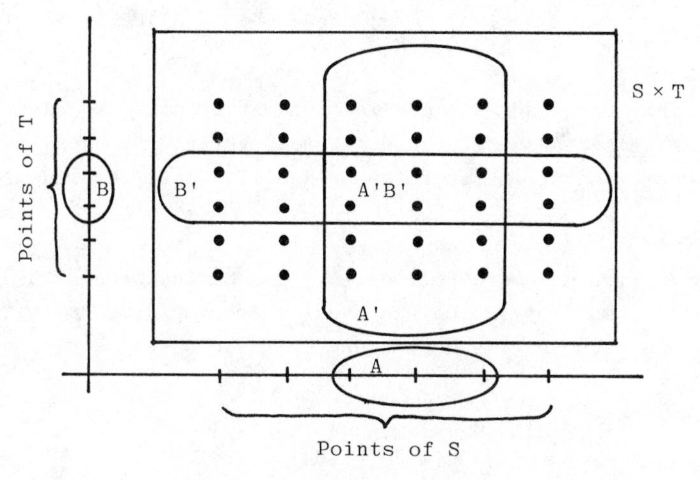

Figure 3.3.1

Events in the Cartesian Product of S and T

We refer to A' as an event which depends only on the first experiment. Since

$$\sum_{i \in A} p_i = P(A) \quad \text{and} \quad \sum_{j \in T} q_j = 1,$$

we find that

$$P(A') = \sum_{(i,j) \in A'} P\{(i,j)\} = \sum_{i \in A,\, j \in T} p_i q_j = \sum_{i \in A} p_i \sum_{j \in T} q_j = P(A).$$

Hence A' has the same probability as A. Since A and A' are defined on different sample spaces, they are strictly speaking different events. However for most purposes we may refer to both events as "event A", and depend upon the context to indicate the appropriate sample space.

We may similarly extend any event B in T to an event

$B' = \{(i,j); i \in S, j \in B\}$ in $S \times T$. Then B' is an event which depends only on the second experiment, and $P(B') = P(B)$.

Event $A'B'$ is now defined on $S \times T$, and consists of all points (i,j) with $i \in A$ and $j \in B$ (see Figure 3.3.1). By (3.3.2) we have

$$P(A'B') = \sum_{(i,j)\in A'B'} P\{(i,j)\} = \sum_{i\in A, j\in B} \sum p_i q_j$$

$$= \sum_{i\in A} p_i \sum_{j\in B} q_j = P(A)P(B) = P(A')P(B')$$

which shows that A' and B' are independent events. We have thus proved that <u>events depending upon different independent experiments are independent events</u>, whatever the probability distributions $\{p_i\}$ and $\{q_j\}$.

<u>Example 3.3.3</u>. Two dice are rolled. Let A be the event "3 or 4 on the first roll", and let B be the event "even outcome on the second roll". Then A and B may be defined on the product sample space as follows:

$$A = \{(3,j),(4,j); \quad 1 \le j \le 6\}$$
$$B = \{(i,2),(i,4),(i,6); \quad 1 \le i \le 6\}.$$

We now have

$$AB = \{(3,2),(3,4),(3,6),(4,2),(4,4),(4,6)\}.$$

Assuming probabilities $\{p_i\}$ for the first die and $\{q_j\}$ for the second, we have by (3.3.2)

$$P(AB) = p_3q_2 + p_3q_4 + p_3q_6 + p_4q_2 + p_4q_4 + p_4q_6$$
$$= (p_3 + p_4)(q_2 + q_4 + q_6) = P(A)P(B).$$

Events A and B are independent no matter what probabilities p_i and q_j are assigned to the faces of the dice. □

<u>Independence of Several Events</u>.

We now wish to extend the definition of independence to a collection of three or more events. Suppose that A,B, and C are events in the same sample space, any two of which are independent:

$$P(AB) = P(A)P(B), \quad P(AC) = P(A)P(C), \quad P(BC) = P(B)P(C). \qquad (3.3.3)$$

One might expect that (3.3.3) would imply that

$$P(ABC) = P(A)P(B)P(C). \tag{3.3.4}$$

The following example illustrates that this is not the case.

Example 3.3.4. A balanced coin is tossed three times. The sample space contains 8 points HHH,HHT,...,TTT, each with probability 1/8. Define A, B, and C to be the events "same outcome on the first and second tosses", "same outcome on the second and third tosses", and "same outcome on the first and third tosses". Then

$$P(A) = P(B) = P(C) = \frac{1}{2}$$

$$P(AB) = P(AC) = P(BC) = \frac{1}{4}.$$

But if two of the events occur, the third must also occur. Hence

$$P(ABC) = \frac{1}{4}$$

so that (3.3.3) is satisfied, but (3.3.4) is not. □

Three events A, B, C for which both (3.3.3) and (3.3.4) hold are called independent, or mutually independent. Three events for which (3.3.3) holds but (3.3.4) possibly does not are called pairwise independent. These definitions may be extended to n events as follows:

Definition. Let $A_1, A_2, \ldots, A_n$ be n events defined on the same sample space. The n events will be called independent, or mutually independent, if the probability that any r of them occur is equal to the product of their r individual probabilities $(r = 2,3,\ldots,n)$. In other words, for any r distinct subscripts $i_1, i_2, \ldots, i_r$, chosen from $\{1,2,\ldots,n\}$, the condition

$$P(A_{i_1} A_{i_2} \ldots A_{i_r}) = P(A_{i_1})P(A_{i_2})\ldots P(A_{i_r}) \tag{3.3.5}$$

must hold. Hence, in order to be mutually independent, events $A_1, A_2, \ldots, A_n$ must satisfy

$$\binom{n}{2} + \binom{n}{3} + \ldots + \binom{n}{n} = 2^n - n - 1$$

different product rules. The n events are called pairwise independent if (3.3.5) holds for $r = 2$ but possibly does not hold for some $r > 2$.

Problems for Section 3.3

1. Let A be the event that a man's left eye is brown, and let B be the event that his right eye is brown. Given that $P(A) = P(B) = 0.7$, can you obtain the probability that both eyes are brown? Why is 0.49 incorrect?
2. A balanced coin is tossed n times. Let A be the event "at least two tails", and let B be the event "one or two heads". Show that A and B are independent events for $n = 3$ but not for $n = 4$.

†3. Let E_1, E_2 and E_3 be events with probabilities $\frac{1}{2}, \frac{1}{4}$ and $\frac{1}{8}$, respectively. Let A be the event that E_1 and E_2 occur; let B be the event that at least one of E_1, E_2, E_3 occurs; and let C be the event that exactly one of E_1, E_2, E_3 occurs. Show these events on Venn diagrams and find their probabilities in each of the following cases:
   (i) E_1, E_2 and E_3 are mutually exclusive;
   (ii) E_1, E_2 and E_3 are mutually independent.

4. The events $A_1, A_2, \ldots, A_n$ are mutually independent and $P(A_i) = p_i$ for $i = 1, 2, \ldots, n$. Find the probability that
   (a) at least one of them occurs;
   (b) exactly one of them occurs.

†5. When A and B play a game, the odds that A wins are 2 to 1. If they play two games, and games are independent, what are the odds that A wins them both?

3.4 Conditional Probability.

Consider an experiment with sample space S and probability distribution $\{p_i\}$. It frequently happens that the sample space contains subsets (i.e. events) within which it is desirable to make separate probability statements. For instance, we may want separate unemployment rates for different sexes, age groups, and regions of the country as well as a single overall unemployment rate for the entire population. We may wish to know the proportions of males and females who are colourblind, as well as the overall fraction of colourblind people.

Conditional probability notation is used to indicate the subset of the sample space within which a probability statement applies. For instance, $P(C|F)$ would be used to denote the probability that a female is colourblind, and $P(C|M)$ the probability that a male is colourblind. The sample space for the first probability statement is F, the set of all females, while the sample space for the second statement is the set of all males. The notation is used to imply that M and F are themselves to be regarded as subsets of a larger sample space $S = M \cup F$ consisting of all humans. In this notation, the probability of colourblindness for the entire population could be denoted by $P(C|S)$. However, we would usually omit reference to S and write merely $P(C)$ unless we were thinking of S as a subset of an even larger sample space.

Probability models are frequently defined by giving several conditional probabilities which are then used to determine the probabilities of events in the entire sample space S. We illustrate such a procedure in the following example.

Example 3.4.1. Unemployment statistics in Canada are collected separately for five regions: British Columbia (10% of the labour force), the Prairies (16%), Ontario (38%), Quebec (27%), and the Atlantic region (9% of the labour force). If the percentages of unemployment in the five regions are 5.3, 2.7, 3.6, 6.4, and 10.3, what is the overall unemployment rate for the country?

Solution. If the size of the labour force in Canada is N, the numbers of workers in the five regions are

$$0.10N,\ 0.16N,\ 0.38N,\ 0.27N,\ 0.09N.$$

The number of unemployed workers in British Columbia is 5.3% of $0.10N$, or $0.0053N$. In this way we find the numbers of unemployed workers in the five regions to be

$$0.00530N,\ 0.00432N,\ 0.01368N.\ 0.01728N,\ .00927N.$$

The total number of unemployed workers in all of Canada is thus

$$(0.00530 + 0.00432 + 0.01368 + 0.01728 + 0.00927)N = 0.04985N$$

and hence the overall unemployment rate is 4.985%.

We now rephrase this solution in terms of conditional probabilities. The sample space is taken to consist of N points corresponding to the N workers in the Canadian labour force, and each is assigned probability $\frac{1}{N}$. Let A_i be the event that a randomly chosen member comes from the ith geographic region $(i = 1,2,\ldots,5)$, and let

U be the event that he is unemployed. Then we are given that

$$P(A_1) = 0.10,\ P(A_2) = 0.16,\ P(A_3) = 0.38,\ P(A_4) = 0.27,\ P(A_5) = 0.09.$$

We are also given the unemployment rate within each region, and these are represented by conditional probabilities:

$$P(U|A_1) = 0.053,\ P(U|A_2) = 0.027,\ P(U|A_3) = 0.036,\ P(U|A_4) = 0.064,$$
$$P(U|A_5) = 0.103.$$

For instance, $P(U|A_1)$ represents the probability that a randomly selected worker is unemployed under the condition that he comes from British Columbia. The overall unemployment rate is found by multiplying these two sets of probabilities together and adding:

$$P(U) = \sum P(U|A_i)P(A_i) = 0.04985. \quad \square$$

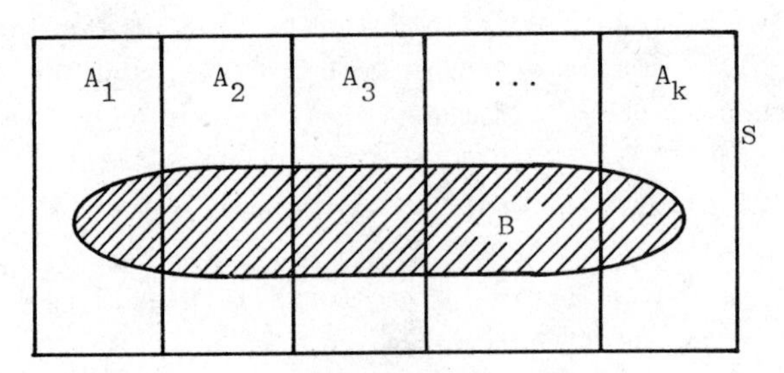

Figure 3.4.1

Partition of S and B into mutually exclusive events
$S = A_1 \cup A_2 \cup \ldots \cup A_k; \quad B = BA_1 \cup BA_2 \cup \ldots \cup BA_k.$

In general, suppose that the sample space S is partitioned into k mutually exclusive subsets $A_1, A_2, \ldots, A_k$, so that each point of S belongs to exactly one of the events A_i (see Figure 3.4.1). Suppose that the (conditional) probability of another event B is known within each of these subsets. Then the (unconditional) probability of event B in the entire sample space is given by

$$P(B) = \sum_{i=1}^{k} P(B|A_i)P(A_i). \tag{3.4.1}$$

If we interpret probabilities as long-run relative frequencies, this result may be established by an obvious generalization of the argument used in Example 3.4.1. □

We have indicated how the conditional probabilities of an event B in subsets $A_1, A_2, \ldots, A_k$ of the sample space may be used to compute the unconditional probability of event B over the entire sample space. Sometimes the opposite procedure is followed; we begin with a probability distribution over the entire sample space and from this deduce conditional probabilities within certain subsets of interest.

Suppose that A and B are two events defined on the same sample space, with $P(A) > 0$. Let us interpret probabilities as long-run relative frequencies, and consider N repetitions of the experiment, where N is very large. Then event A will occur approximately $N \cdot P(A)$ times, and event AB will occur approximately $N \cdot P(AB)$ times. Of those occasions on which A occurs, the fraction on which B also occurs will thus be

$$\frac{N \cdot P(AB)}{N \cdot P(A)} = \frac{P(AB)}{P(A)}.$$

This represents the probability of event B within the subset of the sample space defined by A, and is therefore the conditional probability of B given A, $P(B|A)$. This suggests the following general definition of conditional probability.

Definition. Let A and B be two events defined on the same sample space, with $P(A) > 0$. The conditional probability of B given A is defined by

$$P(B|A) = P(AB)/P(A). \qquad (3.4.2)$$

When $P(A) = 0$, $P(B|A)$ is not defined.

Example 3.4.2. A poker hand of five cards is dealt from a well-shuffled deck. During the deal, one card is exposed and is seen to be the ace of spades. What is the probability that the hand contains exactly two aces?

Solution. As in Example 2.2.1, we take the sample space to consist of $\binom{52}{5}$ equally probable points. Let A be the event that a poker hand contains the ace of spades, and let B be the event that a poker hand contains exactly two aces. The particular hand under consideration is known to contain the ace of spades, and therefore belongs to

A. We wish to know the proportion of such hands which contain exactly two aces; that is, we wish to know the probability of event B within the subset A of the sample space. In conditional probability notation, what we require is $P(B|A)$.

The number of different poker hands which contain the ace of spades is $\binom{51}{4}$. The number of different poker hands which contain the ace of spades and exactly one other ace is $\binom{3}{1}\binom{48}{3}$. Thus

$$P(A) = \binom{51}{4}/\binom{52}{5}, \quad P(AB) = \binom{3}{1}\binom{48}{3}/\binom{52}{5}$$

and by (3.4.2) the required probability is

$$P(B|A) = P(AB)/P(A) = \binom{3}{1}\binom{48}{3}/\binom{51}{4} = 0.208.$$

Given that a poker hand contains the ace of spades, there is about one chance in five that it contains exactly two aces. □

The definition of conditional probability, (3.4.2), is often written as a multiplication rule,

$$P(AB) = P(B|A)P(A). \tag{3.4.3}$$

Since $AB = BA$, we may interchange A and B on the right hand side to obtain

$$P(AB) = P(A|B)P(B). \tag{3.4.4}$$

In Section 3.3, we defined two events A,B to be independent if and only if

$$P(AB) = P(A)P(B).$$

It therefore follows that A and B are independent if and only if

$$P(B|A) = P(B).$$

Equivalently, A and B are independent if and only if

$$P(A|B) = P(A).$$

If A and B are independent, knowledge that one of them has occurred does not change the probability of the other.

For three events, we may apply (3.4.3) twice to obtain:

$$P(ABC) = P(A|BC)P(BC) = P(A|BC)P(B|C)P(C).$$

We can in fact obtain six different formulae for $P(ABC)$ in this way by permuting A,B and C on the right hand side. The generalization to more than three events is straightforward.

We pointed out in the second paragraph of this section that any probability may be written as a conditional probability given the sample space. Consequently, all theorems on probabilities also apply to conditional probabilities. For instance, the probability of the union of two events A_1, A_2 given that a third event B has occurred is given by

$$P(A_1 \cup A_2|B) = P(A_1|B) + P(A_2|B) - P(A_1A_2|B).$$

If we take B to be the entire sample space S, then each of these becomes an unconditional probability, and we obtain

$$P(A_1 \cup A_2) = P(A_1) + P(A_2) - P(A_1A_2),$$

which is formula (3.1.2).

Conditional Probability Models.

Consider an experiment with sample space S and probability distribution $\{p_i\}$, and suppose that we wish to make statements of conditional probability given that an event A has occurred. The conditional probability of any other event B is given by (3.4.2). Let us take B to consist of a single sample point i whose probability is p_i. If i belongs to A, then $AB = \{i\}$, and $P(AB) = p_i$. However if i does not belong to A, then AB is empty, and $P(AB) = 0$. Thus by (3.4.2) we have

$$P(i|A) = \begin{cases} p_i/P(A) & \text{for } i \in A; \\ 0 & \text{otherwise.} \end{cases}$$

The effect of conditioning on event A is to replace the original probability model by a new one which we call a conditional probability model. The new sample space consists only of the points of A; all other points are impossible because we know that A has occurred. The probabilities of all points in A are increased proportionately by the factor $1/P(A)$, which is greater than 1 for $P(A) < 1$, so that

$$\sum_{i \in A} P(i|A) = \sum_{i \in A} p_i/P(A) = P(A)/P(A) = 1.$$

However the ratio of probabilities, or betting odds, for any two points in A remains the same.

Example 3.4.2. (continued). The conditional probability model given event A will be as follows. The sample space will consist of the $\binom{51}{4}$ points of A, corresponding to hands which contain the ace of spades. Since all of these points were equally probable in the original model, they remain equally probable in the conditional model. Each point of A therefore has conditional probability $1/\binom{51}{4}$. The conditional probability of event B is thus found by counting the number of points in A which also belong to B, and dividing by $\binom{51}{4}$. This gives

$$P(B|A) = \binom{3}{1}\binom{48}{3}/\binom{51}{4}$$

as before.

Problems for Section 3.4

†1. Five percent of the television sets from a production line have defective horizontal and vertical controls. In eight percent, only the horizontal control is defective. If a set is found to have a defective horizontal control, what is the probability that the vertical control is also defective?

2. If two balanced dice showed a total score of eight, what is the probability that the first die showed three?

3. Given that a poker hand contains at least one ace, what is the probability that it contains exactly two aces? (Compare with Example 3.4.2.)

†4. Six people are selected at random from a club consisting of ten couples. Find the probability that three couples are selected, given that (a) a particular couple has been chosen;
(b) at least one couple has been chosen.

5. Six balls are randomly distributed into three boxes. What is the probability that at least one box is empty? Given that there is at least one empty box, what is the probability that there are two?

6. Seven balls are randomly distributed into seven boxes. Show that, if exactly two boxes are empty, the odds are 3 to 1 that no box contains three balls.

†7. Each of A, B and C fires one shot at a target. The probability of hitting the target is 0.4 for A, 0.3 for B, and 0.1 for C.
(a) If only one shot hit the target, what is the probability that it was fired by A?

(b) If two shots hit the target, what is the probability that A missed?

8. Each animal in a litter has probability p of being male, independently of other animals in the litter and litter size. The probability that a litter is of size n is a_n $(n = 1,2,\ldots)$.

(a) What is the probability of being able to make at least one male-female mating from a litter of n animals? For what value of p is this probability a maximum?

(b) Show that the proportion of litters from which matings can be made is $G(1) - G(p) - G(1-p)$, where

$$G(x) = a_1 x + a_2 x^2 + a_3 x^3 + \ldots .$$

†9. The probability of a male birth is p, and that of a female birth is $1-p$, with successive births constituting independent trials. The population contains N families, of which n_i have i children $(i = 0,1,2,\ldots;\ n_0 + n_1 + n_2 + \ldots = N)$. Calculate the probability that

(a) a family with i children contains exactly x boys;

(b) a family chosen at random contains exactly i children;

(c) a family chosen at random contains exactly x boys;

(d) a child chosen at random has exactly y brothers and sisters.

10. Let A and B be two mutually exclusive outcomes of an experiment, with probabilities p and q $(p + q \le 1)$. If the experiment is repeated over and over, what is the probability that outcome A occurs before outcome B?

11. In the game of craps, a player rolls two dice. He wins at once if the total on the two dice is 7 or 11; he loses at once if it is 2,3, or 12. Otherwise, he continues rolling the dice until he either wins by throwing his initial total again, or loses by rolling 7. What is the probability that he wins the game?

3.5 Bayes's Theorem.

Let $A_1, A_2, \ldots, A_k$ be a partition of the sample space, and let B be any other event as in Figure 3.4.1. Suppose that we know the probabilities $P(A_i)$, and the conditional probability $P(B|A_i)$ of event B within each of the sets A_i of the partition. The problem is to determine the conditional probabilities $P(A_i|B)$ of the events A_i when B is known to have occurred.

If $P(B) > 0$, the definition (3.4.2) gives

$$P(A_i|B) = P(BA_i)/P(B).$$

But, by (3.4.4), $P(BA_i) = P(B|A_i)P(A_i)$, and hence

$$P(A_i|B) = \frac{P(B|A_i)P(A_i)}{P(B)}. \tag{3.5.1}$$

This result was first given in a paper by the reverend Thomas Bayes in 1763, and is called Bayes's Theorem. Note that the probability $P(B)$ in the denominator can be evaluated using (3.4.1).

The probabilities $P(A_i)$ are called the a priori probabilities, or prior probabilities, of the events A_i. The conditional probabilities $P(A_i|B)$ are called the a posteriori probabilities or posterior probabilities of the events A_i. Bayes's Theorem modifies the prior probabilities to incorporate information provided by the occurrence of event B.

There is no dispute over the mathematical validity of Bayes's Theorem, which is a direct consequence of the definition of conditional probability. It would also be generally agreed that Bayes's Theorem is applicable in the first four examples below, since all of the prior probabilities used can be checked by performing an experiment repeatedly and observing the relative frequencies with which events A_i occur. However, use of Bayes's Theorem is sometimes advocated in situations where the prior probabilities $P(A_i)$ cannot be verified empirically. For instance, they might represent one person's subjective opinion concerning the truth of propositions $A_1, A_2, \ldots, A_k$. There is much controversy over the appropriateness of using Bayes's Theorem in such cases.

Example 3.5.1. The entire output of a factory is produced on three machines which account for 20%, 30%, and 50% of the output, respectively. The fraction of defective items produced is 5% for the first machine, 3% for the second, and 1% for the third.

(a) What fraction of the total output is defective?

(b) If an item is chosen at random from the total output and is found to be defective, what is the probability that it was made by the third machine?

Solution. Let A_i denote the event that a randomly chosen item was made by the ith machine $(i = 1,2,3)$. Let B be the event that a randomly chosen item is defective. Then we are given the following information:

$$P(A_1) = 0.2, \qquad P(A_2) = 0.3, \qquad P(A_3) = 0.5.$$

If the item was made by machine A_1, the probability that it is defective is 0.05; that is $P(B|A_1) = 0.05$. We thus have

$$P(B|A_1) = 0.05, \qquad P(B|A_2) = 0.03, \qquad P(B|A_3) = 0.01.$$

In (a) we are asked to find $P(B)$. This may be obtained from (3.4.1):

$$\begin{aligned} P(B) &= \sum P(B|A_i)P(A_i) \\ &= (0.05)(0.2) + (0.03)(0.3) + (0.01)(0.5) = 0.024. \end{aligned}$$

Hence 2.4% of the total output of the factory is defective.

In (b) we are given that B has occurred, and wish to calculate the conditional probability of A_3. By Bayes's Theorem,

$$P(A_3|B) = \frac{P(B|A_3)P(A_3)}{P(B)} = \frac{(0.01)(0.50)}{0.024} = \frac{5}{24}.$$

Given that the item is defective, the probability that it was made by the third machine is only 5/24. Although machine 3 produces half of the total output, it produces a much smaller fraction of the defective items. Hence the knowledge that the item selected was defective enables us to replace the prior probability $P(A_3) = 0.5$ by the smaller posterior probability $P(A_3|B) = 5/24$.

Example 3.5.2. Two coins, one balanced and one with two tails, are placed in a hat. One coin is selected at random and tossed. You are allowed to see only the up-face, which is tails. What is the probability that the hidden face is also tails?

Solution. Let A be the event that the coin with two tails is selected. Then

$$P(A) = P(\bar{A}) = 0.5.$$

Here A and $\bar{A}$ give a partition of the sample space into $k = 2$ parts. Let B be the event that a tail turns up when the coin is tossed. If the coin has two tails, B is certain to occur. If the coin is balanced, B occurs with probability 0.5. Thus

$$P(B|A) = 1; \quad P(B|\bar{A}) = 0.5.$$

Now (3.4.1) gives

$$P(B) = P(B|A)P(A) + P(B|\bar{A})P(\bar{A}) = 0.75.$$

Given that a tail was obtained, the probability that the two-tailed coin was selected is $P(A|B)$, and Bayes's Theorem gives

$$P(A|B) = \frac{P(B|A)P(A)}{P(B)} = \frac{(1)(0.5)}{(0.75)} = \frac{2}{3}.$$

The probability that the hidden face will also be tails is 2/3.

This result occasionally causes mild surprise. Some people intuitively feel that the probability should be $\frac{1}{2}$. There are, however, three equally probable ways in which tails can come up, and in two of these cases the hidden face will also be tails. If you are not convinced, try performing the experiment yourself several times to check the result. "Coins" of the types required can be made by marking the faces of two poker chips.

Example 3.5.3. Two balls are drawn at random without replacement from an urn containing a white balls and b black balls. What is the probability that the second ball drawn will be white?

Solution. Let W_i be the event that the ith ball drawn is white $(i = 1,2)$. Then for the first ball drawn we have

$$P(W_1) = \frac{a}{a+b}; \qquad P(\bar{W}_1) = \frac{b}{a+b}.$$

If the first ball drawn is white, then at the second draw the urn contains $a-1$ white and b black balls, and the probability of drawing a white ball is

$$P(W_2|W_1) = \frac{a-1}{a+b-1}.$$

Similarly, if the first ball drawn is black, the probability that the second ball will be white is

$$P(W_2|\bar{W}_1) = \frac{a}{a+b-1}.$$

Now we may apply (3.4.1) to obtain

$$\begin{aligned} P(W_2) &= P(W_2|W_1)P(W_1) + P(W_2|\bar{W}_1)P(\bar{W}_1) \\ &= \frac{a-1}{a+b-1} \cdot \frac{a}{a+b} + \frac{a}{a+b-1} \cdot \frac{b}{a+b} = \frac{a}{a+b}. \end{aligned}$$

Thus, if the colour of the first ball drawn is not known, the probability that the second ball will be white is $\frac{a}{a+b}$.

Example 3.5.4. It is known from examining a woman's family tree that she has a 50% chance of being a carrier of the hereditary disease hemophilia. If she is a carrier, there is a 50% chance that any particular son will inherit the disease, with different sons being independent of one another. If she is not a carrier, no son can inherit the disease.

(a) What is the probability that her first son will be normal?

(b) What is the probability that her second son will be normal?

(c) If her first son is normal, what is the probability that her second son will also be normal?

(d) If her first two sons are normal, what is the probability that she is a carrier of the disease?

Solution. Let C be the event that she is a carrier, and let N_i be the event that her ith son is normal $(i = 1,2)$. Then

$$P(C) = P(\bar{C}) = 0.5.$$

If she is a carrier, the probability that her ith son is normal is 0.5, with different sons being independent, so that

$$P(N_1|C) = P(N_2|C) = 0.5;$$

$$P(N_1N_2|C) = P(N_1|C)P(N_2|C) = 0.25.$$

If she is not a carrier, all of her sons must be normal:

$$P(N_1|\bar{C}) = P(N_2|\bar{C}) = P(N_1N_2|\bar{C}) = 1.$$

(a) The probability that her first son will be normal is, by (3.4.1),

$$P(N_1) = P(N_1|C)P(C) + P(N_1|\bar{C})P(\bar{C})$$

$$= (0.5)(0.5) + (1)(0.5) = 0.75.$$

(b) Similarly, the probability that her second son will be normal is

$$P(N_2) = P(N_2|C)P(C) + P(N_2|\bar{C})P(\bar{C}) = 0.75.$$

This probability applies to the situation in which it is not known whether the first son is diseased or healthy.

(c) By definition (3.4.2), the conditional probability of N_2 given that N_1 has occurred is

$$P(N_2|N_1) = P(N_1N_2)/P(N_1).$$

Two normal sons could arise in two ways: from a mother who is a carrier, or from a normal mother. Hence, by (3.4.1),

$$P(N_1N_2) = P(N_1N_2|C)P(C) + P(N_1N_2|\overline{C})P(\overline{C})$$

$$= (0.25)(0.5) + (1)(0.5) = 0.625.$$

The required probability is now

$$P(N_2|N_1) = \frac{0.625}{0.725} = \frac{5}{6}.$$

If the first son is normal, the odds are 5 to 1 that the second son will also be normal. (If the first son is diseased, then the mother is known to be a carrier, and the probability $P(N_2|\overline{N}_1)$ that the second son will be normal is only 50%.)

(d) We now require $P(C|N_1N_2)$, the probability that the woman is a carrier when it is given that her first two sons are normal. By Bayes's Theorem (3.5.1),

$$P(C|N_1N_2) = \frac{P(N_1N_2|C)P(C)}{P(N_1N_2)} = \frac{(0.25)(0.5)}{0.625} = 0.2.$$

Thus if the woman has two normal sons the probability that she is a carrier decreases from 0.5 to 0.2. On the other hand it can be verified that as soon as the woman has a hemophilic son, the probability that she is a carrier increases to one.

<u>Note on independence</u>. Events N_1 and N_2 are independent events in the conditional probability model given C:

$$P(N_1N_2|C) = P(N_1|C)P(N_2|C) = 0.25.$$

They are also independent in the conditional probability model given $\overline{C}$:

$$P(N_1N_2|\overline{C}) = P(N_1|\overline{C})P(N_2|\overline{C}) = 1.$$

However unconditionally they are not independent, for

$$P(N_2|N_1) = \frac{5}{6} > P(N_2).$$

The birth of a normal son increases the probability that the woman is normal, and therefore increases the probability that subsequent sons will be normal.

Example 3.5.5. Suppose that, in Example 3.5.2, the coin need not be selected at random from the hat, but rather an opponent is permitted to select and toss whichever coin he wishes. If the up-face is tails, what are the odds that the down-face is also tails?

Discussion. The conditional probabilities $P(B|A)$ and $P(B|\bar{A})$ in Example 3.5.2 remain unchanged. However the probability of event A is no longer clearly defined. Your opponent might always choose the two-tailed coin; or he might actually choose the coin at random; or he might choose each coin roughly 50% of the time but rarely choose the same coin twice in succession. We may no longer identify the probability of event A with its relative frequency of occurrence in repetitions of a well defined experiment.

An approach which is sometimes suggested is to assign a probability $P(A)$ subjectively. If you thought your opponent was equally likely to select either coin, you would take $P(A) = 0.5$, and then apply Bayes's Theorem as we did in Example 3.5.2. The result would be a posterior probability $P(A|B) = 2/3$, which is numerically the same as that obtained in Example 3.5.2. However the interpretation would be different. If the coin were actually chosen at random, and if you bet 2 to 1 on the two-tailed coin whenever tails came up, the ratio of your winnings to losses would tend to one in a large number of games. There is no such guarantee when $P(A)$ is assigned subjectively. In this case the posterior probability $P(A|B)$ suffers from the same arbitrariness as the prior probability $P(A)$, and there is good reason to doubt its relevance to the betting problem. It is not at all certain that the strategy adopted by your opponent in selecting coins can be described by a simple probability model, and hence there would seem to be no completely satisfactory way of treating this modified game within the framework of probability theory.

Problems for Section 3.5

†1. Suppose that 0.1% of the population is infected with a certain disease. On a medical test for the disease, 98% of those infected give positive results, and 1% of those not infected give positive results. If a randomly chosen person is tested and gives a

positive result, what is the probability that he has the disease?

2. News item (June 19, 1975): "Fifty percent of Canadian men wear coloured underwear, but only twenty percent of American men do". A hotel in Bermuda has four times as many American guests as Canadian guests. The maid has just discovered orange underwear in Room 233. What are the odds that the occupant is Canadian rather than American?

3. A university has four faculties (Arts, Engineering, Mathematics, and Science) with 2000 students each, and two smaller faculties (Environmental Studies and HKLS) with 1000 students each. Fifty-four percent of Arts students are female, and the corresponding percentages for the other faculties are 3%, 29%, 19%, 28% and 59% respectively. What percentage of the university's students are female? What percentage of the female students are in Engineering?

†4. In Example 3.5.4, if the woman has had three normal sons, what is the probability that her fourth son will be normal?

5. The probability that a student knows the correct answer to a question on a multiple choice examination is p. If he doesn't know the correct answer, he chooses one of the k possible answers at random. If the student correctly answered the question, what is the probability that he knew the answer?

6. An automobile insurance company classifies drivers as class A (good risks), class B (medium risks), and class C (poor risks). Class A risks constitute 30% of the drivers who apply to them for insurance, and the probability that such a driver will have one or more accidents in any 12-month period is .01. The corresponding figures for class B are 50% and .03, while those for class C are 20% and .10.
   (a) The company sells Mr. Jones an insurance policy, and within 12 months he has an accident. What is the probability that he is a class A risk? Class B? Class C?
   (b) If a policyholder goes n years without an accident, and years are independent, what is the probability that he is a class A driver?

†7. A speaks the truth 9 times out of 10, and B 7 times out of 8. One ball was drawn from a bag containing 5 white balls and 20 black balls. Both A and B stated that a white ball was drawn. What is the probability that the ball drawn was white?

8. Three convicts A,B, and C know that the warden has randomly chosen one of them to be shot at dawn. Convict A is, naturally enough, curious to know whether he was selected. He realizes that the warden would refuse to answer the direct question "Am I to be

shot?". However, since either B or C is to be spared in any event, he believes that the warden would consent to name one of the other prisoners who was to be spared. He asks, and the warden tells him that C is to be spared. Convict A goes away very unhappy. Before he asked, he had only once chance in three of being shot, but now he knows C is to be spared, and his chance of being shot has risen to 50%. What is the matter with his reasoning?

9. There are 5 urns each containing 5 balls, with the ith urn containing i red balls $(i = 1,2,\ldots,5)$. An urn is selected at random, and from it two balls are chosen at random without replacement. If two red balls are obtained, calculate the probability that the ith urn was selected, for $i = 1,2,3,4,5$.

10. An urn contains 5 red balls and 5 black balls. Three randomly selected balls are transferred to a second urn. Then two additional balls are drawn from the first urn and are found to be red. What is the probability that three black balls were transferred?

†11. A gambler is told that one of two slot machines pays off with probability p_1, while the other pays off with probability $p_2 < p_1$. He selects one slot machine at random and plays that machine n times.

(a) What is the probability that he loses the first time?

(b) What is the probability that he loses all n times?

(c) If he loses all n times, what is the probability that he picked the favourable machine?

*3.6 Union of n Events.

In Section 3.1 we developed formulae for the probability of a union of two or three events. Now suppose that there are n events $A_1, A_2, \ldots, A_n$ defined on the same sample space S, and denote the probability distribution on S by $\{p_i\}$ as usual. We define n sums of probabilities as follows:

$$S_1 = P(A_1) + P(A_2) + \ldots + P(A_n)$$

$$S_2 = P(A_1A_2) + P(A_1A_3) + \ldots + P(A_{n-1}A_n)$$

$$S_3 = P(A_1A_2A_3) + P(A_1A_2A_4) + \ldots + P(A_{n-2}A_{n-1}A_n)$$

$$\ldots$$

$$S_n = P(A_1A_2\ldots A_n).$$

---

* This section may be omitted on first reading; the material which it contains is not used in the sequel.

Note that S_k is a sum of $\binom{n}{k}$ terms $P(A_r A_s \ldots A_t)$ with $r < s < \ldots < t$. Using this notation, the formulae for the probability of a union of 1, 2, or 3 events may be written as follows:

$$P(A_1) = S_1$$
$$P(A_1 \cup A_2) = S_1 - S_2$$
$$P(A_1 \cup A_2 \cup A_3) = S_1 - S_2 + S_3$$

We may now readily anticipate the general result:

<u>Theorem 3.6.1</u>. Let $A_1, A_2, \ldots, A_n$ be n events defined on the same sample space. Then the probability that at least one of them occurs is

$$P(A_1 \cup A_2 \cup \ldots \cup A_n) = S_1 - S_2 + S_3 - + \ldots \pm S_n \qquad (3.6.1)$$

where the sums S_k are as defined above.

<u>Proof</u>. We must show that the right hand side of (3.6.1) is equal to the sum of the probabilities p_i of all points i belonging to one or more of the events $A_1, A_2, \ldots, A_n$.

First, we note that if point i belongs to none of the events $A_1, A_2, \ldots, A_n$, it contributes zero to each of the sums S_k, and therefore contributes zero to the RHS of (3.6.1).

Next, suppose that point i belongs to exactly r of the events $A_1, A_2, \ldots, A_n$, where $0 < r \le n$. Without loss of generality, we may suppose that i belongs to $A_1, A_2, \ldots, A_r$, but not to A_k for any $k > r$. It is necessary to show that the net contribution of point i to the RHS of (3.6.1) is p_i. We note that

(1) i contributes p_i to $P(A_1), P(A_2), \ldots, P(A_r)$, and zero to the remaining terms in S_1. The total contribution of point i to S_1 is therefore $rp_i = \binom{r}{1} p_i$.

(2) i contributes p_i to $P(A_1 A_2), P(A_1 A_3), \ldots, P(A_{r-1} A_r)$, and zero to the remaining terms of S_2. The total contribution of point i to S_2 is therefore $\binom{r}{2} p_i$.

...

(r) i contributes p_i to $P(A_1 A_2 \ldots A_r)$, and zero to the remaining terms of S_r, for a total contribution of $p_i = \binom{r}{r} p_i$.

Since point i belongs to only r of the events, it contributes zero to S_k for every $k > r$. Hence the total contribution of point i to the RHS of (3.6.1) is

$$p_i[\binom{r}{1} - \binom{r}{2} + \binom{r}{3} - + \dots \pm \binom{r}{r}].$$

But, by the Binomial Theorem (2.1.9),

$$0 = (1-1)^r = \binom{r}{0} - \binom{r}{1} + \binom{r}{2} - \binom{r}{3} + - \dots \pm \binom{r}{r}$$

and therefore

$$\binom{r}{1} - \binom{r}{2} + \binom{r}{3} - + \dots \pm \binom{r}{r} = \binom{r}{0} = 1.$$

Hence the total contribution of point i to the RHS of (3.6.1) is p_i as required.

We have thus shown that the RHS of (3.6.1) is equal to the sum of the probabilities p_i of all points i belonging to one or more of the events $A_1, A_2, \dots, A_n$, and the theorem follows. □

Example 3.6.1. Montmort Letter Problem. If n letters are placed at random into n envelopes, what is the probability that at least one letter is placed in the proper envelope?

Solution. Let A_i denote the event that the ith letter is placed in the correct envelope. Then $A_1 \cup A_2 \cup \dots \cup A_n$ is the event that at least one of the letters is correctly placed, and we shall obtain its probability by applying (3.6.1).

There are $n!$ equally probable arrangements of the letters into the envelopes. In exactly $(n-1)!$ of these, the first letter will be placed in its own envelope, so that $P(A_1) = (n-1)!/n!$. Similarly, we have

$$P(A_2) = P(A_3) = \dots = P(A_n) = (n-1)!/n!.$$

We now obtain S_1 as the sum of n equal terms:

$$S_1 = P(A_1) + P(A_2) + \dots + P(A_n) = n\frac{(n-1)!}{n!} = 1.$$

In exactly $(n-2)!$ arrangements, both the first and second letters will be placed in the proper envelopes. Hence

$$P(A_1A_2) = \frac{(n-2)!}{n!} = P(A_iA_j) \quad \text{for} \quad i \neq j.$$

Thus S_2 is the sum of $\binom{n}{2}$ equal terms:

$$S_2 = P(A_1A_2) + P(A_1A_3) + \ldots + P(A_{n-1}A_n) = \binom{n}{2}\frac{(n-2)!}{n!} = \frac{1}{2!}.$$

More generally, a specified set of r letters will be placed in their proper envelopes in $(n-r)!$ of the $n!$ arrangements. Thus

$$P(A_1A_2\ldots A_r) = \frac{(n-r)!}{n!},$$

and S_r is the sum of $\binom{n}{r}$ equal terms:

$$S_r = \binom{n}{r}\frac{(n-r)!}{n!} = \frac{1}{r!}; \qquad r = 1,2,\ldots,n.$$

Now (3.6.1) gives

$$P(A_1 \cup A_2 \cup \ldots \cup A_n) = S_1 - S_2 + S_3 - + \ldots \pm S_n$$

$$= 1 - \frac{1}{2!} + \frac{1}{3!} - + \ldots \pm \frac{1}{n!} \qquad (3.6.2)$$

which is the required probability.

By considering the Taylor's series expansion of e^x, we see that

$$1 - e^{-1} = 1 - \frac{1}{2!} + \frac{1}{3!} - + \ldots$$

Hence, if n is large,

$$P(A_1 \cup A_2 \cup \ldots \cup A_n) \approx 1 - e^{-1} = 0.632\ldots \quad .$$

For large n, the probability of correctly placing at least one letter is very nearly 0.632, whatever the value of n. □

The above problem, which dates back to P.R. Montmort in 1708, can be rephrased and generalized in many ways. For example, if we imagine a hat-check girl who returns n hats to n patrons in a random order, then (3.6.2) gives the probability that at least one hat goes on the right head. It is rather surprising that the probability is practically the same for $n = 10$ as for $n = 10,000$.

The following more general results may be established by counting arguments similar to those used in Theorem 3.6.1.

Theorem 3.6.2. Let $A_1, A_2, \ldots, A_n$ be n events defined on the same sample space. Let $P_{[m]}$ denote the probability that exactly m of the n events will occur, and let P_m denote the probability that at least m of the n events will occur, where $1 \le m \le n$. Then

$$P_{[m]} = S_m - \binom{m+1}{m} S_{m+1} + \binom{m+2}{m} S_{m+2} - + \ldots \pm \binom{n}{m} S_n \qquad (3.6.3)$$

$$P_m = S_m - \binom{m}{m-1} S_{m+1} + \binom{m+1}{m-1} S_{m+2} - + \ldots \pm \binom{n-1}{m-1} S_n \qquad (3.6.4)$$

where $S_1, S_2, \ldots, S_n$ are as defined at the beginning of this section.

Example 3.6.2. In the Montmort Letter Problem, what is the probability that exactly m letters are placed in the correct envelopes?

Solution. In Example 3.6.1 above we showed that $S_r = 1/r!$ $(r = 1,2,\ldots,n)$, and substitution into (3.6.3) gives

$$P_{[m]} = \frac{1}{m!} - \binom{m+1}{m} \frac{1}{(m+1)!} + \binom{m+2}{m} \frac{1}{(m+2)!} - + \ldots \pm \binom{n}{m} \frac{1}{n!}$$

$$= \frac{1}{m!}\left[1 - \frac{1}{1!} + \frac{1}{2!} - + \ldots \pm \frac{1}{(n-m)!}\right].$$

The terms in the square brackets are the initial terms in the Taylor's series expansion of e^{-1}. Hence, if $n - m$ is large,

$$P_{[m]} \approx \frac{1}{m!} e^{-1}.$$

These limiting values sum to one, and are said to form a Poisson probability distribution with mean 1 (see Section 4.3).

Problems for Section 3.6

†1. In the game of bridge, the entire deck of 52 cards is dealt out into four 13-card hands. A "one-suiter" is a hand consisting entirely of cards of one suit. Calculate the probability that at least one of the four hands is a one-suiter.

2. Calculate the probability that exactly m suits are missing from a poker hand $(m = 0,1,2,3)$.

3. A bridge club consists of six married couples. At the beginning of a bridge session, each woman is randomly assigned a male partner for the evening. What is the probability that at least one woman will have her husband for a partner?

†4. N cards numbered $1,2,\ldots,N$ are shuffled and dealt out face up in a row. What is the probability that, for some $i = 1,2,\ldots,N-1$, the card numbered i is followed immediately by the one numbered $i+1$?

5. r balls are distributed at random into n cells, the n^r possible arrangements being equally probable. Let $p_m(r,n)$ be the probability of finding exactly m empty cells.
(a) Without deriving $p_0(r,n)$, show that

$$p_m(r,n) = \binom{n}{m}\left(1 - \frac{m}{n}\right)^r p_0(r,n-m).$$

(b) Use Theorem 1 to show that

$$p_0(r,n) = \sum_{i=0}^{n} (-1)^i \binom{n}{i}\left(1 - \frac{i}{n}\right)^r.$$

(c) Obtain a formula for $p_m(r,n)$ from the results in (a) and (b), and verify that Theorem 2 gives the same formula.

6. There are 10 births over a seven-day period in a small hospital. What is the probability that there were no births on three or more days?

†7. Each month 20 contracts are placed, each one at random, to any of the 10 branches of a company. Find the probability that in one month all branches receive contracts.

*8. Use counting arguments similar to those in Theorem 3.6.1 to establish Theorem 3.6.2.
Hint: use Problem 2.1.5(d).

*9. (a) Derive formula (3.6.3) from formula (3.6.4).
(b) Derive formula (3.6.4) from formula (3.6.3).
Hint: In (b), express P_m as a double sum, change the order of summation, and use (2.1.5).

†*10. A balanced coin is tossed 10 times. What is the probability of obtaining heads at least four times in a row?

*11. In the game of "Stop", each player is dealt 2 cards from a standard deck, and hands are ranked as follows: AA,KK,...,22,AK,AQ, ...,32. For three players, calculate the probability that
(a) at least one hand contains exactly one ace;
(b) at least one hand contains a pair.
Repeat for the case of four players.

Review Problems: Chapter 3

†1. Each question on a multiple choice examination has four possible answers, one of which is chosen at random. What is the probability of achieving 50% or more on a test with four questions of equal value? ten questions of equal value?

2. A drawer contains 4 black socks, 6 green socks, and 2 red socks. Three of the green socks have holes. Two socks are pulled out in the dark.
   (a) What is the probability of obtaining a pair of the same colour?
   (b) If two of the same colour are obtained, what is the probability that they are green?
   (c) If a green pair is obtained, what is the probability that at least one of them has a hole?

†3. Eighty percent of glass jars are produced by Machine 1 and twenty percent by Machine 2. A jar is defective if it has air bubbles or foreign matter or both of these. These two flaws occur independently of one another. With Machine 1, 5% of the output has air bubbles and 2% has foreign matter. With Machine 2, 1% of the output has air bubbles and 3% has foreign matter.
   (a) What is the probability that there are no defectives in ten jars produced on Machine 1?
   (b) What percentage of the total production of jars will be defective?
   (c) If a jar has air bubbles but no foreign matter, what is the probability that it was made by Machine 1?

4. Each of three boxes has two drawers. Each drawer of one box contains a gold coin, each drawer of another box contains a silver coin; and of the third, one drawer contains a gold coin and the other a silver coin. A box is chosen at random, a drawer is opened, and a gold coin is found. What is the probability that the coin in the other drawer of the same box is silver?

†5. Suppose that successive births are independent, and the probability of a male birth is p. Under each of the following conditions, find the probability that there are two boys and two girls in a family of four:
   (i) no further information is available;
   (ii) the eldest child is known to be a son;
   (iii) the eldest and youngest are known to be a son and a daughter, respectively;
   (iv) it is known that there is at least one son and at least one daughter.

6. An airline company has 2-engine and 4-engine planes available for

each flight. For all planes, the probability that a particular engine fails is p, and engines fail independently. A flight cannot be successfully completed unless half of a plane's engines are working.

(a) Find expressions for the probability of a successful flight with a two-engine plane, and with a four-engine plane. For what values of p is the two-engine plane preferable?

(b) On a particular route, two-engine planes are used twice as often as 4-engine planes, and $p = 0.1$. What proportion of flights on this route will be successfully completed? If Aunt Minnie arrived safely on this route, what is the probability that she came on a four-engine plane?

†7. The six members of a family are exposed to an infectious disease with a 7-day incubation period. Each has a 10% chance of contacting the disease, independently of the others. If one or more of them gets the disease, each of the others has a 10% chance of getting it the following week. No individual can get the disease more than once. What is the probability that exactly three members of the family get the disease?

8. Initially, each of urns A and B contains one white ball and one black ball. One ball is drawn simultaneously from each urn and is placed in the other urn. Let p_n, q_n and r_n denote the probabilities that, after n such interchanges, urn A contains two, one, and zero white balls, respectively.

(a) Show that

$$p_n = r_n = \frac{1}{4} q_{n-1}; \quad q_n = p_{n-1} + r_{n-1} + \frac{1}{2} q_{n-1}.$$

(b) Assuming that the limits of these quantities exist as $n \to \infty$, obtain the limiting values.

(c) Show that, for all n, the most probable number of white balls in urn A is 1.

†9. In a knock-out competition, every player gets to play at least one game, and is eliminated as soon as he loses.

(a) Show that there must be $n - 1$ games in a knock-out competition with n players.

(b) If all n players are of equal strength, and the draw for the competition is random, what is the probability that two specified players play each other sometime during the competition?

*10. The probability that a family contains at least one child is α. When a family has i children $(i \geq 1)$, the probability that it grows to $i + 1$ or more is β.

(a) Show that the probability of n children in a completed family is

$$p_n = \alpha(1-\beta)\beta^{n-1} \quad \text{for} \quad n \geq 1.$$

(b) If the probability of a male birth is 0.5, what is the probability that a completed family contains exactly k boys?

†11. An assembly contains three components A,B and C which are each randomly and independently selected from large batches of similar components. The assembly is first tested to see if there is a fault in A; if it passes this test it is then tested to see if there is a fault in B; if it passes this test it is finally tested to see if there is a fault in C. After failing one test, an assembly is never again tested. In a long series of these tests the proportions of all assemblies manufactured that failed the three tests were respectively a,b and c so that a proportion $1-a-b-c$ were satisfactory. What proportions of the original components A, B and C were satisfactory? If these tests were made in the order C,A,B, what proportion would you expect to fail the third test?

12. A lady claims that she can tell by the taste whether the milk was added to the tea or the tea to the milk. In an experiment to test her claim, $2n$ cups of tea are made, n by each method, and are presented for her judgement in a random order. Assume that the lady has no talent, and she merely divides the cups into two groups of n at random.

(a) Determine p, the probability that all cups are classified correctly, as a function of n.

(b) If the experiment is repeated 10 times, what is the probability that all cups will be correctly classified in at most one experiment?

(c) How large should n be in order that the probability in (b) will be at least 0.99?

CHAPTER 4. DISCRETE VARIATES

In most experiments we are interested either in counts of the numbers of times various events occur, or in measurements of quantities such as time, weight, density, etc. Counts and measurements are represented in mathematical probability theory by discrete and continuous random variables, or variates. A variate X is a quantity which is capable of taking on various real values according to chance. A discrete variate has only finitely many or at most countably many possible values. A continuous variate can assume any real value in an interval. In the discrete case, probabilities are found by summing the probability function of X. In the continuous case (to be considered in Chapter 5), probabilities are found by integrating the probability density function of X.

In Section 1 of the present chapter we define discrete variates and probability functions, and introduce notation and terminology. In Sections 2,3, and 4 we introduce two important discrete distributions, the negative binomial and Poisson distributions. In Section 5 we consider the case of two or more variates associated with the same experiment. Finally, in Section 6 the multinomial distribution is derived as a generalization of the binomial distribution.

4.1 Definitions and Notation

A variate X is a real-valued function defined on the sample space S; to each point i in S, X assigns a real number $X(i)$. Variates are also referred to as random variables, chance variables, or stochastic variables.

In this book we shall use a capital letter X,Y,Z, etc. to denote a variate. The corresponding small letter x,y,z, etc. will be used to denote a possible value of the variate. The range of a variate (the set of all its possible values) will be denoted by the corresponding script letter $\mathcal{X},\mathcal{Y},\mathcal{Z}$, etc. In this chapter we continue to require that the sample space S be finite or countable. Then a variate X with domain S can assume at most countably many values, and $\mathcal{X}$ will also be finite or countable. In this case, X is called a discrete variate.

Example 4.1.1. Let X be the total number of heads obtained when a balanced coin is tossed three times. The sample space S for three tosses of a coin contains eight points,

$$S = \{TTT, TTH, THT, HTT, HHT, HTH, THH, HHH\},$$

and they are equally probable. The variate X assigns a real number to each point of S:

$$X(TTT) = 0, \qquad X(TTH) = X(THT) = X(HTT) = 1,$$
$$X(HHT) = X(HTH) = X(THH) = 2, \qquad X(HHH) = 3.$$

Thus the set of possible values of X is

$$\mathcal{X} = \{0,1,2,3\}.$$

The probability distribution of X is the set of probabilities associated with its possible values. The probability that X assumes a particular value x is obtained by summing the probabilities of all the points in S to which the number x is assigned. Here we find

$$P(X = 0) = \frac{1}{8}, \qquad P(X = 1) = \frac{3}{8}, \qquad P(X = 2) = \frac{3}{8}, \qquad P(X = 3) = \frac{1}{8},$$

or equivalently,

$$P(X = x) = \binom{3}{x}/8; \qquad x \in \mathcal{X}.$$

The variate X has a binomial distribution $(n = 3,\ p = \frac{1}{2})$.

Probability Function.

Let X be a discrete variate, and let $x \in \mathcal{X}$ be a possible value of X. The set of all points i in S such that $X(i) = x$ is an event which we shall denote by "$X = x$". For instance, in Example 4.1.1, "$X = 2$" is an event consisting of three sample points,

$$"X = 2" = \{HHT, HTH, THH\}.$$

The probability of event "$X = x$" is the sum of the probabilities of all the points which it contains. Thus, in Example 4.1.1,

$$P(X = 2) = \frac{1}{8} + \frac{1}{8} + \frac{1}{8} = \frac{3}{8}.$$

The function f defined by

$$f(x) = P(X = x); \qquad x \in \mathcal{X} \tag{4.1.1}$$

is called the probability function (p.f. for short) of the discrete variate X. It is clear from (4.1.1) that

$$0 \le f(x) \le 1 \quad \text{for all} \quad x \in \mathcal{X}. \tag{4.1.2}$$

Also, since every point of S is assigned an X-value, the total probability of all X-values must be one:

$$\sum_{x \in \mathcal{X}} f(x) = 1. \tag{4.1.3}$$

Every probability function f must satisfy conditions (4.1.2) and (4.1.3).

A probability function may be specified by giving a table of X-values together with their probabilities. For instance, the probability function for X in Example 4.1.1 is given in the following table:

x	0	1	2	3	Total
f(x)	$\frac{1}{8}$	$\frac{3}{8}$	$\frac{3}{8}$	$\frac{1}{8}$	1

In some instances, particularly when X has many possible values, it may be more convenient to give an algebraic formula for f when this is available.

The probability distribution of a discrete variate X may be represented geometrically by a probability histogram (Section 1.2). We represent the possible values of X by points on a horizontal axis, with which we associate non-overlapping intervals of real numbers. Above the interval corresponding to value x we then construct a rectangle with area f(x). Probabilities thus correspond to areas under the histogram, and the total area of the histogram is 1.

It frequently happens that the individual outcomes of an experiment are of little interest, and that attention is centred on a variate X. It may then be convenient to redefine the probability model for the experiment, taking the sample space to be $\mathcal{X}$, the set of possible X-values, with probabilities given by the probability function f. For instance, in a coin tossing experiment we might be interested only in X, the total number of heads in n tosses. We could then redefine the sample space to be $\mathcal{X} = \{0,1,\ldots,n\}$, with probabilities given by the binomial probability function

$$f(x) = \binom{n}{x} p^x (1-p)^{n-x}, \qquad x \in \mathcal{X}.$$

Cumulative Distribution Function.

There are many situations where one wishes to know the probability that a variate X assumes a value which is less than or equal to some real number t. The <u>cumulative distribution function</u> (c.d.f. for short) is defined as follows:

$$F(t) = P(X \le t) = \sum_{x \le t} f(x), \qquad -\infty \le t \le \infty. \tag{4.1.4}$$

The sum on the right hand side extends over all variate values $x \in X$ such that $x \le t$. We shall generally use a small letter f, g, h, etc. to denote the probability function, and the corresponding capital letter F, G, H, etc. to denote the c.d.f. of a discrete variate.

The cumulative distribution function of X is a non-decreasing function with

$$F(-\infty) = 0; \qquad F(+\infty) = 1. \tag{4.1.5}$$

If X is discrete, F is a step-function. There is a discontinuity, or step, at each variate value $x \in X$, and the height of the step is $f(x)$. The c.d.f. of X in Example 4.1.1 is as follows:

t	$t < 0$	$0 \le t < 1$	$1 \le t < 2$	$2 \le t < 3$	$3 \le t$
$F(t)$	0	$\frac{1}{8}$	$\frac{4}{8}$	$\frac{7}{8}$	1

A graph of this function is shown in Figure 4.1.1. Although $F(t)$ is

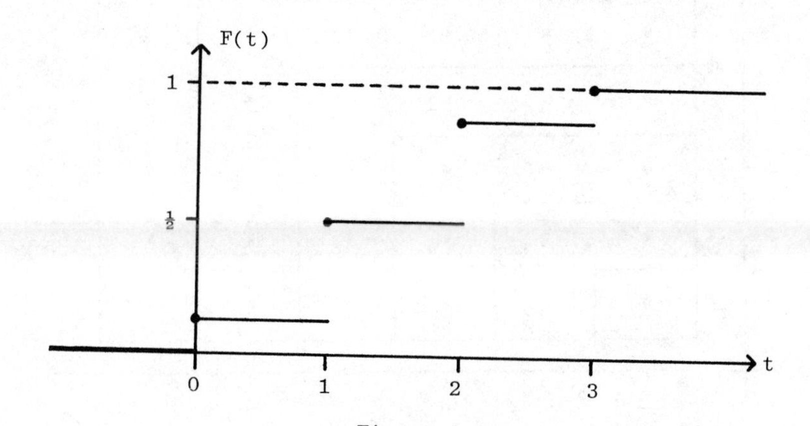

Figure 4.1.1

C.d.f. for X, the number of heads in 3 tosses of a balanced coin.

defined for all real t, it remains constant between successive X-values, and hence F may be completely specified by giving $F(x)$ for each $x \in X$. Thus in Example 4.1.1 we could tabulate the c.d.f. more simply as follows:

x	0	1	2	3
$F(x)$	$\frac{1}{8}$	$\frac{4}{8}$	$\frac{7}{8}$	1

Example 4.1.2. Let X denote the larger of the two numbers obtained when a balanced die is rolled twice. Find the probability function and c.d.f. of X.

Solution. The sample space for two rolls of a die contains the 36 pairs (i,j) with $1 \le i \le 6$ and $1 \le j \le 6$ (see Figure 4.1.2), and these are equally probable. The event "$X = 1$" consists of the single point $(1,1)$, and has probability $f(1) = \frac{1}{36}$. The event "$X = 2$" consists of three points $(1,2),(2,2),(2,1)$ and has probability $f(2) = \frac{3}{36}$.

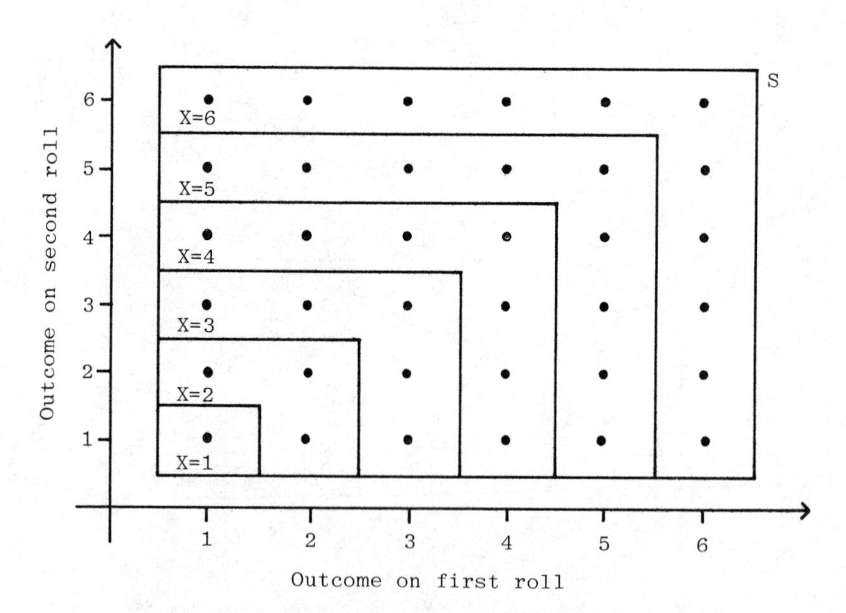

Figure 4.1.2

Partition of sample space by the variate X, where X is the larger outcome in two rolls of a die.

The events corresponding to the other possible X-values are shown in Figure 4.1.2, and their probabilities may be determined in a similar fashion. The probability function and c.d.f. of X are thus as follows:

x	1	2	3	4	5	6
f(x)	1/36	3/36	5/36	7/36	9/36	11/36
F(x)	1/36	4/36	9/36	16/36	25/36	1

In this case, f and F may be given by simple algebraic formulae:

$$f(x) = \frac{2x-1}{36}, \quad F(x) = \frac{x^2}{36}; \quad x = 1,2,\ldots,6.$$

These results may be obtained without listing all of the sample points, by a method which readily extends to more than two rolls. By the definition of F, we have

$$F(x) = P(X \le x) = P(\text{larger number is at most } x)$$
$$= P(\text{both numbers are at most } x).$$

The probability that the outcome of a single roll is at most x is $x/6$, for $x = 1,2,\ldots,6$. Since rolls are independent, the probability that both outcomes are at most x is $(x/6)^2$, and hence

$$F(x) = x^2/36; \quad x = 1,2,\ldots,6.$$

The probability function is now obtained as follows:

$$f(x) = P(X = x) = P(X \le x) - P(X \le x - 1) = F(x) - F(x - 1)$$
$$= \frac{x^2}{36} - \frac{(x-1)^2}{36} = \frac{2x-1}{36}; \quad x = 1,2,\ldots,6.$$

More generally, if X is the largest number obtained in k rolls of a balanced die, then

$$F(x) = \left(\frac{x}{6}\right)^k, \quad f(x) = \frac{x^k - (x-1)^k}{6^k}; \quad x = 1,2,\ldots,6.$$

Example 4.1.3. Let X denote the number of empty boxes when eight balls are distributed at random into six boxes. Find the probability function of X.

Solution. The possible occupancy numbers are listed together with their probabilities in Table 2.4.1. The number of empty boxes ranges from 0 to 5, so that $X = \{0,1,\ldots,5\}$. There are two sets of occupancy numbers such that no boxes are empty, and hence

$$f(0) = P(X = 0) = 0.024005 + 0.090021 = 0.114026.$$

There are three cases in which just one box is empty, so that

$$f(1) = P(X = 1) = 0.030007 + 0.240055 + 0.180041 = 0.450103.$$

The probabilities of the remaining X-values may be obtained in a similar way, and are as follows:

x	0	1	2	3	4	5
f(x)	0.114026	0.450103	0.364584	0.069016	0.002268	0.000004

These results may also be obtained using (3.6.3), where A_i is defined to be the event that the ith box is empty $(i = 1,2,\ldots,6)$.

Example 4.1.4. Discrete Uniform Distribution.

A discrete variate X is said to have a uniform distribution if all of its possible values are equally probable. We thus have

$$f(x) = k; \quad x \in X,$$

where k is a constant. Since f(x) must sum to unity by (4.1.3), it follows that X must contain finitely many points, and $k = \frac{1}{N}$ where N is the number of points in X.

In general, because of (4.1.3) a probability function need be defined only up to a multiplicative constant. The value of the constant may then be determined from the fact that the total probability is 1.

Equality of Variates.

Let X and Y be two variates defined on the same sample space S. If, for each point i in S, the X-value X(i) is equal to the Y-value Y(i), we say that X and Y are identically equal, and write $X \equiv Y$. In this case, the values of X and Y will be equal whatever the outcome of the experiment might be. In particular, $X \equiv c$ means that X is a constant variate (assigns the same value c to each point of S).

A weaker sort of equality can arise if the X-value is equal to the Y-value for some, but not all, of the points in S. Then, if an appropriate outcome occurs, X and Y will be observed equal. By the notation "X = Y" we shall mean the event consisting of all sample points i in S for which $X(i) = Y(i)$.

A third type of "equality" can occur for two variates X,Y not necessarily defined on the same sample space. If X and Y have the same set of possible values with the same probabilities, we say that they have <u>the same distribution</u>, or that they are <u>identically distributed</u>, and write $X \sim Y$.

<u>Example 4.1.5</u>. Let X be the number of heads and Y the number of tails when a coin is tossed 4 times. Then "X = Y" is the event consisting of all sequences of 2 heads and 2 tails. Its probability is

$$P(X = Y) = \binom{4}{2}p^2(1-p)^2.$$

It would be incorrect to write $X \equiv Y$, since this would imply that the number of heads must always equal the number of tails. We can, however, write $X + Y \equiv 4$, because the number of heads plus the number of tails must always be 4.

X and Y clearly have the same set of possible values. Their probability functions f and g are as follows:

$$f(x) = \binom{4}{x}p^x(1-p)^{4-x}; \qquad x = 0,1,\ldots,4$$

$$g(y) = \binom{4}{y}(1-p)^y p^{4-y}; \qquad y = 0,1,\ldots,4.$$

These functions are the same if and only if $p = \frac{1}{2}$. Thus X and Y are identically distributed $(X \sim Y)$ when $p = \frac{1}{2}$, but not otherwise.

<u>Composite Functions</u>.

We recall that a variate X is a real-valued function defined on the sample space S. Suppose that h is another real-valued function defined on $\mathcal{X}$, the range of X. The composite function $h(X)$ associates with each point i in S a real number $h(x)$, where $x = X(i)$. Therefore $h(X)$ is also a variate defined on S. <u>A real-valued function of a variate is itself a variate</u>.

<u>Example 4.1.3</u>. Find the probability function of $(X-7)^2$, where X is the total score in two rolls of a balanced die.

<u>Solution</u>. The possible values of X are $2,3,\ldots,12$; their probabilities $f(x)$ may be obtained as in Example 3.2.1, and are as follows:

x	2	3	4	5	6	7	8	9	10	11	12
f(x)	1/36	2/36	3/36	4/36	5/36	6/36	5/36	4/36	3/36	2/36	1/36
y	25	16	9	4	1	0	1	4	9	16	25

Define $Y \equiv (X-7)^2$, and let g denote the probability function of Y. The Y-value corresponding to each X-value is given in the last row of the table. The event "$Y = 0$" corresponds to only one value of X, and hence

$$g(0) = P(Y=0) = P(X=7) = f(7) = \frac{6}{36}.$$

However, the event "$Y = 1$" corresponds to two X-values, and hence

$$g(1) = P(Y=1) = P(X=6) + P(X=8) = f(6) + f(8) = \frac{10}{36}.$$

Continuing in this way, we find the probability function of Y to be as follows:

y	0	1	4	9	16	25	Total
g(y)	6/36	10/36	8/36	6/36	4/36	2/36	1

In general, if $Y \equiv h(X)$ and g denotes the probability function of Y, we may find $g(y)$ by summing $f(x)$ over all values of x such that $h(x) = y$. In the special case where h is one-to-one, the equation $h(x) = y$ has a unique solution $x = h^{-1}(y)$, and then

$$g(y) = f(h^{-1}(y)).$$

Problems for Section 4.1

†1. Four letters addressed to individuals A,B,C, and D are randomly placed into four addressed envelopes, with one letter in each envelope. Find the probability function of X, the number of letters which go in the proper envelopes.

2. Construct a probability histogram for the number of heads in ten tosses of a balanced coin.

3. A,B, and C fire simultaneously at a target. The probability that A hits the target is 1/2, the probability that B hits the target is 1/3, and the probability that C hits the target is 1/4. Tabulate the probability function for the total number of hits.

†4. Let X be a non-negative discrete variate with cumulative distri-

bution function

$$F(x) = 1 - 2^{-x} \text{ for } x = 0,1,2,\ldots .$$

(a) Find the probability function of X.

(b) What is the probability of the event $X = 5$? $X \geq 5$?

5. An urn contains 4 red balls and 6 black balls. Three balls are selected one at a time, the contents of the urn being thoroughly mixed before each drawing. Let X denote the number of red balls drawn. Tabulate the probability function of X in each of the following cases:
   (a) after a ball has been drawn it is replaced in the urn before the next drawing;
   (b) balls are not replaced;
   (c) after a ball has been drawn it is replaced by one of the opposite colour before the next drawing.

6. Two balls are drawn at random from an urn containing ten balls numbered $0,1,\ldots,9$. Let X be the larger of the numbers on the two balls, and let Y be their total.
   (a) Find the p.f. of X and the p.f. of Y if the sampling is without replacement.
   (b) Repeat (a) for sampling with replacement.

†7. A point starts at the origin and takes unit steps to the left (probability $\frac{1}{2}$) or to the right (probability $\frac{1}{2}$). Assuming that steps are independent, find the probability function of D^2, where D is the distance from the origin after six steps.

8. A and B play a game consisting of n independent rounds. In each round A wins \$1 from B with probability 3/8, or loses \$3 to B with probability 1/8, or breaks even with probability 1/2. Let X represent A's total winnings. Tabulate the probability function of X and draw a probability histogram for each of the cases $n = 1,2,3$.

9. Six people are selected at random from a club consisting of 10 couples. Find the probability function of X, the number of couples selected.

†10. Let X be a discrete variate with probability function

$$f(x) = kx \quad \text{for} \quad x = 1,2,\ldots,n.$$

Evaluate the constant k, and find the cumulative distribution function of X.

11. Let Y denote the smallest number obtained in k rolls of a balanced die. Show that

$$P(Y > y) = (1 - \frac{y}{6})^k,$$

and hence obtain the probability function of Y.

4.2 Waiting Times in Bernoulli Trials

In Section 3.2, we defined Bernoulli trials to be independent repetitions of an experiment with only two possible outcomes, S (success) and F (failure). We considered a fixed number n of Bernoulli trials, and showed that the number of successes in n trials follows a binomial distribution.

We now consider a quite different situation in which the total number of trials is not fixed in advance. Instead we decide to keep repeating the experiment until the rth success has occurred and then stop. We are then interested in the waiting time for the rth success. The number of successes is fixed in advance; the number of failures and the total number of trials are random. We define X to be total number of failures obtained before the rth success occurs. Our aim is to derive expressions for f, the probability function of X.

First let us consider the case $r = 1$. Then $f(x)$ is the probability of obtaining exactly x failures before the first success. This can happen in only one way: we must get failures on the first x trials and a success on the $(x+1)$st trial. But, since trials are independent,

$$P(\underbrace{FF...F}_{x \text{ times}}S) = \underbrace{(1-p)(1-p)...(1-p)}_{x \text{ times}}p$$

where p is the probability of success. Therefore

$$f(x) = p(1-p)^x; \quad x = 0,1,2,... . \tag{4.2.1}$$

This is called a <u>geometric distribution</u> because the probabilities form a geometric series with common ratio $1-p$. Since $0 < p < 1$, the total probability is

$$p[1+(1-p)+(1-p)^2+...] = \frac{p}{1-(1-p)} = 1$$

as it should be.

In the general case, $f(x)$ is the probability of obtaining exactly x failures before the rth success. In order for this to happen, we must obtain a success on the $(r+x)$th trial, preceded by exactly $r-1$ successes and x failures in the first $r+x-1$ trials. The probability of obtaining $r-1$ successes on the first $r+x-1$ trials is given by

$$\binom{r+x-1}{r-1}p^{r-1}(1-p)^{x}.$$

The probability of obtaining a success on trial $r+x$ is p. Because trials are independent, we may multiply these two probabilities to obtain

$$f(x) = \binom{r+x-1}{r-1}p^{r}(1-p)^{x}; \quad x = 0,1,2,\ldots \quad . \tag{4.2.2}$$

For a reason shortly to become apparent, this is called a <u>negative binomial distribution</u>. The ratio of successive terms in (4.2.2) is

$$\frac{f(x)}{f(x-1)} = \frac{r+x-1}{x}(1-p) \tag{4.2.3}$$

and this result can be used for recursive computation of $f(x)$.

It follows from (2.1.8), (2.1.3), and (2.1.1) that

$$\binom{r+x-1}{r-1} = \binom{r+x-1}{x} = (r+x-1)^{(x)}/x!$$

$$= \frac{(r+x-1)(r+x-2)\ldots(r+1)(r)}{x!} .$$

We now multiply each of the x terms in the numerator by -1 to obtain

$$\binom{r+x-1}{r-1} = (-1)^{x}\,\frac{(-r-x+1)(-r-x+2)\ldots(-r-1)(-r)}{x!} .$$

Now reversing the order of multiplication in the numerator and again using (2.1.1) and (2.1.3) gives

$$\binom{r+x-1}{r-1} = (-1)^{x}\,\frac{(-r)(-r-1)(-r-2)\ldots(-r-x+1)}{x!}$$

$$= (-1)^{x}\,\frac{(-r)^{(x)}}{x!} = (-1)^{x}\binom{-r}{x}.$$

Upon substitution in (4.2.2), we obtain another formula for the probability function of a negative binomial distribution:

$$f(x) = \binom{-r}{x}p^{r}(p-1)^{x}; \quad x = 0,1,2,\ldots \quad . \tag{4.2.4}$$

The binomial theorem (2.1.9) with negative index, $n = -r$, gives

$$\sum_{x=0}^{\infty} f(x) = p^r \sum_{x=0}^{\infty} \binom{-r}{x}(p-1)^x = p^r(1+p-1)^{-r}$$

and hence the total probability is 1.

Now let Y be the total number of trials required to obtain the rth success, and let g be the probability function of Y. Then Y is identically equal to the number of successes plus the number of failures: $Y \equiv r + X$. It follows that

$$g(y) = P(Y = y) = P(X = y - r) = f(y - r); \quad y = r, r+1, r+2, \ldots .$$

Two different expressions for $g(y)$ can be now obtained from (4.2.2) and (4.2.4).

In the above derivation, we have taken r to be a positive integer. In fact, (4.2.4) defines a proper probability distribution for all real positive r. Although negative binomial distributions with fractional values of r do not arise as distributions of waiting times in Bernoulli trials, they can be useful in other situations.

Example 4.2.1. A typist has an 83% probability of correctly typing a page. Pages containing errors must be retyped. Find the probability distribution for the total number of pages which must be typed in order to complete a 10-page report.

Solution. We assume that successive pages are Bernoulli trials with probability of success $p = 0.83$. Trials are to be continued until the 10th success occurs. Let X be the number of failures (pages with errors) before the 10th success. Taking $r = 10$ in (4.2.2), we find that X has probability function

$$f(x) = \binom{9+x}{9}p^{10}(1-p)^x; \quad x = 0, 1, 2, \ldots .$$

Let Y be the total number of pages which must be typed, with probability function g. Then $Y \equiv 10 + X$, and

$$g(y) = P(Y = y) = P(X = 10 - y) = f(y - 10) = \binom{y-1}{9}p^{10}(1-p)^{y-10};$$
$$y = 10, 11, 12, \ldots .$$

These probabilities may be calculated recursively, and the first few of them are given in the following table:

y	10	11	12	13	14	15	≥16
g(y)	0.155	0.264	0.247	0.168	0.093	0.044	0.029

Although the typist makes errors on only about one page in six, there is a fairly large probability that 14 or more pages will need to be typed in order to complete a 10-page report. A histogram of these probabilities would show a long tail to the right, and this is typical of most waiting-time distributions.

Example 4.2.2. An urn contains a white balls and b black balls. Balls are drawn at random one by one until the rth white ball has been obtained. Let X denote the number of black balls obtained before the rth white ball is drawn. Find the probability function of X for (a) sampling with replacement; (b) sampling without replacement.

Solution. (a) We remarked in Section 3.2 that, under sampling with replacement, successive draws are Bernoulli trials. The probability function of X is thus given by (4.2.2) with $p = \frac{a}{a+b}$.

(b) Under sampling without replacement, successive draws are no longer independent, and the results obtained above for Bernoulli trials are not applicable. We therefore obtain an expression for $f(x)$ from first principles. In order that there be exactly x black balls prior to the rth white ball, there must be exactly $r-1$ white and x black balls in the first $r+x-1$ draws, followed by a white ball on the $(r+x)$th draw. The probability of obtaining $r-1$ white and x black balls in the first $r+x-1$ draws is

$$\binom{a}{r-1}\binom{b}{x}\Big/\binom{a+b}{r+x-1}.$$

There will then be $a+b-(r+x-1)$ balls left in the urn, of which $a-(r-1)$ are white, so that the probability of a white ball on the $(r+x)$th draw is $(a-r+1)/(a+b-r-x+1)$. It follows that

$$f(x) = \frac{a-r+1}{a+b-r-x+1}\binom{a}{r-1}\binom{b}{x}\Big/\binom{a+b}{r+x-1}; \quad x = 0,1,2,\ldots,b.$$

In this case it is necessary to have $r \le a$, because no white ball can be drawn more than once. Also, since there are only b black balls which can be drawn, $f(x) = 0$ for $x > b$.

Problems for Section 4.2

1. A baseball player has a 20% chance of hitting the ball each time at bat, with successive times at bat being independent. What is the probability that
   (a) he gets his first hit on his fifth time at bat?
   (b) he gets his second hit on his tenth time at bat?
   (c) he gets his first and second hits on his fifth and tenth times at bat?

†2. A batch of g good light bulbs has become mixed up with a batch of b bad ones. Bulbs are randomly selected one at a time without replacement and tested until all of the bad ones have been found. Find the probability function of X, the total number of bulbs tested.

3. A drawer contains 4 black socks and 6 red socks. Socks are selected at random without replacement until a pair of red socks has been obtained. Find the probability function of X, the number of black socks left in the drawer.

4. Find the value of x for which the negative binomial probability (4.2.2) is maximized. Show that the maximum is unique unless $r-1$ is a multiple of p.

5. Let X be the number of failures before the rth success in repeated Bernoulli trials. Show that

$$P(X \le x) = \sum_{i=0}^{x} \binom{x+r}{i} p^{x+r-i}(1-p)^{i},$$

and hence obtain the probability function of X.

†6. A drawer contains 20 socks of five different colours, there being four of each colour. Socks are selected at random without replacement until two of the same colour have been obtained. Find the probability function of the number of socks left in the drawer.

†7. Balls are randomly distributed one at a time into n cells until some cell contains two balls. Find the probability that more than r balls are required, and hence obtain the probability function of the total number of balls distributed.

8. A very large population is divided into n age groups each containing the same number of individuals. Individuals are randomly selected one at a time until all age groups are represented. Let X_n be the number of individuals selected, with probability function f_n. Show that

   (a) $f_2(x) = 1/2^{x-1}$ for $x = 2,3,\ldots;$

(b) $f_3(x) = 2(2^{x-2} - 1)/3^{x-1}$ for $x = 3,4,\ldots;$

(c) $f_4(x) = 3(3^{x-2} - 2^{x-1} + 1)/4^{x-1}$ for $x = 4,5,\ldots$.

†*9. Balls are randomly distributed one at a time into n cells until all the cells are occupied. Find the probability function of the total number of balls required. (Hint: see problem 3.6.5.)

*10. A balanced coin is tossed until two consecutive heads are obtained. Let p_n be the probability that n tosses are required. Show that

$$p_n = f_{n-1}/2^n \quad \text{for} \quad n = 1,2,3,\ldots$$

where f_n is the nth Fibonacci number:

$$f_0 = 0, \quad f_1 = 1, \quad f_n = f_{n-1} + f_{n-2} \quad \text{for} \quad n \geq 2.$$

4.3 The Poisson Distribution

Let μ be a positive real number. A discrete variate X with probability function

$$f(x) = \mu^x e^{-\mu}/x!; \qquad x = 0,1,2,\ldots \tag{4.3.1}$$

is said to have a Poisson distribution. The parameter μ is called the mean of the distribution. If one interprets probabilities as long-run relative frequencies, then μ is the average or mean of the X-values that one would obtain in a very large number of repetitions of the experiment. The mean is the central value of the probability distribution in the same sense that the centre of mass is the central point of a mass distribution. See Chapter 5 for further details.

The total probability in (4.3.1) is

$$\sum_{x=0}^{\infty} f(x) = \sum_{x=0}^{\infty} \mu^x e^{-\mu}/x! = e^{-\mu}\left[1 + \mu + \frac{\mu^2}{2!} + \frac{\mu^3}{3!} + \ldots\right].$$

The series in square brackets is the Taylor's series expansion of e^{μ}, and hence the total probability is $e^{-\mu}e^{\mu} = 1$ for all μ. The ratio of successive terms in (4.3.1) is

$$\frac{f(x)}{f(x-1)} = \frac{\mu}{x}, \tag{4.3.2}$$

and this result may be used for recursive calculation of $f(x)$.

The Poisson distribution has a great many applications. Some of these are considered below and in the next section.

Poisson approximation to the binomial distribution.

If n is large in comparison with np, then the binomial distribution with parameters n and p may be approximated by a Poisson distribution with mean $\mu = np$; that is

$$\binom{n}{x} p^x (1-p)^{n-x} \sim \mu^x e^{-\mu}/x! \qquad (4.3.3)$$

Proof. We substitute $p = \mu/n$ in the left hand side of (4.3.3) to obtain

$$\binom{n}{x} p^x (1-p)^{n-x} = \frac{n(n-1)\dots(n-x+1)}{x!} \left(\frac{\mu}{n}\right)^x \left(1-\frac{\mu}{n}\right)^{n-x}$$

$$= \frac{\mu^x}{x!}\left[\frac{n}{n} \cdot \frac{n-1}{n} \cdot \dots \cdot \frac{n-x+1}{n}\right]\left(1-\frac{\mu}{n}\right)^{-x}\left(1-\frac{\mu}{n}\right)^{n}.$$

Now, keeping x fixed, we let $n \to \infty$ and $p \to 0$ in such a way that $\mu = np$ remains constant. The expression within the square brackets tends to 1 because it is a product of finitely many terms, each of which tends to 1. For the same reason, $(1-\frac{\mu}{n})^{-x} \to 1$. Finally, by a well-known result from calculus,

$$\lim_{n\to\infty} \left(1-\frac{\mu}{n}\right)^n = e^{-\mu},$$

and combining these results gives (4.3.3).

Remarks. (i) Because of (4.3.3), the Poisson distribution is sometimes called the law of rare events. In the approximation, one equates the expected frequency of success np with the mean μ of the Poisson distribution. The approximation applies when np is small in comparison with n; that is, when successes are rare events.

(ii) We may also obtain a Poisson approximation to the binomial distribution for the case when n is large and $n(1-p)$ is small. Letting $q = 1-p$ and $y = n-x$, we have

$$\binom{n}{x} p^x (1-p)^{n-x} = \binom{n}{y} q^y (1-q)^{n-y} \sim \mu^y e^{-\mu}/y!$$

where $\mu = nq$ is the expected number of failures.

(iii) To obtain a Poisson approximation to the hypergeometric distribution, we first apply the binomial approximation (2.5.2), followed by the Poisson approximation (4.3.3).

Example 4.3.1. The probability function of a binomial distribution with parameters $n = 100$ and $p = .02$ is

$$f(x) = \binom{100}{x}(.02)^x(.98)^{100-x}; \quad x=0,1,2,\ldots \ .$$

Since $\mu = np = 2$, the Poisson approximation has probability function

$$g(x) = 2^x e^{-2}/x!; \quad x = 0,1,2,\ldots \ .$$

We compare these functions for small values of x:

x	0	1	2	3	4	5	6
f(x)	0.1326	0.2707	0.2734	0.1823	0.0902	0.0353	0.0114
g(x)	0.1353	0.2707	0.2707	0.1804	0.0902	0.0361	0.0120

The agreement is very close (ratio near 1) for small values of x, but it gets progressively worse as x increases beyond 4.

Example 4.3.2. A rare type of blood occurs in the population with relative frequency 0.001. If n people are tested, what is the probability that at least two of them have the rare blood type? How large must n be if this probability is to exceed 95%?

Solution: The probability that x of n people tested have the rare blood type is given by a binomial distribution,

$$f(x) = \binom{n}{x}(.001)^x(.999)^{n-x}; \quad x=0,1,2,\ldots,n.$$

The probability that at least two have the rare blood type is then

$$\begin{aligned} 1-f(0)-f(1) &= 1-(.999)^n - n(.001)(.999)^{n-1} \\ &= 1-(.999)^{n-1}(.999+.001n). \end{aligned}$$

In order for this probability to exceed 95%, it is necessary to have

$$(.999 + .001n)(.999)^{n-1} < 0.05.$$

Using an electronic calculator, we may show by trial and error that this inequality holds only for $n \ge 4742$. It is necessary to test at least 4742 people in order to be 95% sure of obtaining two or more with the rare blood type.

The calculations are slightly simpler if we approximate $f(x)$ by the Poisson probability function

$$g(x) = \mu^x e^{-\mu}/x!; \quad x = 0,1,2,\ldots$$

where $\mu = .001n$. The probability that at least two have the rare blood type is now approximated by

$$1 - g(0) - g(1) = 1 - (1+\mu)e^{-\mu}.$$

In order for this to exceed 95%, we require that

$$(1+\mu)e^{-\mu} < 0.05.$$

Trial and error shows that this inequality is satisfied for $\mu \geq 4.744$. We thus have $n = 1000\mu \geq 4744$, which agrees closely with the exact result obtained above.

Example 4.3.3. Suppose that n organisms are randomly distributed throughout a volume V of fluid, so that the probability that a specific organism is located in a given drop of volume D is D/V. The placement of the n organisms either in or outside the drop may be thought of as n independent trials with constant probability D/V of success at each trial. Hence the number X of organisms in a drop of volume D has a binomial distribution with probability function

$$f(x) = \binom{n}{x}\left(\frac{D}{V}\right)^x\left(1 - \frac{D}{V}\right)^{n-x}; \quad x = 0,1,\ldots,n.$$

In practice, n will usually be very large while D/V is quite small. Hence the binomial distribution will be well approximated by a Poisson distribution with mean $\mu = \frac{nD}{V} = \lambda D$, where $\lambda = \frac{n}{V}$ is the number of organisms per unit volume of solution:

$$f(x) \sim (\lambda D)^x e^{-\lambda D}/x!; \quad x = 0,1,2,\ldots .$$

This result can also be obtained directly from the arguments in the next section.

Problems for Section 4.3

1. A book of 250 pages contains 200 misprints. Assume that each misprint was assigned to a page at random, independently of other misprints. Give an expression for the probability that
   (a) a particular page contains fewer than two misprints;
   (b) a four-page section contains more than five misprints.
   Use the Poisson approximation to evaluate these probabilities.

†2. A sociologist decides that it would be useful, though not essential, to include in his experimental group at least two persons whose an-

nual income exceeds $1,000,000. The large population from which the group is to be randomly selected contains 1% of such people.

(a) How large a group must he select in order to be 95% sure of including at least two millionaires?

(b) Considerations of time and expense subsequently force him to limit the group size to 300. How sure can he now be about including two millionaires?

3. Of 4,000 voters in an election, only 50 voted for candidate A. If a sample of size 50 is selected at random, what is the probability that at most 2 voted for candidate A?

4. A gasoline company runs a contest. There are 500,000 tickets, of which 500 are winners, and a motorist receives one ticket with each fill-up.

   (a) If a motorist has ten fill-ups during the contest, what is the probability that he wins at least one prize?

   (b) A particular dealer distributed 2000 tickets. What is the probability that he had at least one winner?

4.4 Random Events in Time and Space

We have encountered the Poisson distribution as an approximation to the binomial distribution (Section 4.3) and in the Montmort letter problem (Example 3.6.2). The following derivation justifies some of the many other uses of the Poisson distribution.

Consider a system which is subject to instantaneous changes due to the occurrence of random events - for example, the splitting of physical particles, or the arrival of telephone calls at an exchange. Suppose that the following conditions hold:

(1) The numbers of changes in disjoint time intervals are independent.

(2) If h is sufficiently small, the probability of one change in $(t,t+h)$ is λh where λ is constant with respect to t; the probability of more than one change in $(t,t+h)$ is negligible (i.e. events occur singly and not in pairs or groups).

These assumptions define a Poisson process, and are sufficient to determine $f_t(x)$, the probability of x changes during a time interval of length t.

We first obtain an expression for $f_{t+h}(x)$. If h is sufficiently small, the probability of more than one change in $(t,t+h)$ is negligible. Then x changes in $(0,t+h)$ can occur in only two ways: either there must be x changes in $(0,t)$ and none in $(t,t+h)$, or else there must be $x-1$ changes in $(0,t)$ and one in $(t,t+h)$.

Therefore, if h is sufficiently small,

$$f_{t+h}(x) = (1-\lambda h)f_t(x) + \lambda h f_t(x-1).$$

After rearranging and dividing by h we obtain

$$\frac{f_{t+h}(x) - f_t(x)}{h} = \lambda f_t(x-1) - \lambda f_t(x).$$

Now letting h tend to zero gives

$$\frac{d}{dt} f_t(x) = \lambda f_t(x-1) - \lambda f_t(x). \tag{4.4.1}$$

In the above argument we have assumed $x \geq 1$. However (4.4.1) also holds for $x = 0$ if we define $f_t(x) \equiv 0$ for $x < 0$.

Since there must be zero changes in zero time, we have $f_0(0) = 1$, and $f_0(x) = 0$ for $x > 0$. Subject to these boundary conditions, the unique solution of (4.4.1) is

$$f_t(x) = (\lambda t)^x e^{-\lambda t}/x!; \quad x = 0,1,2,\ldots \quad . \tag{4.4.2}$$

This can be proved by induction, or by using probability generating functions (Chapter 8). Note that (4.4.2) defines the probability function of a Poisson distribution with mean $\mu = \lambda t$. We shall show in Section 5.3 that λt is the average or expected number of changes in time t. Hence λ represents the average number of changes per unit time, and is sometimes called the intensity parameter.

The derivation of equation (4.4.1) did not use any special properties of time, and thus applies equally well to random occurrences on a line, in the plane, or in volumes of any dimension. For example, if we replace time intervals by regions in the plane, the assumptions become

(1) The numbers of occurrences in non-overlapping regions are independent.

(2) For regions of sufficiently small area, the probability of exactly one occurrence in a region is proportional to its area, and the probability of more than one occurrence is negligible.

We then conclude that the probability of x occurrences in a region of area A is

$$f_A(x) = (\lambda A)^x e^{-\lambda A}/x!; \quad x = 0,1,2,\ldots,$$

where λ is the average number of occurrences per unit area. For example, the distribution of flying-bomb hits on South London during World War II, the distribution of dwellings in rural areas of Japan, and the distribution of bacterial colonies over a glass plate, are all well described by Poisson distributions.

Example 4.4.1. A switchboard serving a large number of telephones receives calls at the approximately constant rate of 3 per minute during a certain period (e.g. 2:00 p.m. to 4:00 p.m. on Tuesday). What is the probability of receiving more than 6 calls in a particular one-minute interval? What is the probability of receiving at most 2 calls in a particular two-minute interval?

Solution. Because the switchboard serves a large number of telephones, assumptions (1) and (2) above seem reasonable, at least as a first approximation. If we assume that arrivals of telephone calls are "random events in time", the number of arrivals in an interval of t minutes will have a Poisson distribution with mean λt, where $\lambda = 3$ is the average number of arrivals per minute. The probability of x arrivals in one minute is thus

$$f_1(x) = 3^x e^{-3}/x!; \quad x = 0,1,2,\ldots .$$

Using (4.3.2), we may compute $f_1(x)$ recursively for $x \le 6$. The probability of more than 6 arrivals in one minute is then

$$1 - f_1(0) - f_1(1) - \ldots - f_1(6) = 0.0335.$$

Let Y be the number of arrivals in two minutes. Then Y has a Poisson distribution with mean $2\lambda = 6$, and probability function

$$f_2(y) = 6^y e^{-6}/y!; \quad y = 0,1,2,\ldots .$$

The probability of at most two arrivals in two minutes is thus

$$f_2(0) + f_2(1) + f_2(2) = e^{-6}(1 + 6 + 6^2/2) = 0.0620.$$

Example 4.4.2. A radioactive source consists of a very large number of atoms. When the atoms decay, α-particles are emitted, and a proportion of these are recorded by a counter. If the period of observation is short in comparison with the half-life of the element, the number of atoms in the source will remain nearly constant. Then λ, the average number of particles recorded per second, should remain very nearly constant. Assumptions (1) and (2) above seem plausible, and one might expect that X, the number of particles recorded in t seconds, would have a Poisson distribution with mean $\mu = \lambda t$.

The numbers of particles reaching a counter in each of 2608 intervals of 7.5 seconds were recorded. Table 4.4.1 gives 0_x, the number of time periods in which x particles were recorded ($x = 0,1,\ldots,9$). There were 16 periods in which more than 9 particles were recorded, and the total number of particles recorded in all 2608 intervals was 10,094. The average number of particles recorded per time interval was thus $\frac{10094}{2608} = 3.870$.

Table 4.4.1

Observed and Expected Frequencies for Example 4.4.2

x	0_x	e_x	x	0_x	e_x
0	57	54.4	6	273	253.8
1	203	210.5	7	139	140.3
2	383	407.4	8	45	67.9
3	525	525.5	9	27	29.2
4	532	508.4	≥10	16	17.1
5	408	393.5			
			Total	2608	2608.0

Rutherford, Chadwick, and Ellis, *Radiations from radioactive sources*, Cambridge (1920), p. 172. The table is taken from H. Cramer. *Mathematical Methods of Statistics*, Uppsala and Princeton (1945), p. 436.

If X has a Poisson distribution with mean $\mu = 3.870$, the probability of recording x particles in an interval is

$$f(x) = (3.870)^x e^{-3.870}/x!; \quad x = 0,1,2,\ldots .$$

The expected number of intervals in which x particles are recorded is then

$$e_x = 2608f(x).$$

We first compute e_0, and then obtain $e_1, e_2, \ldots, e_9$ recursively using

$$e_x = \frac{3.870}{x} e_{x-1}.$$

The final expected frequency e_{10} is then obtained by subtraction:

$$e_{10} = 2608 - e_0 - e_1 - \ldots - e_9.$$

These are listed in Table 4.4.1 where they may be compared with the observed frequencies 0_x. The overall agreement is good, and the Poisson distribution seems to fit the data quite well. The observed and expected frequencies are not equal, but some differences are bound to occur owing to random variation. The question which arises is whether the observed differences are too great to be accounted for purely by chance. We shall consider such questions in the second part of the book.

Example 4.4.3. During World War II, the city of London was subjected to flying-bomb attacks. The technology at that time did not allow bombs to be aimed at particular targets, but only at the general vicinity. Consequently, the points of impact of flying bombs should have the appearance of "random points in the plane". The number of hits in a region of area A should have a Poisson distribution with mean $\mu = \lambda A$, where λ is the expected number of hits per unit area.

The number of flying-bomb hits was determined for each of 576 regions of equal area in south London. There were 537 hits altogether, and thus the average number of hits per region was $\frac{537}{576} = 0.9323$. Table 4.4.2 gives 0_x, the observed number of regions with x hits, and e_x, the number expected under a Poisson distribution with mean $\mu = 0.9323$. The fit of the Poisson distribution to these data is extremely good.

Table 4.4.2

Observed and Expected Frequencies of Flying-bomb Hits

x	0	1	2	3	4	≥ 5
0_x	229	211	93	35	7	1
e_x	226.74	211.39	98.54	30.62	7.14	1.57

This example was taken from Feller*, who comments as follows: "It is interesting to note that most people believed in the tendency of points of impact to cluster. If this were true, there would be a higher frequency of areas with many hits or no hit and a deficiency in the intermediate classes. Table 4.4.2 indicates perfect randomness and homogeneity of the area; we have here an instructive illustration of the established fact that to the untrained eye randomness appears as regularity or tendency to cluster."

* W. Feller, *An Introduction to Probability Theory and Its Applications, vol. 1*, 3rd edition, Wiley (1968), p. 160.

Problems for Section 4.4

†1. Flaws in the plating of large sheets of metal occur at random. On the average there are 2.56 flaws per 100 square feet. Calculate the probability that a sheet 4 feet by 8 feet will have no flaws. Out of 100 such sheets, how many can be expected to have two or more flaws?

2. Coliform bacteria are randomly distributed throughout river water at the average concentration of one per twenty cubic centimeters of water.

   (a) What is the probability of finding exactly two coliform bacteria in a test tube containing 10 cc of river water?

   (b) What is the probability of finding at least one coliform bacterium in a test tube containing 1 cc of river water?

3. Accidents occur in a large manufacturing plant at the average rate of 2.3 per day, and there are 200 working days in a year. Determine the expected number of days in a year on which x accidents will occur $(x=0,1,\ldots,5)$, and the expected number of days on which more than 5 accidents will occur.

†4. A cookie manufacturer prepares a large vat of batter from which to make raisin cookies. He wants 95% of the cookies produced to contain at least two raisins. How many raisins per cookie should he allow in mixing the batter?

4.5 Bivariate Distributions

It is often necessary to simultaneously consider two or more variates defined on the same sample space. We shall take up the case of two variates (bivariate distributions) in some detail. The extension to three or more variates (multivariate distributions) is straightforward.

Let X and Y be discrete variates defined on the same sample space with ranges $\mathcal{X}$ and $\mathcal{Y}$. The set of all sample points i such that $X(i) = x$ and $Y(i) = y$ is an event which we shall denote by "$X = x,\ Y = y$". The joint probability function of X and Y is a function of two variables:

$$f(x,y) = P(X=x,\ Y=y);\quad x \in \mathcal{X},\ y \in \mathcal{Y}. \tag{4.5.1}$$

Similarly, the joint cumulative distribution function of X and Y is defined as follows:

$$F(s,t) = P(X \le s,\ Y \le t);\quad -\infty \le s \le \infty,\ -\infty \le t \le \infty. \tag{4.5.2}$$

The probability function and c.d.f. for the first variate X will be denoted by f_1 and F_1; those for the second variate Y will be denoted by f_2 and F_2.

To obtain statements of probability for X alone, we let Y range over all its possible values. Thus

$$f_1(x) = P(X=x) = \sum_{y \in \mathcal{Y}} P(X=x, Y=y) = \sum_{y \in \mathcal{Y}} f(x,y); \tag{4.5.3}$$

$$F_1(s) = P(X \le s) = P(X \le s, Y \le \infty) = F(s,\infty). \tag{4.5.4}$$

We may obtain f_2 and F_2 in a similar way. If we arrange the probabilities $f(x,y)$ in a two-way table, then $f_1(x)$ and $f_2(y)$ can be obtained as marginal totals (row and column totals) in the table. For this reason, when a variate X is considered as one of a set of two or more variates $X, Y, \ldots$, it is usual to refer to f_1 as the <u>marginal probability function</u> of X, or as the probability function for the <u>marginal distribution</u> of X.

<u>Example 4.5.1</u>. A poker hand is dealt from a well-shuffled deck. Derive the joint and marginal probability functions of X, the number of aces, and Y, the number of kings in the poker hand, and find the probability that $X = Y$.

<u>Solution</u>. There are $\binom{52}{5}$ equally probable poker hands. We first find $f(x,y)$, the probability that a poker hand contains x aces and y kings. The x aces can be chosen in $\binom{4}{x}$ ways, the y kings in $\binom{4}{y}$ ways, and the other $5-x-y$ cards in $\binom{44}{5-x-y}$ ways. The number of poker hands containing x aces and y kings is therefore $\binom{4}{x}\binom{4}{y}\binom{44}{5-x-y}$, and hence

$$f(x,y) = \binom{4}{x}\binom{4}{y}\binom{44}{5-x-y} \Big/ \binom{52}{5}; \quad x = 0,1,\ldots; \quad y = 0,1,\ldots .$$

Note that $f(x,y) = 0$ when $x > 4$, $y > 4$, or $x+y > 5$.

By (4.5.3), the marginal probability function of X is obtained by summing the joint probability function over all values of Y:

$$f_1(x) = \sum_{y=0}^{\infty} f(x,y) = \binom{4}{x}\Big[\sum_{y=0}^{\infty} \binom{4}{y}\binom{44}{5-x-y}\Big] \Big/ \binom{52}{5}.$$

The sum in square brackets may be evaluated by the hypergeometric identity (2.1.10) to give

$$f_1(x) = \binom{4}{x}\binom{48}{5-x}/\binom{52}{5}; \quad x = 0,1,\ldots .$$

A similar argument gives

$$f_2(y) = \binom{4}{y}\binom{48}{5-y}/\binom{52}{5}; \quad y = 0,1,\ldots .$$

These results can also be obtained directly by counting arguments.

The joint probabilities $f(x,y)$ are given in the body of Table 4.5.1. The marginal probabilities $f_1(x)$ and $f_2(y)$ appear as row totals and column totals in the table. A special feature of this

Table 4.5.1

Joint and Marginal Probability Functions for the Number of Aces (X) and Kings (Y) in a Poker hand

	y=0	y=1	y=2	y=3	y=4	$f_1(x)$
x=0	0.417863	0.208931	0.030575	0.001456	0.000017	0.658842
1	0.208931	0.081534	0.008736	0.000271	0.000002	0.299474
2	0.030575	0.008736	0.000609	0.000009	0	0.039929
3	0.001456	0.000271	0.000009	0	0	0.001736
4	0.000017	0.000002	0	0	0	0.000019
$f_2(y)$	0.658842	0.299474	0.039929	0.001736	0.000019	1.000000

example is the symmetry: $f(x,y) = f(y,x)$. As a result, the marginal probability functions of X and Y are identical. Since X and Y have the same possible values with the same probabilities, they are identically distributed, and we may write $X \sim Y$ (see Section 4.1).

The probability of the event "X = Y" is obtained by summing $f(x,y)$ over all pairs (x,y) with $x = y$. This amounts to adding up the terms on the main diagonal in Table 4.5.1:

$$P(X = Y) = f(0,0) + f(1,1) + f(2,2) = 0.500006.$$

The probability that a poker hand contains equal numbers of aces and kings is almost exactly one-half.

Distribution of a function of X and Y.

Frequently one wishes to obtain the probability distribution of some function of X and Y, such as $X+Y$, XY, or X^2+Y^2. If

h is any real-valued function with suitable domain of definition, we may define a new variate $Z \equiv h(X,Y)$. The probability of the event $Z = z$ may be calculated by adding $f(x,y)$ over all pairs (x,y) such that $h(x,y) = z$, and in this way the probability function of Z is obtained.

Example 4.5.2. Find the probability function of $X + Y$ in Example 4.5.1.

Solution. Define $Z \equiv X+Y$, and denote its probability function by g. We obtain $g(z)$ by summing $f(x,y)$ over all pairs (x,y) with $x+y = z$; that is, we sum over all pairs $(x,z-x)$:

$$g(z) = \sum_{x=0}^{z} f(x,z-x) = \left[\sum_{x=0}^{z} \binom{4}{x}\binom{4}{z-x}\right]\binom{44}{5-z}\Big/\binom{52}{5}.$$

The sum in square brackets may be evaluated using (2.1.10) to give

$$g(z) = \binom{8}{z}\binom{44}{5-z}\Big/\binom{52}{5}; \quad z = 0,1,2,\ldots \quad .$$

This result can also be obtained by counting the number of hands which contain z aces and kings and $5-z$ other cards, and then dividing by $\binom{52}{5}$. The probabilities $g(z)$ can be obtained numerically from Table 4.5.1 by summing along the appropriate diagonals. For example

$$g(3) = P(X+Y=3) = f(0,3)+f(1,2)+f(2,1)+f(3,0) = 0.020384.$$

Conditional distributions.

By the definition (3.4.2), the conditional probability of event B given event A is

$$P(B|A) = P(AB)/P(A),$$

provided that $P(A) > 0$. Taking A to be the event "$X = x$" and B to be the event "$Y = y$", we find that

$$P(Y=y|X=x) = \frac{P(X=x, Y=y)}{P(X=x)} = \frac{f(x,y)}{f_1(x)}$$

provided that $f_1(x) > 0$. Accordingly, we define the conditional probability function of Y given that $X = x$ as follows:

$$f_2(y|x) = P(Y=y|X=x) = \frac{f(x,y)}{f_1(x)}; \quad y \in \mathcal{Y}. \qquad (4.5.5)$$

The conditional p.f. is not defined if $f_1(x) = 0$. Note that

$$\sum_{y\in\mathcal{Y}} f_2(y|x) = [\sum_{y\in\mathcal{Y}} f(x,y)]/f_1(x) = f_1(x)/f_1(x)$$

by (4.5.3). Hence the total probability is 1, and (4.5.5) defines a proper probability distribution on $\mathcal{Y}$. Sometimes (4.5.5) is used in the equivalent form

$$f(x,y) = f_1(x)f_2(y|x). \tag{4.5.6}$$

The conditional probabilities $f_2(y|x)$ can be obtained from a two-way table of joint probabilities by dividing all entries $f(x,y)$ in the xth row by their total $f_1(x)$.

Example 4.5.3. Find the probability distribution of the number of aces in poker hands containing one king.

Solution. We require $f_1(x|1)$, the conditional p.f. of X given that Y = 1. By the definition (4.5.5), we have

$$f_1(x|1) = \frac{f(x,1)}{f_2(1)} = \frac{\binom{4}{x}\binom{4}{1}\binom{44}{4-x}/\binom{52}{5}}{\binom{4}{1}\binom{48}{4}/\binom{52}{5}} = \frac{\binom{4}{x}\binom{44}{4-x}}{\binom{48}{4}}; \quad x = 0,1,\ldots .$$

This result can also be obtained by counting arguments. The conditional probabilities $f_1(x|1)$ can be obtained numerically by dividing the entries in column $y = 1$ of Table 4.5.1 by the column total.

Example 4.5.4. The number of α-particles emitted by a radioactive source in a fixed time period has a Poisson distribution with mean μ. Each particle emitted has the same probability p of being recorded, independently of other particles. (The value of p is determined by the solid angle subtended by the counter at the centre of the source.) Find the probability distribution of the number of particles recorded by the counter.

Solution. Let X denote the number of particles emitted, and let Y be the number recorded. Since X has a Poisson distribution with mean μ, its (marginal) probability function is

$$f_1(x) = \mu^x e^{-\mu}/x!; \quad x = 0,1,2,\ldots .$$

If x particles are emitted, we then have x independent trials with probability of success p at each trial. Given that x particles are emitted, the probability that y are recorded will therefore be given by a binomial distribution:

$$f_2(y|x) = \binom{x}{y}p^y(1-p)^{x-y}; \qquad y = 0,1,\ldots,x.$$

By (4.5.6), the joint probability function of X and Y is

$$\begin{aligned} f(x,y) &= f_1(x)f_2(y|x) \\ &= \binom{x}{y}p^y(1-p)^{x-y}\mu^x e^{-\mu}/x!; \quad 0 \le y \le x. \end{aligned}$$

We now obtain the marginal probability function of y by summing over all possible values of x:

$$f_2(y) = \sum_{x=y}^{\infty} f(x,y) = \sum_{x=y}^{\infty} \frac{1}{y!(x-y)!}(\mu p)^y\{\mu(1-p)\}^{x-y}e^{-\mu}$$

$$= (\mu p)^y e^{-\mu}\left[\sum_{x=y}^{\infty} \frac{\{\mu(1-p)\}^{x-y}}{(x-y)!}\right]\frac{1}{y!}.$$

The sum in square brackets is the Taylor's series expansion of $e^{\mu(1-p)}$, and therefore

$$f_2(y) = (\mu p)^y e^{-\mu}e^{\mu-\mu p}/y! = (\mu p)^y e^{-\mu p}/y!; \qquad y = 0,1,2,\ldots .$$

The number of particles recorded by the counter has a Poisson distribution with mean μp.

Independence

By the definition (3.3.1), the two events "$X = x$" and "$Y = y$" are independent if and only if

$$P(X = x,\; Y = y) = P(X = x)P(Y = y);$$

that is, if and only if

$$f(x,y) = f_1(x)f_2(y). \tag{4.5.7}$$

X and Y are called <u>independent variates</u> if and only if (4.5.7) holds for all $x \in \mathcal{X}$ and $y \in \mathcal{Y}$. Equivalently, X and Y are independent variates if and only if their joint c.d.f. factors,

$$F(x,y) = F_1(x)F_2(y), \tag{4.5.8}$$

for all variate values x and y. Upon comparing (4.5.7) and (4.5.6), we see that yet another necessary and sufficient condition for X and Y to be independent variates is that

$$f_2(y|x) = f_2(y)$$

for all variate values x and y.

In Section 3.3 we showed that events depending upon different independent experiments are independent events. It follows that <u>two variates X and Y which depend upon different independent experiments are independent variates</u>, and their joint probability function can be obtained from (4.5.7).

By far the most important application of this result is to repeated counts or measurements of the same quantity in independent repetitions of an experiment. For instance, consider the situation described in Example 4.4.2, where a counter records the numbers of particles in each of n disjoint time intervals of equal length t. Let X_i denote the number of particles recorded in the ith interval $(i = 1,2,\ldots,n)$. Then $X_1, X_2, \ldots, X_n$ are independent variates, and each of them has a Poisson distribution with mean $\mu = \lambda t$, where λ is the intensity parameter. Variates which are independent and have the same distribution are called <u>i.i.d.</u> (independent and identically distributed). In this instance, the probability function of X_i is

$$f_i(x_i) = \mu^{x_i} e^{-\mu}/x_i!; \quad x_i = 0,1,2,\ldots \; .$$

The joint probability function of $X_1, X_2, \ldots, X_n$ is obtained as a product

$$f(x_1,x_2,\ldots,x_n) = f_1(x_1)f_2(x_2)\ldots f_n(x_n) = \prod_{i=1}^{n} \mu^{x_i} e^{-\mu}/x_i!.$$

Because the X_i's are identically distributed, the functions $f_1, f_2, \ldots, f_n$ are identical; that is,

$$f_1(x) = f_2(x) = \ldots = f_n(x) = \mu^x e^{-\mu}/x!; \quad x = 0,1,2,\ldots \; .$$

Example 4.5.5. Let X and Y be independent Poisson variates with means μ and ν, respectively. Show that $X + Y$ has a Poisson distribution with mean $\mu + \nu$.

Solution. Consider a Poisson process of intensity λ. Then we can find t_1 and t_2 such that $\mu = \lambda t_1$ and $\nu = \lambda t_2$. We can interpret X as the number of events occurring in a time interval of length t_1, and Y as the number of events occurring in a disjoint time interval of length t_2. Then $X + Y$ represents the number of events occurring in a time interval of length $t_1 + t_2$. Consequently, $X + Y$ will have a Poisson distribution with mean $\lambda(t_1 + t_2) = \mu + \nu$.

Now we give an algebraic proof. The probability functions of X and Y are

$$f_1(x) = \mu^x e^{-\mu}/x!; \quad x = 0,1,2,\ldots ,$$

$$f_2(y) = \nu^y e^{-\nu}/y!; \quad y = 0,1,2,\ldots .$$

Since X and Y are independent, their joint probability function is

$$f(x,y) = f_1(x)f_2(y) = \mu^x \nu^y e^{-(\mu+\nu)}/x!y!.$$

Define $Z \equiv X + Y$, and let g denote the probability function of Z. Then $g(z)$ is obtained by summing $f(x,y)$ over all pairs (x,y) such that $x + y = z$; that is, over all pairs $(x, z - x)$. Hence

$$g(z) = \sum_{x=0}^{z} f(x, z - x) = e^{-(\mu+\nu)} \sum_{x=0}^{z} \mu^x \nu^{z-x}/x!(z - x)!.$$

To put the sum in a recognizable form, we note that

$$\frac{1}{x!(z-x)!} = \frac{1}{z!}\,\frac{z!}{x!(z-x)!} = \frac{1}{z!}\binom{z}{x}.$$

Upon making this substitution and removing the constant factor $\nu^z/z!$, we see that

$$g(z) = e^{-(\mu+\nu)}\nu^z\Big[\sum_{x=0}^{z}\binom{z}{x}\Big(\frac{\mu}{\nu}\Big)^x\Big]/z!.$$

The sum in square brackets may be evaluated using the binomial theorem (2.1.9) to give

$$g(z) = e^{-(\mu+\nu)}\nu^z(1 + \frac{\mu}{\nu})^z/z! = e^{-(\mu+\nu)}(\mu + \nu)^z/z!; \quad z = 0,1,2,\ldots$$

which is the probability function of a Poisson distribution with mean $\mu + \nu$.

Corollary. Let $X_1, X_2, \ldots, X_n$ be independent Poisson variates with means $\mu_1, \mu_2, \ldots, \mu_n$. Then their total $X_1 + X_2 + \ldots + X_n$ has a Poisson distribution with mean $\mu_1 + \mu_2 + \ldots + \mu_n$.

Proof. By induction on n, using Example 4.5.5.

Problems for Section 4.5

†1. Let X and Y be discrete variates with joint probability function

$$f(x,y) = k(\frac{1}{x} + \frac{1}{y}) \quad \text{for} \quad x = 1,2,3; \quad y = 2,3.$$

Evaluate the constant k, and determine $P(X \geq Y)$.

2. Two balanced dice are rolled simultaneously. Let X be the number of sixes obtained and let Y be the total score on the two dice. Tabulate the joint probability function of X and Y, and the conditional p.f. of Y given that $X = 0$.

3. Let X be the largest outcome and Y the smallest when three balanced dice are rolled. The following table gives $216 \cdot P(X = x, Y = y)$.

	x=1	2	3	4	5	6
y=1	1	6	12	18	24	30
2	0	1	6	12	18	24
3	0	0	1	6	12	18
4	0	0	0	1	6	12
5	0	0	0	0	1	6
6	0	0	0	0	0	1

(a) Tabulate the marginal p.f. of Y and the conditional p.f. of Y given that $X = 5$.

(b) Tabulate the probability function of the range, $R \equiv X - Y$.

(c) Show that $7 - Y$ and X are identically distributed.

†4. The joint distribution of variates X and Y is given in the following table:

	x=0	1	2	3
y=0	.00	.05	.10	.05
1	.05	.10	.10	.10
2	.10	.10	.10	.05
3	.05	.05	.00	.00

(a) Compute the following:

$$P(X > Y), \quad P(X = 2, Y < 2), \quad P(Y < 2 | X = 2).$$

(b) If (X_1, Y_1) and (X_2, Y_2) are independent, each pair having the above bivariate distribution, what is the probability that $Y_1 + Y_2 = 6(X_1 + X_2)$?

5. Let the joint distribution of X and Y be as in the preceding problem, and define $U \equiv XY$, $V \equiv X + Y$. Find the joint probability function of U and V.

6. Let X and Y be the numbers of Liberals and Conservatives on a committee of 8 selected at random from a group containing 10 Liberals, 20 Conservatives, and 10 Independents. Give expressions for

(a) the joint probability function of X and Y;

(b) the marginal probability function of X;

(c) the conditional probability function of Y given that $X = 2$;

(d) the probability function of $X + Y$.

†7. Ten mice are randomly selected and divided into two groups of five. Each mouse in the first group is given a dose of a certain poison which is sufficient to kill one in ten; each mouse in the second group is given a dose sufficient to kill three in ten. What is the probability that there will be more deaths in the first group than in the second?

8. In Example 4.5.4, find the conditional distribution of X, the number of particles emitted, given that y were recorded.

9. Suppose that X and Y are independent variates having binomial distributions with parameters (n,p) and (m,p), respectively. Show that $X + Y$ has a binomial distribution with parameters $(n + m, p)$.

†10. Suppose that flash bulbs are purchased in batches of n, and that the number of defectives in a batch has a binomial distribution with parameters (n,p). From a batch, m bulbs are selected at random without replacement. Derive the marginal distribution of the number of defectives in the sample.

11. Let X and Y be independent and identically distributed with probability function (4.2.1).

(a) Show that $X + Y$ has a negative binomial distribution $(r = 2)$.

(b) Show that the conditional distribution of X given that $X + Y = t$ is uniform on $0, 1, \ldots, t$.

(c) Find the joint p.f. of X and Z where $Z \equiv Y - X$. Are X and Z independent?

*12. (a) Show that the number of ways to partition $2N$ distinguishable objects into N pairs is $(2N)! \div (N!2^N)$.

(b) An urn contains R red balls and W white balls, where $R+W=2N$ is even. Pairs of balls are removed from the urn at random and are used to fill N boxes, each of which will hold two balls. The numbers of boxes having two red balls, two white balls, and one ball of each colour are, respectively, $X, Y,$ and $n-X-Y$. Find the p.f. of X.

†*13. An urn contains N balls numbered $1,2,\ldots,N$. A sample of n balls is chosen with replacement. Let X and Y represent the largest and smallest numbers drawn. Give expressions for the following probabilities:

$$P(X \le x), \quad P(Y > y), \quad P(X \le x, \; Y > y).$$

Hence obtain the joint p.f. and marginal p.f.'s of X and Y.

*14. Suppose that X and Y are independent variates with probability functions

$$f_1(x) = \alpha^x(1-\alpha) \quad \text{for} \quad x = 0,1,2,\ldots \; ;$$

$$f_2(y) = \beta^y(1-\beta) \quad \text{for} \quad y = 0,1,2,\ldots \; .$$

Define $Z \equiv Y - X$. Show that Z has probability function

$$f(z) = \begin{cases} k\beta^z & \text{for} \quad z = 0,1,2,\ldots \\ k\alpha^{-z} & \text{for} \quad z = -1,-2,-3,\ldots \end{cases}$$

where $k = (1-\alpha)(1-\beta)/(1-\alpha\beta)$.

*15. Let $X_1, X_2, \ldots, X_n$ be independent and identically distributed with probability function

$$f(x) = \theta^x(1-\theta) \quad \text{for} \quad x = 0,1,2,\ldots \; .$$

(a) Show that

$$P(X_1 \le X_2 \le \ldots \le X_n) = \frac{(1-\theta)^n}{(1-\theta)(1-\theta^2)\ldots(1-\theta^n)}.$$

Hint: Express the probability as an n-fold sum with all of the upper limits being $+\infty$.

(b) Show that the smallest of the X_i's has probability function

$$\theta^{xn}(1-\theta^n) \quad \text{for} \quad x = 0,1,2,\ldots \; .$$

4.6 The Multinomial Distribution

The binomial and multinomial distribution are of fundamental importance in statistical applications because they arise as probability distributions of frequencies in independent repetitions of an experiment. Suppose that S, the sample space of the experiemnt, is partitioned into k mutually exclusive events, $S = A_1 \cup A_2 \cup \ldots \cup A_k$. Let p_i be the probability of event A_i, and define X_i to be the frequency with which event A_i occurs in n independent repetitions of the experiment $(i = 1,2,\ldots,k)$. The notation is summarized in the following table:

Event	A_1	A_2	A_3	...	A_k	Total
Probability	p_1	p_2	p_3	...	p_k	1
Frequency	X_1	X_2	X_3	...	X_k	n

Because exactly one of the events A_i must occur at each repetition, the probabilities p_i sum to 1, and the frequencies X_i sum to n:

$$p_1 + p_2 + \ldots + p_k = 1; \qquad X_1 + X_2 + \ldots + X_k \equiv n. \tag{4.6.1}$$

We shall derive the joint distribution for the k frequencies $X_1, X_2, \ldots, X_k$.

First consider the case $k = 2$. Each outcome belongs either to A_1 (a success) or A_2 (a failure), and $p_2 = 1 - p_1$. Then X_1 is the number of successes in n independent repetitions, and $X_2 \equiv n - X_1$ is the number of failures. We showed in Section 3.2 that the probability of obtaining x_1 successes and $n - x_1$ failures in n independent repetitions is given by a binomial distribution,

$$\binom{n}{x_1} p_1^{x_1}(1 - p_1)^{n-x_1} = \frac{n!}{x_1! x_2!} p_1^{x_1} p_2^{x_2}$$

where $x_1 = 0,1,\ldots,n$ and $x_2 = n - x_1$.

In the general case, the result of n independent repetitions will be a sequence of n A_i's. For instance, with $n = 7$ we might observe

$$A_4\ A_1\ A_1\ A_2\ A_1\ A_2\ A_1.$$

Since repetitions are independent, the probability of this sequence is

$$p_4\ p_1\ p_1\ p_2\ p_1\ p_2\ p_1 = p_1^4\ p_2^2\ p_4.$$

In general, the probability of a particular sequence in which A_1 appears x_1 times, A_2 appears x_2 times,..., and A_k appears x_k times is

$$p_1^{x_1} p_2^{x_2} \cdots p_k^{x_k}.$$

From Section 2.1, the number of different sequences of n things of k different kinds in which there are x_i things of the ith kind $(i = 1, 2, \ldots, k)$ is

$$\binom{n}{x_1 x_2 \ldots x_k} = \frac{n!}{x_1! x_2! \ldots x_k!}.$$

Hence the probability that $A_1, A_2, \ldots, A_k$ respectively occur $x_1, x_2, \ldots, x_k$ times in any order is

$$f(x_1, x_2, \ldots, x_k) = \binom{n}{x_1 x_2 \ldots x_k} p_1^{x_1} p_2^{x_2} \cdots p_k^{x_k}, \qquad (4.6.2)$$

where we must have $\sum p_i = 1$ and $\sum x_i = n$.

It follows by the Multinomial Theorem (2.1.12) that the sum of (4.6.2) over all non-negative integers $x_1, x_2, \ldots, x_k$ with $\sum x_i = n$ is equal to

$$(p_1 + p_2 + \ldots + p_k)^n = 1^n = 1.$$

The probabilities defined by (4.6.2) are said to form a <u>multinomial distribution</u> with index n and probability parameters $p_1, p_2, \ldots, p_k$.

We could rewrite (4.6.2) as a function of $k-1$ variables by substituting $x_k = n - x_1 - \ldots - x_{k-1}$. This is usually done in the case $k = 2$ (binomial distribution). However, in the general case it is usually convenient to retain one redundant variable in order to preserve the symmetry.

Marginal Distributions

Suppose that we are interested only in the occurrence or non-occurrence of one of the events A_i. If we define the occurrence of A_i to be a success (probability p_i), and its non-occurrence to be a failure (probability $1 - p_i$), then we have n Bernoulli trials, and the probability of x_i successes is

$$f_i(x_i) = \binom{n}{x_i} p_i^{x_i} (1 - p_i)^{n - x_i}; \quad x_i = 0, 1, \ldots, n.$$

Hence the marginal distribution of a single variable X_i is binomial with index n and probability parameter p_i. This result can also be proved algebraically, by summing all of the other variables out of the joint probability function (4.6.2).

Example 4.6.1. A balanced die is rolled twelve times. Find the probability that each face comes up twice.

Solution. We have $n = 12$ independent repetitions of an experiment with $k = 6$ possible outcomes, each of which has probability $\frac{1}{6}$. Let X_i be the number of times the ith face comes up $(i = 1,2,\ldots,6)$. Their joint probability function is multinomial:

$$f(x_1,x_2,\ldots,x_6) = \binom{12}{x_1 x_2 \ldots x_6}\left(\frac{1}{6}\right)^{x_1}\left(\frac{1}{6}\right)^{x_2}\ldots\left(\frac{1}{6}\right)^{x_6}$$

$$= \frac{12!}{x_1!x_2!\ldots x_6!}\, 6^{-12}.$$

The required probability is

$$f(2,2,\ldots,2) = 12!/2^6 6^{12} = 0.0034.$$

This result was also obtained in Example 2.4.2, but the method given there can be used only if the outcomes are equally probable, and is therefore of less general applicability.

Example 4.6.2. Three coins are tossed and the total number of heads obtained is recorded. This experiment is repeated ten times. What is the probability of obtaining the following frequency table?

Number of heads	0	1	2	3	Total
Frequency observed	1	3	4	2	10

Solution. The probability distribution of frequency tables in independent repetitions of an experiment is multinomial. Let A_i be the event that i heads occur when three coins are tossed, and let p_i be the probability of event A_i $(i = 0,1,2,3)$. Let X_i be the number of times A_i occurs in $n = 10$ independent repetitions of the experiment. Then, by (4.6.2),

$$f(x_0,x_1,x_2,x_3) = \binom{10}{x_0 x_1 x_2 x_3} p_0^{x_0} p_1^{x_1} p_2^{x_2} p_3^{x_3},$$

and the required probability is $f(1,3,4,2)$. For balanced coins, we have $p_i = \binom{3}{i}/8$, and hence

$$f(1,3,4,2) = \frac{10!}{1!3!4!2!}(\tfrac{1}{8})^1(\tfrac{3}{8})^3(\tfrac{3}{8})^4(\tfrac{1}{8})^2 = 0.0257.$$

More generally, if the probability of heads is θ (the same for all three coins), then $p_i = \binom{3}{i}\theta^i(1-\theta)^{3-i}$, and the probability of the observed frequencies can be obtained as a function of θ:

$$f(1,3,4,2) = \binom{10}{1\ 3\ 4\ 2}[(1-\theta)^3]^1[3\theta(1-\theta)^2]^3[3\theta^2(1-\theta)]^4[\theta^3]^2$$

$$= \frac{10!}{1!\,3!\,4!\,2!}3^7\theta^{17}(1-\theta)^{13}; \qquad 0 < \theta < 1.$$

The observed frequencies have the greatest probability when $\theta = 17/30$, in which case $f(1,3,4,2) = 0.0335$. (See Chapter 9.)

Example 4.6.3. An urn contains N balls of k different colours, there being a_i balls of the ith colour $(a_1 + a_2 + \ldots + a_k = N)$. Suppose that n balls are drawn randomly one at a time, and let X_i be the number of balls drawn which are of the ith colour $(X_1 + X_2 + \ldots + X_k \equiv n)$. Find the joint distribution of $X_1, X_2, \ldots, X_k$ under (a) sampling with replacement, and (b) sampling without replacement.

Solution. (a) The probability that the first ball drawn has colour i is a_i/N $(i = 1,2,\ldots,k)$. Under sampling with replacement these probabilities remain constant, and we have n independent repetitions of an experiment with k possible outcomes. The joint distribution of $X_1, X_2, \ldots, X_k$ is therefore multinomial; their joint probability function is given by (4.6.2) with $p_i = a_i/N$.

(b) Under sampling without replacement, the composition of the urn changes from one trial to the next. We no longer have independent repetitions, and the multinomial distribution (4.6.2) does not apply.

The joint probability function of $X_1, X_2, \ldots, X_k$ can be obtained by counting arguments. The number of ways to select x_1 balls of colour 1, x_2 of colour 2, ..., and x_k of colour k is

$$\binom{a_1}{x_1}\binom{a_2}{x_2}\cdots\binom{a_k}{x_k}.$$

In total, there are $\binom{N}{n}$ equally probable ways to select n balls from the urn, and dividing gives

$$f(x_1,x_2,\ldots,x_k) = \binom{a_1}{x_1}\binom{a_2}{x_2}\cdots\binom{a_k}{x_k} \Big/ \binom{N}{n},$$

where we must have $\sum a_i = N$ and $\sum x_i = n$. This is a generalization of the hypergeometric distribution (2.3.1). We previously made use of the special case $k = 3$ in Example 4.5.1.

Problems for Section 4.6

†1. The probability that a lightbulb lasts fewer than 500 hours is 0.5, and the probability that it lasts longer than 800 hours is 0.2. Find the probability that, out of 10 bulbs, exactly 4 burn out before 500 hours, and exactly 2 last longer than 800 hours.

2. According to Mendel's theory of heredity, if pea plants having RY (round yellow) seeds are crossed with pea plants having WG (wrinkled green) seeds, then $\frac{9}{16}$ of the offspring will have RY seeds, $\frac{3}{16}$ will have WY seeds, $\frac{3}{16}$ will have RG seeds, and $\frac{1}{16}$ will have WG seeds. Find the probability that in 10 offspring there will be
   (a) exactly 8 with RY seeds;
   (b) 5 with RY, 3 with WY, 1 with RG, and 1 with WG;
   (c) exactly 5 with RY and exactly 3 with WY.

†3. The Hardy-Weiberg law implies that, under certain conditions, the relative frequencies with which three genotypes AA, Aa, and aa occur in the population will be

$$\theta^2, \quad 2\theta(1-\theta), \quad \text{and} \quad (1-\theta)^2,$$

respectively, where $0 < \theta < 1$. Eight members are randomly selected from the population and their genotypes are determined.
   (a) Determine, as a function of θ, the probability that
      (i) there are no AA's in the sample;
      (ii) there are two AA's, four Aa's, and two aa's.
   (b) For what value of θ is the probability in (a)(ii) a maximum?

4. Suppose that $X_1,X_2,\ldots,X_k$ have a multinomial distribution (4.6.2), and define $Y \equiv X_1 + X_2 + \ldots + X_r$ where $r < k$. Show that Y has a binomial distribution.

5. Let $X_1,X_2,\ldots,X_k$ be independent Poisson-distributed variates with means $\mu_1,\mu_2,\ldots,\mu_k$. Define $Y \equiv X_1 + X_2 + \ldots + X_k$. Show that the

conditional distribution of $X_1, X_2, \ldots, X_k$ given that $Y=y$ is multinomial with index y and probability parameters $p_i = \mu_i / \sum \mu_i$ $(i=1,2,\ldots,k)$.

Review Problems: Chapter 4

†1. Suppose that oil spills occur off the British Columbia coast at the average rate of 3 per annum.

(a) Find the probability of having two or more spills in a given one-month period.

(b) A one-month cleanup period is required after each spill. What is the probability that a cleanup will be completed before the next spill occurs?

2. The probability that a patient will recover from a certain rare disease is .25. To test the effectiveness of a new drug, it is given to ten patients with the disease. If at least four patients recover, the drug is judged to be effective in the treatment of the disease.

(a) If the drug raises the recovery rate to .35, what is the probability that it will be judged ineffective?

(b) If the drug has no effect (recovery rate still .25), what is the probability that it will be judged effective?

†3. Flaws in the plating of standard sized sheets of metal occur at random over the entire area being plated, and 67% of the sheets produced have no flaws.

(a) Ten sheets are taken from the assembly line. What is the probability that exactly 3 of them contain flaws?

(b) Sheets are taken from the assembly line until 7 without flaws have been obtained. What is the probability that ten sheets will be selected?

(c) What percentage of the sheets plated will have more than one flaw?

4. Sixty percent of the customers at a store pay cash, thirty percent use charge cards, and ten percent write personal cheques.

(a) What is the probability that, out of 10 customers, 4 will pay cash?

(b) What is the probability that the 10th customer served will be the fourth one to pay cash?

(c) What is the probability that, out of 10 customers, 5 pay cash, 2 use charge cards, and 3 write cheques?

†5. Cassettes are produced by cutting lengths from a large reel of magnetic recording tape, in which defects occur randomly at the aver-

age rate of 1.5 per thousand linear feet. Cassettes with two or more defects are designated as "standard quality", while those with fewer than two defects are designated as "finest quality".

(a) What fraction of 300-foot cassettes will be of standard quality?

(b) What fraction of 600-foot cassettes will be of finest quality?

(c) If a 600-foot cassette is of finest quality, what is the probability that it contains a defect?

(d) If a box contains six 600-foot cassettes, what is the probability that one contains no defects, two contain one defect, and three contain two or more defects?

(e) What length of tape should be put in each cassette in order that 50% of cassettes produced will have no defects?

6. The probability that an individual dies from a certain type of accident during a one-year period is .00004, independently of other individuals. Give an expression for the probability that there will be at least two deaths in one year among the 30,000 people covered by a particular insurance company. Use a suitable approximation to evaluate this probability.

7. An urn contains 2 red, 2 white, and 2 blue balls. One ball is chosen at random and removed from the urn. Five balls are then selected at random with replacement, and there are 2 red, 2 white, and 1 blue. What is the probability that the ball transferred from the urn was blue?

†8. Suppose that 40% of the population are in favour of a certain government action. What is the probability that, of five people selected at random, a majority are in favour? If five groups of five people are selected, what is the probability that a majority of groups show a majority in favour? What is the probability that a majority of the 25 people contacted are in favour?

9. A certain low-budget alphabet soup contains only the letters A,B, C,D. A large vat of soup contains equal proportions of the four letters randomly distributed throughout the volume, with a total of two letters per ounce.

(a) What is the probability that an eight ounce serving contains exactly 12 letters?

(b) If an eight-ounce serving contains 12 letters, what is the probability that it contains three letters of each type?

(c) Three letters are selected randomly without replacement from a bowl containing three letters of each type. What is the probability that

(i) the three letters are all the same?

(ii) the three letters are all different?

(iii) the three letters can be arranged to spell the work "DAD"?

10. A shipment of 100 lightbulbs includes 40 forty watt bulbs and 60 sixty watt bulbs. Bulbs are selected at random without replacement until three forty watt bulbs have been obtained. Find the probability function of the total number of bulbs selected.

†11. Two people toss a balanced coin n times each. What is the probability that they obtain the same number of heads?

12. Suppose that the number of eggs laid by a female robin follows a Poisson distribution with mean μ, and that each egg has probability p of hatching independently of other eggs. Find the probability distribution of the number of offspring.

13. A biassed coin with an unknown probability p of heads is tossed repeatedly. Based on the outcomes of previous tosses, the outcome of the next toss is to be predicted. At the first toss, and on any subsequent toss preceded by equal numbers of heads and tails, the prediction is made at random (probability $\frac{1}{2}$ for heads and $\frac{1}{2}$ for tails). If x heads and y tails have already occurred, the prediction is "heads" if $x > y$ and "tails" if $x < y$. Show that the probability of a correct prediction at trial n tends to $\frac{1}{2} + |\frac{1}{2} - p|$ as $n \to \infty$.

†*14. In Problem 4.6.2, what is the probability that among the ten offspring there will be at least one having each seed type?

15. Corn borers are distributed at random throughout a corn field, with 22 borers per 100 cobs of corn on the average. Six cobs of corn are selected at random. What is the probability that

(a) two corn borers will be found in the six cobs?

(b) two of the six cobs will be infected with corn borers?

CHAPTER 5. MEAN AND VARIANCE

Let X be a discrete variate with range $\mathcal{X}$ and probability function f, and let h be a real-valued function defined on $\mathcal{X}$. The expected value of the variate $h(X)$ is a real number given by

$$E\{h(X)\} = \sum_{x \in \mathcal{X}} h(x)f(x).$$

If probabilities are interpreted as long-run relative frequencies, then $E\{h(X)\}$ represents the average value of $h(X)$ in infinitely many repetitions of the experiment (Section 1).

The expected value of X is called the mean and is usually denoted by μ; it indicates the location of the "centre" of the probability distribution in the sense of a centre of gravity. The expected value of $(X-\mu)^2$ is called the variance, and is denoted by σ^2. Its positive square root σ is called the standard deviation. For most common distributions, the interval $(\mu - 3\sigma, \mu + 3\sigma)$ contains almost all of the probability (Section 2). Means and variances are derived for several discrete distributions in Section 3.

Although various other measures of location and spread are possible, the mean and variance have the advantage that they are easily calculated for sums and linear combinations of random variables (Section 4). Section 5 shows how one can sometimes use indicator variables to simplify probabilistic arguments, and Section 6 considers expected values with respect to conditional distributions.

Although the discussion in this chapter is restricted to discrete distributions, one need merely replace sums by integrals to adapt the definitions and results to the continuous case; see Chapter 6.

5.1 Mathematical Expectation

Let X be a discrete variate with range $\mathcal{X} = \{x_1, x_2, \ldots\}$ and probability function f. The <u>mathematical expectation</u> of X is a number defined by

$$E(X) = \sum_{x \in \mathcal{X}} xf(x) = x_1 f(x_1) + x_2 f(x_2) + \ldots, \tag{5.1.1}$$

provided that this series converges absolutely (see discussion below). $E(X)$ is also called the expected value, average value, or mean value of X; also the mean or first moment of the distribution. It is often denoted by μ, μ_X, or $\langle X \rangle$.

When probabilities are interpreted as long-run relative frequencies, $E(X)$ represents the average value of X in infinitely many repetitions of the experiment.

Example 5.1.1. A gambler rolls two balanced dice, and wins X dollars, where X is the total score obtained on the two dice. If he plays this game a large number of times, what will be his average winnings per game?

Solution. Suppose that the gambler plays n games, and let n_i be the number of times that a total i is obtained $(i = 2,3,\ldots,12)$. On n_2 occasions he wins \$2 for a total gain of $\$2n_2$; on n_3 occasions he wins \$3 for a total gain of $\$3n_3$; and so on. His total gain in all n games is

$$2n_2 + 3n_3 + \ldots + 12n_{12}.$$

Dividing this by n gives his average gain per game:

$$2\frac{n_2}{n} + 3\frac{n_3}{n} + \ldots + 12\frac{n_{12}}{n}.$$

Now let $n \to \infty$. Then n_i/n tends to $f(i)$, the probability of the event $X = i$, and hence the gambler's average gain per game tends to

$$2f(2) + 3f(3) + \ldots + 12f(12) = E(X).$$

Using the table of f in Example 4.1.6, we obtain

$$E(X) = 2(\tfrac{1}{36}) + 3(\tfrac{2}{36}) + \ldots + 12(\tfrac{1}{36}) = 7.$$

On the average, the gambler wins \$7 per game. □

In general, suppose that in n repetitions, the value x_i occurs n_i times $(i = 1,2,\ldots)$, so that $n_1 + n_2 + \ldots = n$. Then the total of the n observed X-values is

$$\underbrace{x_1 + x_1 + \ldots}_{n_1 \text{ times}} + \underbrace{x_2 + x_2 + \ldots}_{n_2 \text{ times}} + \ldots = x_1n_1 + x_2n_2 + \ldots,$$

and hence their average value is

$$\frac{1}{n}(x_1n_1 + x_2n_2 + \ldots) = x_1\frac{n_1}{n} + x_2\frac{n_2}{n} + \ldots .$$

As $n \to \infty$, n_i/n tends to $f(x_i)$, the probability of the event $X = x_i$, and hence the average of the X-values tends to

$$x_1 f(x_1) + x_2 f(x_2) + \ldots = E(X).$$

Thus $E(X)$ is the average of the X-values that would be obtained in infinitely many repetitions of the experiment.

Example 5.1.2. Crown and Anchor

The game of crown and anchor is played with three dice and a betting board divided into six sections. The sections of the board and the six faces of each die are labelled Club, Diamond, Heart, Spade, Crown, and Anchor. To play the game, you place a bet on a section of the betting board, and the owner of the board rolls the three dice. If your section fails to come up on any of the dice, you lose your stake. If your section comes up on i of the dice, you get back your stake plus i times your stake $(i = 1,2,3)$. What is your expected gain from (a) a bet of \$6 on Hearts? (b) simultaneous bets of \$1 on each section?

Solution. (a) Let X denote your net gain from a bet of \$6 on Hearts. We wish to find $E(X)$, the average or expected value of X in a large number of plays.

The probability that Hearts comes up on a particular die is $\frac{1}{6}$, and the three dice are independent. The probability that Hearts comes up on exactly i of the dice is therefore given by a binomial distribution,

$$P(\text{Hearts on } i \text{ dice}) = \binom{3}{i}\left(\frac{1}{6}\right)^i\left(\frac{5}{6}\right)^{3-i} \quad \text{for } i = 0,1,2,3.$$

If $i = 0$ you lose \$6; if $i = 1,2$, or 3 you win \$6, \$12, or \$18. Hence the probability function of X is as follows:

x	-6	6	12	18	Total
$f(x)$	$\frac{125}{216}$	$\frac{75}{216}$	$\frac{15}{216}$	$\frac{1}{216}$	1

Now, by (5.1.1), the expected gain is

$$E(X) = \sum x f(x)$$

$$= -6\left(\frac{125}{126}\right) + 6\left(\frac{75}{216}\right) + 12\left(\frac{15}{216}\right) + 18\left(\frac{1}{216}\right) = -\frac{17}{36}.$$

Hence, on the average, you will lose about 47¢ per game.

(b) We now wish to find $E(Y)$, where Y is the net gain from a bet of \$1 on each of the six sections. The probability that the three dice show the same face is $\frac{6}{216}$; in this case you collect \$3 on one section and lose \$1 on each of the other five sections, for a net loss of \$2. The probability that the three dice show different faces is $\frac{6\times5\times4}{216} = \frac{120}{216}$; in this case you win \$1 on three sections and lose \$1 on three sections, for a net gain of \$0. Finally, the probability that just two dice show the same face is $1 - \frac{6}{216} - \frac{120}{216} = \frac{90}{216}$; in this case you win \$2 on one section, \$1 on another section, and lose \$1 on four sections, for a net loss of \$1. Hence the probability function of Y is as follows:

y	-2	-1	0	Total
$g(y)$	$\frac{6}{216}$	$\frac{90}{216}$	$\frac{120}{216}$	1

The expected gain is

$$E(Y) = \sum yg(y)$$

$$= -2\left(\frac{6}{216}\right) - 1\left(\frac{90}{216}\right) + 0\left(\frac{120}{216}\right) = -\frac{17}{36},$$

which is the same as in (a). The expected loss per game is not affected by the betting strategy adopted.

Convergence

If X has only finitely many possible values, as in the preceding two examples, then $\sum xf(x)$ is a sum of finitely many terms. In this case $E(X)$ is finite and uniquely defined. However, if X has infinitely many possible values, then $\sum xf(x)$ may diverge, or its value may depend upon the order in which the summation is performed. In order that $E(X)$ be well defined, it is necessary that the series $\sum xf(x)$ be convergent to the same finite value regardless of the order of summation. By a standard result from calculus, all rearrangements of a series are convergent to the same value if and only if the series is absolutely convergent. Thus, in order that $E(X)$ be well defined, the series of absolute values

$$\sum |xf(x)| = \sum |x| \cdot f(x)$$

must be convergent. If this is not so, we say that X has no finite expectation.

Expectation of a Function of X

Let $U \equiv h(X)$, where h is a real-valued function defined on $\mathcal{X}$. Then U is a variate, with range $\mathcal{U}$ and probability function g, say. By (5.1.1), the expected value of U is

$$E(U) = \sum_{u \in \mathcal{U}} ug(u). \tag{5.1.2}$$

We shall show that

$$E(U) = E\{h(X)\} = \sum_{x \in \mathcal{X}} h(x)f(x) \tag{5.1.3}$$

provided that this series is absolutely convergent. The advantage of this result is that it permits us to find the expected value of U without having to derive the probability function g.

To prove (5.1.3), we note that the sum over $\mathcal{X}$ can be performed in two stages. First we sum over all x such that $h(x)$ has a constant value u, and then we sum over u:

$$\sum_{x \in \mathcal{X}} h(x)f(x) = \sum_{u} \sum_{h(x)=u} h(x)f(x) = \sum_{u} u \sum_{h(x)=u} f(x).$$

The absolute convergence of (5.1.3) makes this rearrangement of the series permissible. But, as we noted in Section 4.1, summing $f(x)$ over all x such that $h(x) = u$ gives $g(u)$:

$$g(u) = P\{h(X) = u\} = \sum_{h(x)=u} f(x).$$

It follows that

$$\sum_{x \in \mathcal{X}} h(x)f(x) = \sum_{u \in \mathcal{U}} ug(u) = E(U)$$

as claimed.

A similar result holds when U is a function of two or more variates. Suppose that $U \equiv h(X,Y)$, where X and Y are discrete variates with joint probability function f. We could find $E(U)$ by deriving g, the probability function of U, as in Section 4.5, and then using (5.1.2). However it is usually easier to obtain $E(U)$ directly from the joint distribution of X and Y as follows:

$$E(U) = E\{h(X,Y)\} = \sum\sum h(x,y)f(x,y) \tag{5.1.4}$$

where the sum is taken over all pairs of variate values (x,y). In particular $E(X)$ can be determined either from the marginal distribution of X, or from the joint distribution of X and Y:

$$E(X) = \sum\sum xf(x,y) = \sum_{x \in \mathcal{X}} x \sum_{y \in \mathcal{Y}} f(x,y) = \sum_{x \in \mathcal{X}} xf_1(x).$$

Example 5.1.1 (continued). Find the gambler's expected winnings per game if his payoff is $(X-7)^2$, where X is the total on the two dice.

Solution. By (5.1.3), his expected gain is

$$\begin{aligned} E\{(X-7)^2\} &= \sum_{x=2}^{12} (x-7)^2 f(x) \\ &= 25f(2) + 16f(3) + 9f(4) + 4f(5) + 1f(6) + 0f(7) \\ &\qquad + 1f(8) + 4f(9) + 9f(10) + 16f(11) + 25f(12) \\ &= 35/6. \end{aligned}$$

Alternatively, let $U \equiv (X-7)^2$. The probability function of U is tabulated in Example 4.1.6, and (5.1.2) gives

$$\begin{aligned} E(U) &= 0g(0) + 1g(1) + 4g(4) + 9g(9) + 16g(16) + 25g(25) \\ &= 35/6. \end{aligned}$$

Since $g(0) = f(7)$, $g(1) = f(6) + f(8)$, $g(4) = f(5) + f(9)$, etc., the second sum is merely a rearrangement of the first. □

Expectation of Linear Functions

If a and b are any constants, then

$$\sum(ax+b)f(x) = a\sum xf(x) + b\sum f(x).$$

Since $\sum f(x) = 1$, it follows by (5.1.3) that

$$E(aX+b) = aE(X) + b \tag{5.1.5}$$

provided that $E(X)$ exists. Taking $a=0$, we see that $E(b) = b$ for any constant b.

More generally, since

$$\sum\sum(ax+by)f(x,y) = a\sum\sum xf(x,y) + b\sum\sum yf(x,y),$$

it follows from (5.1.4) that

$$E(aX+bY) = aE(X) + bE(Y) \tag{5.1.6}$$

whenever $E(X)$ and $E(Y)$ exist. This result holds quite generally for any two variates X,Y defined on the same sample space. For instance, Y may be a function of X, or X and Y may themselves be functions of one or more other variates $U,V,\ldots$.

Example 5.1.1 (continued). Using (5.1.5) and (5.1.6) we have

$$E\{(X-7)^2\} = E(X^2 - 14X + 49) = E(X^2) - 14E(X) + 49.$$

We previously showed that $E(X) = 7$. By (5.1.3), the expected value of X^2 is

$$E(X^2) = \sum x^2 f(x).$$

Using the table of f in Example 4.1.6, we obtain

$$E(X^2) = 2^2\cdot\frac{1}{36} + 3^2\cdot\frac{2}{36} + 4^2\cdot\frac{3}{36} + \ldots + 11^2\cdot\frac{2}{36} + 12^2\cdot\frac{1}{36} = \frac{1974}{36},$$

and hence

$$E\{(X-7)^2\} = \frac{1974}{36} - 98 + 49 = \frac{35}{6}. \quad \square$$

We conclude this section with an example in which the expected value of X fails to exist.

Example 5.1.3. A balanced coin is tossed repeatedly until a tie (equal numbers of heads and tails) occurs. Let X be the number of tosses required to obtain the first tie, with probability function f. Since ties can occur only on even-numbered trials, $f(x) = 0$ for x odd. The probability of a tie at trial 2 is

$$f(2) = P\{HT, TH\} = \frac{2}{4};$$

the probability of a tie at trial 4 but not at trial 2 is

$$f(4) = P\{HHTT, TTHH\} = \frac{2}{16};$$

the probability of a tie at trial 6 but not trial 2 or 4 is

$$f(6) = P\{HHHTTT, HHTHTT, TTHTHH, TTTHHH\} = \frac{4}{64}.$$

By Theorem 2.7.1, the general result is

$$f(2x) = \frac{2}{x}\binom{2x-2}{x-1}2^{-2x} \quad \text{for} \quad x = 1,2,\ldots .$$

If $E(X)$ exists, it is given by

$$E(X) = \sum xf(x) = \sum_{x=1}^{\infty} (2x)f(2x)$$

$$= 4\sum_{x=1}^{\infty} \binom{2x-2}{x-1} 2^{-2x} = \sum_{n=0}^{\infty} \binom{2n}{n} 2^{-2n}.$$

Now, using Problem 2.1.5(c), we obtain

$$E(X) = \sum_{n=0}^{\infty} \binom{-1/2}{n} (-1)^n.$$

This series represents the binomial expansion of $(1-1)^{-1/2}$, and is divergent to $+\infty$. The expected waiting time for the first tie is infinite (does not exist). As we noted in Section 2.7, there is an appreciably large probability of waiting an enormous length of time for the first tie to occur, and it is for this reason that the expected waiting time is infinite.

Problems for Section 5.1

†1. The following table gives the probability distribution of the number of people per passenger vehicle crossing a toll bridge:

No. of people	1	2	3	4	5	6
Probability	.05	.43	.27	.12	.09	.04

Would more money be raised by charging \$1.00 per car, or by charging 30¢ per person, or by charging 50¢ for the car and driver and 25¢ for each passenger?

2. In Problem 4.1.3, what is the expected number of hits on the target?

3. Find the expected value of the range R in Problem 4.5.3.

4. Suppose that there are n players in a game, and that the winner collects c dollars. The ith player contributes c_i dollars and has probability p_i of winning $(i = 1,2,\ldots,n)$, where $\sum c_i = c$ and $\sum p_i = 1$. The game is said to be fair if the expected gain is zero for each player. Show that the game is fair if and only if $c_i = p_i c$ for $i = 1,2,\ldots,n$.

†5. In the game described in Problem 3.2.5, each player contributes money to a pot which is taken by the winner. If B contributes \$3.00, how much should A and C contribute in order that the game will be fair?

6. Suppose that n people take a blood test, and that each has proba-

bility p of being infected with a disease, independently of the others. To save time, blood samples from k people are pooled and analysed together. If this analysis gives a negative result, then none of the k people is infected, and the one test suffices for all k people. If the pooled test gives a positive result, then at least one of the k people is infected, and they must be tested separately. In this case $k+1$ tests are necessary for the k people.

(a) Show that, on the average, the number of tests required for a group of k people is $k+1-k(1-p)^k$.

(b) What is the expected number of tests required for the n/k groups of k people?

(c) Show that, if p is small, the expected number of tests in (b) is approximately $n(kp + k^{-1})$, and is minimized for $k \approx 1/\sqrt{p}$.

7. A man waits for a bus at a stop where buses pass every ten minutes. There is a probability q_i that the ith bus to reach the stop will be full and unable to take him. When he reached the stop he had just missed a bus $(q_0 = 1)$. Show that the mean number of hours that he must wait is

$$\frac{1}{6} \sum_{r=1}^{\infty} \left(\prod_{i=0}^{r-1} q_i \right).$$

8. A slot machine works on the insertion of a penny. If the player wins, the penny is returned with an additional penny. If he loses, the original penny is lost. The probability of winning is arranged to be $1/2$ if the preceding play resulted in a loss, and to be $p < 1/2$ if the preceding play resulted in a win.

(a) Show that the probability of winning at trial n approaches $(3-2p)^{-1}$ as $n \to \infty$.

(b) The cost to the owner of maintaining the machine is c pennies per play. Show that, in order to make a profit in the long run, the owner must arrange that $p < (1-3c)/2(1-c)$.

*9. A balanced coin is tossed repeatedly until two successive heads are obtained (Problem 4.2.10). Show that, on the average, six tosses will be required.

†*10. The basic premium for an insurance policy is P, and this amount is paid for the first year's insurance. The insurance is renewed annually, and if any claim was made in the preceding year the basic premium P must be paid. If no claim was made, the premium is to be λ times the preceding year's premium, where $0 < \lambda < 1$. Thus, after r consecutive claim-free years, the premium for the next

year is $\lambda^r P$. The probability of no claim in a year is θ, and years are independent.

(a) Find the probability distribution of the premium paid in the nth year of the policy, and hence find the expected premium payable.

(b) Show that, if this expected premium must exceed kP for all n, then $\lambda \geq (k+\theta-1)/k\theta$.

5.2 Moments; the Mean and Variance

Let X be a discrete variate with range $\mathcal{X}$ and probability function f, and let r be a non-negative integer. The rth moment of X (or of the distribution of X) is defined by

$$m_r = E(X^r) = \sum_{x \in \mathcal{X}} x^r f(x) \tag{5.2.1}$$

provided that the series converges absolutely.

Moments are numbers which describe the probability distribution, and the first two moments are the most useful. Many practical situations call for a comparison of two or more probability distributions - for instance, the distributions of yield for two varieties of wheat, or the distributions of survival times for cancer patients under two different treatments. It often happens that the two distributions to be compared have nearly the same shape, and then the comparison can be made in terms of the first two moments only. Somewhat less frequently, a comparison of third and fourth moments may also be useful.

The first moment is the mean of X (or of the distribution),

$$m_1 = E(X),$$

and it is often denoted by μ_X or μ. The mean specifies the location of the "centre" of the distribution in the sense of a centre of gravity or balance point in mechanics. Suppose that at points $x_1, x_2, \ldots$ along a rigid horizontal axis we place weights $f(x_1), f(x_2), \ldots$, where $\sum f(x_i) = 1$. Then the centre of gravity or balance point of the system is located at the point $\sum x_i f(x_i) = \mu_X$.

If X has mean μ, then by (5.1.5),

$$E(X-\mu) = E(X) - \mu = \mu - \mu = 0,$$

so that the variate $X-\mu$ has mean 0. The second moment of $X-\mu$ is called the variance of X.

Definition. The variance of X (or of the distribution) is a non-negative number defined by

$$\mathrm{var}(X) = E\{(X-\mu)^2\} = \sum_{x \in X} (x-\mu)^2 f(x). \qquad (5.2.2)$$

The positive square root of the variance is called the standard deviation of X, and is often denoted by σ_X or σ. The variance is then denoted by σ_X^2 or σ^2.

Since $f(x) \geq 0$, $\mathrm{var}(X)$ is clearly non-negative. Furthermore, $\mathrm{var}(X) > 0$ unless $(x-\mu)^2 f(x) = 0$ for all $x \in X$; that is, unless $f(x) = 0$ whenever $x \neq \mu$. If $\mathrm{var}(X) = 0$, then the distribution of X is singular, with all of the probability being concentrated at the single point μ.

The standard deviation measures the spread or dispersion of the probability distribution. If the distribution is tightly clustered about the mean, then $f(x)$ will be very small whenever $(x-\mu)^2$ is large, and hence σ will be small. On the other hand, σ will be large whenever the probability is spread out over a wide interval.

For most commonly used distributions, almost all of the probability lies within three standard deviations of the mean. The probability content of the interval $(\mu - 3\sigma, \mu + 3\sigma)$ is always at least $8/9$ by Chebyshev's Inequality (see below).

Since μ is a constant, (5.1.5) and (5.1.6) give

$$E\{(X-\mu)^2\} + E\{X^2 - 2\mu X + \mu^2\} = E(X^2) - 2\mu E(X) + \mu^2,$$

and it follows that

$$\sigma^2 = E(X^2) - \mu^2. \qquad (5.2.3)$$

The variance is equal to the mean square minus the squared mean. Formula (5.2.3) is often used in deriving σ^2.

Example 5.2.1. Find the standard deviation of the net gain under the two betting strategies described for crown and anchor in Example 5.1.2.

Solution. Let X and Y denote the net gains under the two strategies, as in Example 5.1.2. We previously tabulated the probability functions of X and Y, and showed that $\mu_X = \mu_Y = -\frac{17}{36}$. To find the variances, we first compute $E(X^2)$ and $E(Y^2)$:

$$\begin{aligned} E(X^2) &= \sum x^2 f(x) \\ &= (-6)^2 \cdot \frac{125}{216} + (6)^2 \cdot \frac{75}{216} + (12)^2 \cdot \frac{15}{216} + (18)^2 \cdot \frac{1}{216} = \frac{9684}{216}\,; \end{aligned}$$

$$E(Y^2) = \sum y^2 g(y)$$

$$= (-2)^2 \cdot \frac{6}{216} + (-1)^2 \cdot \frac{90}{216} + (0)^2 \cdot \frac{120}{216} = \frac{114}{216}.$$

Now from (5.2.3) we obtain

$$\text{var}(X) = \frac{9684}{216} - (-\frac{17}{36})^2 = \frac{57815}{1296} = 44.61$$

$$\text{var}(Y) = \frac{114}{216} - (-\frac{17}{36})^2 = \frac{395}{1296} = 0.3048.$$

Finally, we take square roots to obtain the standard deviations:

$$\sigma_X = 6.68; \quad \sigma_Y = 0.552.$$

The standard deviation of X is much greater than the standard deviation of Y, owing to the fact that the distribution of X is spread out over a much wider interval.

Variance of a Linear Function

Suppose that $Y \equiv aX + b$ where a and b are constants. By (5.1.5), the mean of Y is

$$\mu_Y = E(aX + b) = aE(X) + b = a\mu_X + b.$$

Thus $Y - \mu_Y = a(X - \mu_X)$, and the variance of Y is

$$E\{(Y - \mu_Y)^2\} = E\{a^2(X - \mu_X)^2\} = a^2 E\{(X - \mu_X)^2\}.$$

It follows that, for any constants a and b,

$$\text{var}(aX + b) = a^2 \text{var}(X). \tag{5.2.4}$$

Taking positive square roots now gives

$$\sigma_{aX+b} = |a| \cdot \sigma_X. \tag{5.2.5}$$

Putting $a = 1$ in (5.2.4), we see that $\text{var}(X + b) = \text{var}(X)$ for any constant b. Addition of a constant b to X merely shifts the probability distribution to the right by b units without changing either its spread or its shape. Hence the variance remains constant.

It is important to note that, while μ and σ have the same units as X, σ^2 does not. For instance, if X denotes net gain in dollars (as in the preceding example), then the units for μ and σ will also be dollars, but the units for σ^2 will be "squared dollars".

If we change units from dollars to cents, the net gain is $100X$ with mean 100μ and standard deviation 100σ, but with variance $10000\sigma^2$.

Standard Form

If X has mean μ and variance σ^2, the standard form of X is the variate X^* defined by

$$X^* \equiv \frac{X-\mu}{\sigma} . \tag{5.2.6}$$

Since μ and σ are constants, (5.1.5) and (5.2.4) give

$$E(X^*) = \frac{1}{\sigma} E(X-\mu) = \frac{1}{\sigma}(\mu-\mu) = 0;$$

$$\mathrm{var}(X^*) = \mathrm{var}(\frac{1}{\sigma} X - \frac{\mu}{\sigma}) = \frac{1}{\sigma^2}\mathrm{var}(X) = 1.$$

A variate which has mean 0 and variance 1 is called a standardized variate. We have just shown that the standard form of X is a standardized variate.

Chebyshev's Inequality

Let X be a variate with finite standard deviation σ and mean μ. Then, for any $t > 0$,

$$P\{|X-\mu| \geq t\sigma\} \leq 1/t^2. \tag{5.2.7}$$

Proof. Let $Z \equiv \frac{X-\mu}{\sigma}$ be the standard form of X, and let g be the probability function of Z. Then

$$P\{|X-\mu| \geq t\sigma\} = P\{|Z| \geq t\} = \sum_{|z|\geq t} g(z).$$

Since $z^2/t^2 \geq 1$ over the range of summation, we have

$$\sum_{|z|\geq t} g(z) \leq \sum_{|z|\geq t} g(z)z^2/t^2 = t^{-2} \sum_{|z|\geq t} z^2 g(z).$$

Next we note that

$$\sum_{|z|\geq t} z^2 g(z) \leq \sum_{\text{all } z} z^2 g(z)$$

because the inclusion of extra non-negative terms cannot decrease the sum. Finally, since Z has mean 0 and variance 1, we have

$$\sum z^2 g(z) = E(Z^2) = \mathrm{var}(Z) = 1.$$

Together, these results give

$$P\{|X - \mu| \geq t\sigma\} \leq t^{-2}$$

as required. □

Putting $t = 3$ in Chebyshev's Inequality gives

$$P\{|X - \mu| \geq 3\sigma\} \leq 1/9.$$

Hence, for any distribution with finite μ and σ, the interval $(\mu - 3\sigma,\ \mu + 3\sigma)$ contains at least $8/9$ of the probability. The exact probability content of this interval depends upon the shape of the distribution, and is generally greater than $8/9$. For instance, in Example 5.2.1 we have

$$(\mu_X - 3\sigma_X,\ \mu_X + 3\sigma_X) = (-20.51,\ 19.57);$$

$$(\mu_Y - 3\sigma_Y,\ \mu_Y + 3\sigma_Y) = (-2.128,\ 1.184)$$

and these intervals include all of the probability.

Skewness and Kurtosis

The first moment of X is its mean μ, which specifies the location of the centre of the distribution. The second moment of $X - \mu$ is the variance σ^2, which measures the spread or dispersion of the distribution. The third and fourth moments of the standardized variate $X^* \equiv \frac{X - \mu}{\sigma}$ give indications of the shape of the distribution.

The coefficient of skewness is the third moment of X^*, and is usually denoted by γ_1:

$$\gamma_1 = E(X^{*3}) = \sigma^{-3}E\{(X - \mu)^3\}. \qquad (5.2.8)$$

If the distribution of X is symmetrical about its mean (e.g. uniform distribution, or binomial distribution with $p = \frac{1}{2}$), then $\gamma_1 = 0$. If the distribution of X has a long tail to the right (e.g. geometric or Poisson distribution) then $\gamma_1 > 0$ and the distribution is said to be positively skewed. A long tail to the left gives $\gamma_1 < 0$, and the distribution is said to be negatively skewed.

The coefficient of kurtosis or excess, usually denoted by γ_2, is 3 less than the 4th moment of X^*:

$$\gamma_2 = E(X^{*4}) - 3 = \sigma^{-4}E\{(X - \mu)^4\} - 3. \qquad (5.2.9)$$

The 4th moment of X^* is decreased by 3 in order that $\gamma_2 = 0$ for a normal distribution (see Section 6.6). A distribution with thicker tails than the normal distribution (e.g. Student's distribution, Section 6.9) will have $\gamma_2 > 0$. A distribution with thinner tails (e.g. uniform distribution) will have $\gamma_2 < 0$.

<u>Note on Convergence</u>

The rth moment of X was defined as

$$m_r = E(X^r) = \sum x^r f(x),$$

provided that the series converges absolutely. Note that

$$m_0 = \sum f(x) = 1$$

for all probability distributions. However, some or all of the higher moments $m_1, m_2, \ldots$ may fail to exist, depending upon the shape of the distribution.

<u>Example 5.2.2</u>. The series $\sum_{x=1}^{\infty} x^{-n}$ is known to converge for $n > 1$ and diverge for $n \le 1$. Choose $n > 1$ and let $c_n = 1/\sum x^{-n}$. Then

$$f(x) = c_n x^{-n} \quad \text{for} \quad x = 1,2,\ldots$$

is a probability function. The rth moment of this distribution is

$$m_r = E(X^r) = c_n \sum x^r x^{-n} = c_n \sum x^{-(n-r)},$$

which converges for $n - r > 1$ but diverges for $n - r \le 1$. Thus, for each $n \ge 2$ we have a distribution such that the first $n - 2$ moments exist and the remaining moments do not exist. □

In general, the existence of the rth moment implies the existence of all moments of lower order. To see this, take $r \ge 1$, and note that $|x| \le 1$ implies that $|x^{r-1}| \le 1$. Also, if $|x| > 1$, then $|x^{r-1}| < |x^r|$. Hence we have

$$\begin{aligned}\sum |x^{r-1}| \cdot f(x) &= \sum_{|x| \le 1} |x^{r-1}| \cdot f(x) + \sum_{|x|>1} |x^{r-1}| \cdot f(x) \\ &\le \sum_{|x| \le 1} f(x) + \sum_{|x|>1} |x^r| \cdot f(x) \\ &\le 1 + \sum |x^r| \cdot f(x).\end{aligned}$$

If m_r exists, then $\sum|x^r|\cdot f(x)$ converges. Hence $\sum|x^{r-1}|\cdot f(x)$ also converges, and m_{r-1} exists.

Problems for Section 5.2

†1. Find the mean and variance of X, the number of white balls in the sample in Example 2.5.1, under sampling with replacement and without replacement. Why would one expect the variance to be smaller in the latter case?

2. Under controlled driving conditions, the Mercedes Diesel gives an average of 24.0 miles per U.S. gallon, with variance 1.44. What are the mean and variance if mileage is measured per Imperial gallon? (5 U.S. gallons = 4 Imperial gallons)

†3. Find the mean and variance of the distribution with c.d.f.

$$F(x) = 1 - 2^{-x} \quad \text{for} \quad x = 0,1,2,\ldots .$$

4. Using formulae for a sum of squares and a sum of cubes, find the mean and variance of X in Problem 4.1.10.

†5. Let X have a discrete uniform distribution on the N values $(2i-1)/2N$ $(i=1,2,\ldots,N)$. Find the mean and variance of X, and their limiting values as $N \to \infty$.

6. If X is a variate with mean μ, the rth moment about the mean is defined by

$$m_r(\mu) = E\{(X-\mu)^r\}.$$

Show that, if the probability function of X is symmetrical about μ, then all odd moments about the mean which exist are zero.

7. Markov's Inequality. Let X be a positive-valued variate with mean μ. Show that, for any $k > 0$,

$$P\{X > k\mu\} < \frac{1}{k} .$$

*8. (a) Show that, for $n > 0$,

$$\int_1^N (x-1)^n dx \le \sum_{x=1}^{N} (x-1)^n \le \int_0^N x^n dx$$

and hence that

$$\sum_{x=1}^{N} (x-1)^n \sim \frac{1}{n+1} N^{n+1}.$$

(The single approximation symbol "$\sim$" means that the ratio of the two quantities tends to 1 as $N \to \infty$.)

(b) An urn contains N balls numbered $1,2,\ldots,N$. A sample of n

balls is chosen at random with replacement. Let X represent the largest number drawn. Show that, as $N \to \infty$,

$$E(X) \sim \frac{nN}{n+1}\ ; \quad \mathrm{var}(X) \sim \frac{nN^2}{(n+1)^2(n+2)}.$$

5.3 Some Examples

In this section, we work out the means and variances for several of the more common discrete distributions. In these derivations, we shall use factorial moments instead of the moments m_r defined in Section 5.2. The rth factorial moment g_r is defined as follows:

$$g_r = E\{X^{(r)}\} = E\{X(X-1)\ldots(X-r+1)\} = \sum x^{(r)} f(x) \tag{5.3.1}$$

provided that the sum converges absolutely.

The first two factorial moments are

$$g_1 = E(X) = m_1 = \mu;$$

$$g_2 = E\{X(X-1)\} = E(X^2 - X) = E(X^2) - E(X) = m_2 - \mu.$$

Now (5.2.3) gives

$$\sigma^2 = m_2 - \mu^2 = g_2 + \mu - \mu^2. \tag{5.3.2}$$

The mean and variance can thus be found from the first two factorial moments.

Poisson Distribution

The probability function for a Poisson distribution is given by (4.3.1), and the rth factorial moment is

$$g_r = E\{X^{(r)}\} = \sum_{x \in X} x^{(r)} f(x) = \sum_{x=0}^{\infty} x^{(r)} \mu^x e^{-\mu}/x!.$$

Since $x^{(r)} = 0$ for $x < r$, we may change the lower limit of summation from $x = 0$ to $x = r$. Also, for $x \geq r$ we have

$$x! = x(x-1)\ldots(x-r+1)(x-r)\ldots(2)(1) = x^{(r)}(x-r)!.$$

It follows that

$$g_r = \sum_{x=r}^{\infty} \frac{\mu^x e^{-\mu}}{(x-r)!} = \mu^r \sum_{x=r}^{\infty} \frac{\mu^{x-r} e^{-\mu}}{(x-r)!} = \mu^r \sum_{y=0}^{\infty} \frac{\mu^y e^{-\mu}}{y!},$$

where $y = x - r$. This sum represents the total probability in a Poisson distribution, and therefore equals 1. Hence $g_r = \mu^r$, and

$$E(X) = g_1 = \mu;$$
$$\operatorname{var}(X) = g_2 + \mu - \mu^2 = \mu.$$

The variance of a Poisson distribution equals the mean.

Binomial Distribution

The probability function of a binomial distribution is given by (2.5.1), and the rth factorial moment is

$$g_r = E\{X^{(r)}\} = \sum x^{(r)}\binom{n}{x}p^x(1-p)^{n-x}.$$

Because $\binom{n}{x}$ is defined to be zero for $x < 0$ or $x > n$, the sum may be taken over all integers x. The cancellation formula (2.1.7) gives

$$x^{(r)}\binom{n}{x} = n^{(r)}\binom{n-r}{x-r}$$

for all x, and hence

$$g_r = \sum n^{(r)}\binom{n-r}{x-r}p^x(1-p)^{n-x}$$
$$= n^{(r)}p^r\sum\binom{n-r}{x-r}p^{x-r}(1-p)^{n-x}$$
$$= n^{(r)}p^r\sum\binom{n-r}{y}p^y(1-p)^{n-r-y}$$

where the sum is over all integers y. This sum represents the total probability in a binomial distribution with index $n - r$, and is therefore equal to 1. It follows that

$$g_r = n^{(r)}p^r$$

from which we obtain

$$E(X) = g_1 = np;$$
$$\operatorname{var}(X) = g_2 + \mu - \mu^2 = n(n-1)p^2 + np - n^2p^2 = np(1-p).$$

The binomial distribution has mean np and variance $np(1-p)$.

Note. In approximating the binomial distribution using (4.3.3), we are replacing it by a Poisson distribution having the same mean. In Section 6.8 we shall see that when both np and $n(1-p)$ are large, the binomial distribution can be approximated by a normal distribution having the same mean $\mu = np$, and the same variance $\sigma^2 = np(1-p)$.

Negative Binomial Distribution

Two expressions for the probability function of a negative binomial distribution are given by (4.2.2) and (4.2.4). If the second of these is used, it may be shown that

$$\mu = r(1-p)/p; \quad \sigma^2 = r(1-p)/p^2$$

by essentially the same argument that was used above for the binomial distribution. The mean and variance of the geometric distribution (4.2.1) are obtaining by setting $r = 1$.

Hypergeometric Distribution

The rth factorial moment for the hypergeometric distribution (2.3.1) is given by

$$g_r = E\{X^{(r)}\} = \sum x^{(r)}\binom{a}{x}\binom{b}{n-x}/\binom{a+b}{n},$$

where once again the sum may be taken over all integers x. When we use the cancellation rule (2.1.7) this becomes

$$a^{(r)}\sum\binom{a-r}{x-r}\binom{b}{n-x}/\binom{a+b}{n} = a^{(r)}\left[\sum\binom{a-r}{y}\binom{b}{n-r-y}\right]/\binom{a+b}{n}.$$

The sum in square brackets may be evaluated using the hypergeometric identity (2.1.10) to give

$$g_r = a^{(r)}\binom{a+b-r}{n-r}/\binom{a+b}{n}.$$

This may be simplified using (2.1.3) or (2.1.7) to give

$$g_r = a^{(r)}n^{(r)}/(a+b)^{(r)}.$$

The first two factorial moments are

$$g_1 = \frac{an}{a+b}; \quad g_2 = \frac{a(a-1)n(n-1)}{(a+b)(a+b-1)}.$$

One can use (5.3.2) and a little algebra to show that

$$\mu = np; \quad \sigma^2 = np(1-p)\cdot\frac{a+b-n}{a+b-1} \tag{5.3.3}$$

where $p = \dfrac{a}{a+b}$.

Discussion.

Suppose that n balls are drawn at random from an urn containing a white and b black balls, and let X be the number of white balls in the sample. If balls are drawn without replacement, the distribution of X is hypergeometric (Section 2.3), and its mean and variance are

$$\mu_1 = np; \qquad \sigma_1^2 = np(1-p) \cdot \frac{a+b-n}{a+b-1},$$

where $p = \frac{a}{a+b}$. If balls are drawn with replacement, the distribution of X is binomial (Section 2.5), and its mean and variance are

$$\mu_2 = np; \qquad \sigma_2^2 = np(1-p).$$

The expected number of white balls in the sample is the same for both sampling methods. The variances are equal when $n = 1$, in which case the two methods of sampling are equivalent. However, for $n > 1$ we have $\frac{a+b-n}{a+b-1} < 1$, and the variance is always smaller for sampling without replacement. This shows up clearly in Example 2.5.1, where the two sampling methods are compared when $a = 6$, $b = 14$ and $n = 8$. Then $p = 0.3$, and the mean of both distributions is 2.4. The binomial distribution f_2 has variance $\sigma_2^2 = 1.68$; the hypergeometric distribution f_1 is more concentrated about the mean, and thus has a smaller variance, $\sigma_1^2 = \frac{12}{19}\sigma_2^2 = 1.06$. The reason for the larger tail probabilities under sampling with replacement is explained in Example 2.5.1.

If n is much smaller than $a+b$, then $\frac{a+b-n}{a+b-1} \approx 1$, and hence $\sigma_1^2 \approx \sigma_2^2$. It is precisely in this case that (2.5.2) applies and the hypergeometric can be approximated by the binomial distribution.

Problems for Section 5.3

1. Show that the kth factorial moment of the negative binomial distribution (4.2.4) is given by

$$g_k = (r+k-1)^{(k)}(1-p)^k/p^k \quad \text{for } k = 1,2,\ldots .$$

Hence derive the mean and variance of the negative binomial and geometric distributions.

†2. A secretary makes an average of two typing errors per page. Pages with more than two errors must be retyped. How many pages can she expect to type in order that all 100 pages of a report shall be acceptable?

3. The probability that x particles will be emitted in an experiment is $\theta^x(1-\theta)$ for $x=0,1,2,\ldots$, where $0<\theta<1$. The lifetime of a system of x particles is xk^x where $0<k<1/\theta$. Show that the expected lifetime of a system arising from a performance of the experiment is $k\theta(1-\theta)/(1-k\theta)^2$.

4. Show that the coefficient of skewness for a binomial distribution with parameters (n,p) is

$$\gamma_1=(1-2p)/\sqrt{np(1-p)}.$$

What can be concluded about the shape of the distribution for various values of p? for fixed p and increasing n?

†5. A point starts at the origin and takes unit steps to the right with probability p, or the left with probability $1-p$. Assuming that steps are independent, find the expected squared distance from the origin after n steps.

6. Let Y be the fraction of successes in n Bernoulli trials. Use Chebyshev's Inequality to show that, for any $\epsilon>0$,

$$\lim_{n\to\infty} P\{|Y-p| > \epsilon\}=0.$$

With probability 1, the relative frequency Y tends to the probability p as $n\to\infty$ (the law of large numbers).

*7. An urn contains a white balls and b black balls. Balls are drawn at random without replacement until the rth white ball has been obtained. Let X denote the number of black balls obtained before the rth white ball is drawn. Show that the kth factorial moment of X is

$$g_k=b^{(k)}\binom{r+k-1}{k}/\binom{a+k}{k}.$$

Hence find the mean and variance, and show that they are approximately equal to the mean and variance of a negative binomial distribution when $a+b$ is large.

5.4 Variance of a Sum; Covariance and Correlation

In many problems, the quantity of interest can be written as the sum of n variates,

$$S_n \equiv X_1+X_2+\ldots+X_n.$$

Sometimes the distribution of S_n can be obtained fairly easily using generating functions (Chapter 8). Otherwise, derivation of the distribution of S_n requires the evaluation of an $(n-1)$-fold sum (or integral in the continuous case), and this may prove to be very difficult. However, the mean and variance of S_n are relatively easy to find from the formulae to be given below. These indicate the most important features of the distribution, its location and spread, although they give no information about its shape.

The distribution of S_n for large n will be considered in Section 6.7. We shall see that, under quite general conditions, S_n has approximately a normal distribution when n is sufficiently large. Then the mean and variance of S_n completely determine its distribution, and probabilities can be computed using normal distribution tables.

Sum of Two Variates

Let X and Y be discrete or continuous variates with finite means and variances. By (5.1.6) the mean of $X+Y$ is $\mu_X+\mu_Y$. Hence, by the definition (5.2.2), the variance of $X+Y$ is

$$\text{var}(X+Y) = E\{(X+Y-\mu_X-\mu_Y)^2\}.$$

We regroup the terms inside the round brackets and expand:

$$\begin{aligned}(X+Y-\mu_X-\mu_Y)^2 &\equiv [(X-\mu_X)+(Y-\mu_Y)]^2 \\ &\equiv (X-\mu_X)^2+(Y-\mu_Y)^2+2(X-\mu_X)(Y-\mu_Y).\end{aligned}$$

Now we take expected values and use (5.1.6) to obtain

$$\text{var}(X+Y) = E\{(X-\mu_X)^2\} + E\{(Y-\mu_Y)^2\} + 2E\{(X-\mu_X)(Y-\mu_Y)\}.$$

We therefore have

$$\text{var}(X+Y) = \text{var}(X)+\text{var}(Y)+2\text{cov}(X,Y) \tag{5.4.1}$$

where $\text{cov}(X,Y)$ is the covariance of X and Y (see below). The variance of $X+Y$ is finite whenever X and Y have finite variances.

Covariance

The covariance of X and Y is defined by

$$\text{cov}(X,Y) = E\{(X-\mu_X)(Y-\mu_Y)\}. \tag{5.4.2}$$

It exists whenever X and Y have finite variances. Note that

$$E\{(X-\mu_X)(Y-\mu_Y)\} = E\{XY - X\mu_Y - \mu_X Y + \mu_X\mu_Y\}$$

$$= E(XY) - \mu_Y E(X) - \mu_X E(Y) + \mu_X \mu_Y$$

by (5.1.6). Hence the following is an equivalent definition of covariance:

$$\text{cov}(X,Y) = E(XY) - \mu_X \mu_Y. \tag{5.4.3}$$

Using (5.4.2) or (5.4.3), it is easy to show that

$$\text{cov}(aX + b,\ cY + d) = ac\ \text{cov}(X,Y) \tag{5.4.4}$$

for any constants a,b,c,d.

The covariance of X with itself is the variance of X:

$$\text{cov}(X,X) = E\{(X - \mu_X)^2\} = \text{var}(X).$$

Formula (5.2.3) is thus a special case of (5.4.3).

A positive covariance indicates that, on the average, $(X - \mu_X)(Y - \mu_Y)$ is positive, so that large values of X tend to occur with large values of Y, and small values of X with small values of Y. For instance, let X be the height and Y the weight of an adult male. Since tall men generally weigh more than short men, X and Y will have a positive covariance. On the other hand, a negative covariance indicates that large values of one variate tend to occur with small values of the other. For instance, let X be the number of hearts and Y the number of spades in a bridge hand. If a hand contains many hearts, it will probably contain only a few spades, and hence the covariance of X and Y will be negative.

Correlation Coefficient

The correlation coefficient ρ of two variates X and Y is the covariance of their standard forms:

$$\rho(X,Y) = \text{cov}(X^*,Y^*). \tag{5.4.5}$$

Since $E(X^*) = E(Y^*) = 0$, we have

$$\rho(X,Y) = E(X^*Y^*) = \frac{E\{(X-\mu_X)(Y-\mu_Y)\}}{\sigma_X \sigma_Y} = \frac{\text{cov}(X,Y)}{\sigma_X \sigma_Y}. \tag{5.4.6}$$

It follows from (5.4.6), (5.4.4) and (5.2.5) that

$$\rho(aX + b, cY + d) = \begin{cases} \rho(X,Y) & \text{for} \quad ac > 0 \\ -\rho(X,Y) & \text{for} \quad ac < 0 \end{cases}$$

for any constants a,b,c,d. Covariance depends upon the scale of measurement used, but the correlation coefficient does not. For instance, if X and Y represent temperatures, $cov(X,Y)$ will depend upon whether the measurements are in degrees fahrenheit or centigrade, but $\rho(X,Y)$ will not.

A convenient property of the correlation coefficient is that $-1 \le \rho \le +1$. To prove this, we note that, by (5.4.1),

$$var(X^* \pm Y^*) = var(X^*) + var(Y^*) \pm 2\ cov(X^*,Y^*)$$

where $var(X^*) = var(Y^*) = 1$. Hence

$$var(X^* \pm Y^*) = 2(1 \pm \rho) \ge 0$$

because variance is non-negative. It follows that $-1 \le \rho \le 1$.

If $\rho = 1$, then $var(X^* - Y^*) = 0$, and hence the variate $X^* - Y^*$ assumes a single value with probability 1. Since

$$E(X^* - Y^*) = E(X^*) - E(Y^*) = 0,$$

it follows that $X^* - Y^* \equiv 0$; that is

$$\frac{X - \mu_X}{\sigma_X} - \frac{Y - \mu_Y}{\sigma_Y} \equiv 0.$$

Hence X may be expressed as a linear function of Y,

$$X \equiv aY + b,$$

where $a = \sigma_X/\sigma_Y$ and $b = \mu_X - a\mu_Y$. A similar result is obtained for $\rho = -1$, but with $a = -\sigma_X/\sigma_Y$. A correlation coefficient of ± 1 thus indicates that one variate is a linear function of the other.

Uncorrelated Variates

Two variates X and Y such that $cov(X,Y) = 0$ are said to be uncorrelated. If X and Y are uncorrelated, then (5.4.1) gives

$$var(X + Y) = var(X) + var(Y). \tag{5.4.7}$$

<u>The mean of a sum always equals the sum of the means, but the variance of a sum is equal to the sum of the variances only when the variates are uncorrelated</u>.

If X and Y are independent variates, their joint probability function factors:

$$f(x,y) = f_1(x)f_2(y) \quad \text{for all } x,y.$$

It follows that

$$E(XY) = \sum\sum xyf(x,y) = \sum xf_1(x)\cdot\sum yf_2(y) = E(X)\cdot E(Y).$$

Hence $\text{cov}(X,Y) = 0$ by (5.4.3). If X and Y are independent variates, then they are uncorrelated, and (5.4.7) applies.

Two variates X and Y can be uncorrelated without being independent. For example, let X be a discrete variate taking values $-1,0,1$ with probabilities $\frac{1}{3}, \frac{1}{3}, \frac{1}{3}$, and define $Y \equiv X^2$. Then X and Y are certainly not independent. However, they are uncorrelated, for

$$E(X) = (-1)\cdot\frac{1}{3} + (0)\cdot\frac{1}{3} + (1)\cdot\frac{1}{3} = 0$$

$$E(XY) = E(X^3) = (-1)^3\cdot\frac{1}{3} + (0)^3\cdot\frac{1}{3} + (1)^3\cdot\frac{1}{3} = 0$$

and hence $\text{cov}(X,Y) = 0$ by (5.4.3).

Sum of n variates.

Now consider a sum of n variates,

$$S_n \equiv X_1 + X_2 + \ldots + X_n,$$

where X_i has finite variance σ_i^2 and mean μ_i $(i = 1,2,\ldots,n)$. By (5.1.6), the mean of the sum is always equal to the sum of the means:

$$E(S_n) = \sum E(X_i) = \sum\mu_i. \tag{5.4.8}$$

We thus have

$$S_n - E(S_n) = \sum X_i - \sum\mu_i = \sum(X_i - \mu_i),$$

and by the definition (5.2.2), the variance of S_n is

$$\text{var}(S_n) = E\{[\sum(X_i - \mu_i)]^2\}.$$

Now we use the result

$$(\sum a_i)^2 = \sum a_i^2 + 2\sum\sum_{i<j} a_i a_j$$

together with (5.1.6) to obtain

$$\text{var}(S_n) = \sum E\{(X_i - \mu)^2\} + 2\sum\sum_{i<j} E\{(X_i - \mu_i)(X_j - \mu_j)\}.$$

It follows by (5.2.2) and (5.4.2) that

$$\text{var}(S_n) = \sum \text{var}(X_i) + 2\sum\sum_{i<j} \text{cov}(X_i, X_j). \tag{5.4.9}$$

The variance of a sum is equal to the sum of the variances plus twice the sum of the covariances.

If all of the covariances are zero, the variance of the sum is equal to the sum of the variances:

$$\mathrm{var}(S_n) = \sum \mathrm{var}(X_i). \tag{5.4.10}$$

In particular, (5.4.10) applies when $X_1, X_2, \ldots, X_n$ are independent.

Linear Combinations

The above results for a sum S_n may be adapted for a linear combination $\sum a_i X_i$, where $a_1, a_2, \ldots, a_n$ are constants. By (5.1.5), (5.2.4) and (5.4.4) we have

$$E(a_i X_i) = a_i \mu_i; \quad \mathrm{var}(a_i X_i) = a_i^2\, \mathrm{var}(X_i);$$

$$\mathrm{cov}(a_i X_i, a_j X_j) = a_i a_j \mathrm{cov}(X_i, X_j).$$

Combining these results with (5.4.8) and (5.4.9), we obtain

$$E(\sum a_i X_i) = \sum a_i \mu_i; \tag{5.4.11}$$

$$\mathrm{var}(\sum a_i X_i) = \sum a_i^2 \mathrm{var}(X_i) + 2 \sum_{i<j}\sum a_i a_j \mathrm{cov}(X_i, X_j). \tag{5.4.12}$$

Example 5.4.1. Find the mean and variance of the total score on n balanced dice.

Solution. The total score on n dice may be written

$$S_n \equiv X_1 + X_2 + \ldots + X_n$$

where X_i is the score on the ith die $(i = 1, 2, \ldots, n)$. Since X_i has possible values $1, 2, \ldots, 6$, each with probability $\frac{1}{6}$, we have

$$\mu_i = E(X_i) = \frac{1}{6}(1 + 2 + \ldots + 6) = \frac{7}{2};$$

$$E(X_i^2) = \frac{1}{6}(1^2 + 2^2 + \ldots + 6^2) = \frac{91}{6};$$

$$\mathrm{var}(X_i) = E(X_i^2) - \mu_i^2 = \frac{91}{6} - (\frac{7}{2})^2 = \frac{35}{12}.$$

The covariances are zero because the X_i's are independent. Hence (5.4.8) and (5.4.10) give

$$E(S_n) = \sum \mu_i = \frac{7n}{2}; \quad var(S_n) = \sum \sigma_i^2 = \frac{35n}{12}.$$

Alternatively, we could work out the probability function of S_n and obtain the mean and variance from it. This was done for the case $n = 2$ in Example 5.1.1. However, unless n is quite small, it is much easier to use formulae (5.4.8) and (5.4.10).

Example 5.4.2. Find the mean and variance of the total net gain in n games of crown and anchor if, in each game, a bet of \$6 is placed on "hearts".

Solution. The total net gain in n games is

$$S_n \equiv X_1 + X_2 + \ldots + X_n$$

where X_i is the net gain in the ith game. From Examples 5.1.2 and 5.2.1, X_i has mean $\mu_i = -\frac{17}{36}$ and variance $\sigma_i^2 = 44.61$. Since the X_i's depend on different games, they are independent. Hence, by (5.4.8) and (5.4.10),

$$E(S_n) = \sum \mu_i = -\frac{17n}{36}; \quad var(S_n) = \sum \sigma_i^2 = 44.61n.$$

We can use these results to determine the probable magnitude of the total gain in n games. For instance, in 36 games the expected gain is $\mu = -17$, with standard deviation

$$\sigma = \sqrt{44.61(36)} = 40.07.$$

By Chebyshev's Inequality (5.2.7), the interval

$$(\mu - 3\sigma, \mu + 3\sigma) = (-137.22, 103.22)$$

contains at least $\frac{8}{9}$ of the probability. In fact, the Central Limit Theorem (Section 6.7) shows that this interval will contain more than 99% of the probability. It is therefore very improbable that one would lose more than \$137 or gain more than \$103 in 36 games.

Example 5.4.3. Consider a single game of crown and anchor in which a bet of \$6 is placed on hearts and another bet of \$6 is placed on spades. Let X_1 be the net gain on the first bet and X_2 the net gain on the second bet. Find the covariance of X_1 and X_2.

Solution. In this case X_1 and X_2 are not independent because they refer to the same game. A large win on one section will be accompanied by a small win or loss on the other. Large values of one variate tend to occur with small values of the other, and hence they will have a negative covariance.

From Example 5.1.2 we have

$$E(X_1) = E(X_2) = -\frac{17}{36}.$$

We shall evaluate $E(X_1X_2)$ and then obtain the covariance from (5.4.3).

We may think of the rolls of the dice as independent repetitions of an experiment with three possible outcomes: "heart", "spade", or "other". From Section 4.6, the probability of obtaining i hearts, j spades and $3-i-j$ others is given by the multinomial distribution:

$$\binom{3}{i\ j\ 3-i-j}\left(\frac{1}{6}\right)^i\left(\frac{1}{6}\right)^j\left(\frac{4}{6}\right)^{3-i-j} = \frac{3!4^{3-i-j}}{i!j!(3-i-j)!6^3}.$$

We have $x_1 = -6$ for $i=0$ and $x_2 = -6$ for $j=0$; otherwise $x_1 = 6i$ and $x_2 = 6j$. Hence we may obtain the following table of possible values (x_1, x_2) and probabilities $f(x_1, x_2)$:

ij	00	01	02	03	10	11	12	20	21	30
x_1	-6	-6	-6	-6	6	6	6	12	12	18
x_2	-6	6	12	18	-6	6	12	-6	6	-6
$6^3f(x_1,x_2)$	64	48	12	1	48	24	3	12	3	1

Now, by (5.1.4), the expected value of X_1X_2 is

$$E(X_1X_2) = \sum\sum x_1x_2f(x_1,x_2)$$

$$= \frac{1}{216}[(-6)(-6)(64) + (-6)(6)(48) + \ldots + (18)(-6)(1)] = -\frac{50}{6},$$

and (5.4.3) gives

$$\text{cov}(X_1,X_2) = -\frac{50}{6} - \left(-\frac{17}{36}\right)^2 = -\frac{11089}{1296}.$$

Example 5.4.4. Consider a single game of crown and anchor in which a total bet of \$6 is spread over the six sections. Let a_i be the fraction of the bet which is placed on the ith section, so that $0 \le a_i \le 1$ and $a_1 + a_2 + \ldots + a_6 = 1$. Find the mean and variance of the gain under this strategy.

Solution. Let Y denote the net gain. Then

$$Y \equiv a_1X_1 + a_2X_2 + \ldots + a_6X_6$$

where X_i is the net gain from a bet of \$6 on the ith section. The X_i's refer to the same game and will not be independent. By the preceding example and the symmetry of the X_i's, we have

$$\text{cov}(X_i, X_j) = -\frac{11089}{1296} = c, \quad \text{say.}$$

Also from Examples 5.1.2 and 5.2.1 we have

$$E(X_i) = -\frac{17}{36} = m; \quad \text{var}(X_i) = \frac{57815}{1296} = v.$$

By (5.4.11), the mean of Y is

$$E(Y) = \sum a_i E(X_i) = m\sum a_i = m.$$

The expected gain is the same no matter how the bets are distributed. By (5.4.12), the variance of Y is

$$\text{var}(Y) = \sum a_i^2 \text{var}(X_i) + 2 \sum\sum_{i<j} a_i a_j \text{cov}(X_i, X_j)$$

$$= v\sum a_i^2 + 2c \sum\sum_{i<j} a_i a_j.$$

But since $\sum a_i = 1$, we have

$$1 = (\sum a_i)^2 = \sum a_i^2 + 2 \sum\sum_{i<j} a_i a_j,$$

and therefore

$$\text{var}(Y) = v\sum a_i^2 + c(1 - \sum a_i^2) = c + (v-c)\sum a_i^2.$$

Since $v > 0$ and $c < 0$, $v - c$ is positive.

To maximize $\sum a_i^2$ subject to $\sum a_i = 1$ and $0 \le a_i \le 1$, we must take $a_i = 1$ for one value of i and $a_i = 0$ otherwise. Hence the variance of Y is maximized by placing the entire bet on a single section. If this strategy is followed over several games (see Example 5.4.2), there is some chance of making a profit, but there is also a good chance of sustaining a big loss.

To minimize $\sum a_i^2$ subject to $\sum a_i = 1$, we must take the a_i's to be equal. Hence the variance of Y is minimized by distributing the bet evenly over the six sections. In this case, a large loss is improbable, but there is no chance of making a profit either.

Problems for Section 5.4

†1. Six balanced coins and two balanced dice are tossed together. Let Z be the sum of the scores on the dice and the number of heads. Find the mean and variance of Z.

2. In Problem 4.5.3, compute

(a) the means, variances, and covariance of X and Y;

(b) the mean and variance of the range, $R \equiv X - Y$.

Why would one anticipate that $\text{cov}(X,Y) > 0$ in this example?

3. In Problem 4.1.8, compute the mean and variance of X in n rounds of the game. Use Chebyshev's Inequality to obtain an interval containing at least 8/9 of the probability.

†4. Each of A,B, and C fires 20 shots at a target. The probability that a particular shot hits the target is 0.4 for A, 0.3 for B, and 0.1 for C, and shots are mutually independent. Let X be the total number of hits on the target.

(a) Express X as a sum of three independent variates, and hence find its mean and variance.

(b) Use Chebyshev's Inequality to obtain an interval which contains the total number of hits with probability at least 8/9.

5. Show that if X and Y have finite second moments, then $\text{cov}(X,Y)$ exists. Hint: First show that $|xy| \le \frac{1}{2}(x^2 + y^2)$ for all x,y.

6. Find the covariance of X and Y in Problem 4.5.6. Why would one anticipate a negative covariance in this case?

†7. Let $X_1, X_2, \ldots, X_n$ be independent variates with the same mean μ, but with possibly different variances $\sigma_1^2, \sigma_2^2, \ldots, \sigma_n^2$. Define $Y \equiv \sum c_i X_i$ where $c_1, c_2, \ldots, c_n$ are constants. Find the mean and variance of Y. What values of the constants c_i should be chosen in order that $E(Y) = \mu$ and $\text{var}(Y)$ is minimized? Specialize this result to the case in which all of the variances are equal.

8. Let $X_1, X_2, \ldots, X_n$ be independent variates having the same mean μ and variance σ^2. The sample mean and variance are defined by

$$\overline{X} \equiv \frac{1}{n} \sum X_i \; ; \quad S^2 \equiv \frac{1}{n-1} \sum (X_i - \overline{X})^2 .$$

(a) Show that $E(\overline{X}) = \mu$ and $\text{var}(\overline{X}) = \sigma^2/n$. Using Chebyshev's Inequality, prove that, for every $\epsilon > 0$,

$$\lim_{n\to\infty} P\{|\overline{X} - \mu| > \epsilon\} = 0.$$

(b) Show that $\sum(X_i - \overline{X})^2 \equiv \sum X_i^2 - n\overline{X}^2$, and prove that $E(S^2) = \sigma^2$.

9. Let $Y_1, Y_2, \ldots, Y_n$ be uncorrelated variates with zero means and common variance σ^2. New variates are defined as follows:

$$X_1 \equiv Y_1 \; ; \quad X_i \equiv \alpha X_{i-1} + Y_i \quad \text{for } i = 1,2,\ldots,n$$

where α is a constant. Find the mean and variance of X_n, and the covariance of X_1 and X_n.

†10. A machine works for a time X_1 until it breaks down. It is then repaired, which takes time Y_1. It then works for a further time X_2 until it breaks down again. The new repair time is Y_2; and so on. The X_i are independent variates with common mean μ and variance σ^2. The value of Y_i depends upon how "old" the machine is, where age is measured in terms of the time it has been working, $X_1 + X_2 + \ldots + X_i$. It may be assumed that

$$Y_i \equiv \alpha + \beta(X_1 + X_2 + \ldots + X_i)$$

with $\beta > 0$ and variation negligible compared with σ^2. When the machine breaks down for the nth time it is immediately scrapped. Find the mean and variance of the time between the installation of a new machine and its scrapping.

*5.5 Indicator Variables

An indicator variable for an event A is a variate X which takes the value 1 if A occurs, and the value 0 if A does not occur:

$$P(X=1) = P(A); \quad P(X=0) = P(\overline{A}) = 1 - P(A).$$

The expected values of X and X^2 are

$$E(X) = \sum x f(x) = 1 \cdot P(A) + 0 \cdot P(\overline{A}) = P(A); \tag{5.5.1}$$

$$E(X^2) = \sum x^2 f(x) = 1^2 \cdot P(A) + 0^2 \cdot P(\overline{A}) = P(A). \tag{5.5.2}$$

If X is an indicator variable for event A, then $1 - X$ is an indicator variable for the event $\overline{A}$. If Y is an indicator variable for another event B defined on the same sample space, then $XY = 1$ if both A and B occur, and $XY = 0$ otherwise. Therefore XY is an indicator variable for the event AB, and

* This section may be omitted on first reading.

$$E(XY) = P(AB). \tag{5.5.3}$$

Similarly, $(1-X)(1-Y)$ is an indicator variable for the event $\bar{A}\bar{B}$.

Probabilistic arguments can sometimes be simplified through the use of indicator variables and the formulae of Section 5.4. This is illustrated in the following examples.

Example 5.5.1. Suppose that n balls are drawn one at a time from an urn containing a white and b black balls. Let A_i be the event that the ith ball drawn is white, and let X_i be an indicator variable for A_i $(i=1,2,\ldots,n)$. The total of the X_i's is

$$S_n \equiv X_1 + X_2 + \ldots + X_n,$$

which represents the total number of white balls drawn. S_n has a binomial distribution under sampling with replacement and a hypergeometric distribution under sampling without replacement. The means and variances of these distributions were found in Section 5.3. The following alternate derivation has the advantage that no complicated sums are involved.

(a) Sampling with replacement. The probability that the ith ball drawn will be white is $P(A_i) = \frac{a}{a+b} = p$, say. Then

$$E(X_i) = E(X_i^2) = p;$$

$$\text{var}(X_i) = E(X_i^2) - [E(X_i)]^2 = p(1-p)$$

by (5.5.1),(5.5.2), and (5.2.3). In this case, the X_i's are independent, and thus

$$E(S_n) = \sum E(X_i) = np; \qquad \text{var}(S_n) = \sum \text{var}(X_i) = np(1-p)$$

by (5.4.8) and (5.4.10).

(b) Sampling without replacement. There are $(a+b)^{(n)}$ possible ordered sequences of n balls. The number of ordered sequences in which the ith ball is white is $a(a+b-1)^{(n-1)}$. Hence the probability that the ith ball drawn will be white is

$$P(A_i) = \frac{a(a+b-1)^{(n-1)}}{(a+b)^{(n)}} = \frac{a}{a+b} = p.$$

(See Example 3.5.3 for another proof.) Hence as in (a) we have

$$E(X_i) = E(X_i^2) = p; \quad var(X_i) = p(1-p);$$

$$E(S_n) = \sum E(X_i) = np.$$

However, in this case the X_i's are not independent. We must therefore determine the covariances and use the more general formula (5.4.9).

The number of ordered sequences in which the ith and jth balls are white $(i \neq j)$ is $a(a-1)(a+b-2)^{(n-2)}$. Hence the probability that both the ith and jth balls are white is

$$P(A_i A_j) = \frac{a(a-1)(a+b-2)^{(n-2)}}{(a+b)^{(n)}} = \frac{a(a-1)}{(a+b)(a+b-1)}.$$

Hence by (5.5.3) we have

$$E(X_i X_j) = \frac{a(a-1)}{(a+b)(a+b-1)}.$$

Now by (5.4.3), the covariance of X_i and X_j is found to be

$$cov(X_i, X_j) = E(X_i X_j) - E(X_i)E(X_j) = -\frac{p(1-p)}{a+b-1}.$$

The covariance is negative because drawing white on the ith ball reduces the chance that the jth ball will be white. Finally, since there are n equal variances and $\binom{n}{2}$ equal covariances on the right hand side of (5.4.9), we get

$$var(S_n) = n\,var(X_i) + 2\binom{n}{2}cov(X_i, X_j) = np(1-p)\cdot\frac{a+b-n}{a+b-1}$$

as in Section 5.3.

Example 5.5.2. Montmort Letter Problem

Suppose that n letters are placed at random into n envelopes. Let A_i be the event that the ith letter is placed in the correct envelope, and let X_i be an indicator variable for A_i $(i = 1,2,\ldots,n)$. Then the total number of correctly placed letters is given by the sum

$$S_n \equiv X_1 + X_2 + \ldots + X_n.$$

We shall find the mean and variance of S_n.

There are $n!$ possible arrangements of the letters. There are $(n-1)!$ arrangements in which the ith letter is correctly placed, and $(n-2)!$ arrangements in which the ith and jth letters are correctly placed $(i \neq j)$. Hence

$$P(A_i) = \frac{(n-1)!}{n!} = \frac{1}{n} ; \quad P(A_i A_j) = \frac{(n-2)!}{n!} = \frac{1}{n(n-1)} .$$

Now by (5.5.1),(5.5.2) and (5.5.3), we have

$$E(X_i) = E(X_i^{\,2}) = P(A_i) = \frac{1}{n} ;$$

$$E(X_i X_j) = \frac{1}{n(n-1)} \quad \text{for } i \neq j.$$

It follows that

$$\text{var}(X_i) = \frac{n-1}{n^2} ; \quad \text{cov}(X_i, X_j) = \frac{1}{n^2(n-1)} .$$

Substitution in (5.4.8) gives

$$E(S_n) = \sum E(X_i) = \frac{n}{n} = 1.$$

Whatever the value of n, the expected number of correctly placed letters is 1.

The variance of S_n may be obtained from (5.4.9). Since there are n equal variances and $\binom{n}{2}$ equal covariances on the right hand side of (5.4.9), we get

$$\text{var}(S_n) = n\text{var}(X_i) + 2\binom{n}{2}\text{cov}(X_i, X_j) = 1,$$

which is also constant with respect to n.

Alternatively, the distribution of S_n can be derived as in Section 3.6, and the mean and variance obtained from it. The argument based on indicator variables is much easier since it avoids both the derivation of the distribution of S_n and the evaluation of difficult sums.

Example 5.5.3. Union of n Events

Let $A_1, A_2, \ldots, A_n$ be n events defined on the same sample space, and let B be their union,

$$B = A_1 \cup A_2 \cup \ldots \cup A_n .$$

In Section 3.6 we "anticipated" a formula for $P(B)$, and then proved it using a rather complicated counting argument and the binomial theorem. The same result can be obtained directly and much more simply through the use of indicator variables.

Let X_i be an indicator variable for A_i $(i=1,2,\ldots,n)$. Then as we noted at the beginning of this section, $1-X_1$ is an indicator variable for $\bar{A}_1$, and $(1-X_1)(1-X_2)$ is an indicator variable for $\bar{A}_1\bar{A}_2$. In general, the product $(1-X_1)(1-X_2)\ldots(1-X_n)$ takes the value 1 if none of the events occur ($X_i=0$ for all i), and the value 0 otherwise. Hence the variate Y defined by

$$Y \equiv 1-(1-X_1)(1-X_2)\ldots(1-X_n)$$

takes the value 1 if at least one event A_i occurs, and the value 0 otherwise; that is, Y is an indicator variable for event B. Therefore, by (5.5.1) we have

$$\begin{aligned} P(B) &= E(Y) = E\{1-(1-X_1)(1-X_2)\ldots(1-X_n)\} \\ &= E\{\sum X_i - \sum\sum X_iX_j + \sum\sum\sum X_iX_jX_k - + \ldots\} \\ &= \sum E(X_i) - \sum\sum E(X_iX_j) + \sum\sum\sum E(X_iX_jX_k) - + \ldots . \end{aligned}$$

Finally, note that $E(X_i)=P(A_i)$ by (5.5.1), $E(X_iX_j)=P(A_iA_j)$ by (5.5.3), and similarly $E(X_iX_jX_k) = P(A_iA_jA_k)$, etc. It follows that

$$P(B) = \sum P(A_i) - \sum\sum P(A_iA_j) + \sum\sum\sum P(A_iA_jA_k) - + \ldots ,$$

which is the result obtained previously in Section 3.6.

Problems for Section 5.5

1. A deck of N cards numbered 1,2,...,N is shuffled and the cards are then turned over one by one. A match is said to occur whenever the ith card turned over is numbered i. Let X be the total number of matches. Show that X has mean and variance 1.

†2. There are n well-spaced homes in a residential area. The probability that the fire department will be summoned to the ith house at least once during a one-year period is p_i. Let X be the number of different homes visited by the fire department over a one year period. Find the mean and variance of X. Show that, for $\sum p_i$ fixed, the variance of X is maximized when all of the p_i's are equal.

*3. In a row of $n+1$ points, any point has probability p of being black and $q=1-p$ of being white. The colour is determined independently for each point. Show that the number of unlike pairs among the n pairs of adjacent points has mean $2npq$ and variance $2(2n-1)pq - 4(3n-2)p^2q^2$.

Hint: Define indicator variables $X_i = 1$ if points i and $i+1$ have different colours $(i = 1,2,\ldots,n)$.

4. Suppose that A and B are mutually exclusive events with indicator variables X and Y. Show that $X + Y$ is an indicator variable for $A \cup B$.

*5. Let $A_1, A_2, \ldots, A_n$ be n events defined on the same sample space, with indicator variables $X_1, X_2, \ldots, X_n$, respectively. Let B be the event that exactly m of the A_i's occur.

(a) Find an indicator variable for the event

$$A_1 A_2 \ldots A_m \bar{A}_{m+1} \bar{A}_{m+2} \ldots \bar{A}_n.$$

(b) Express B as a union of such events, and hence obtain an indicator variable for B.

(c) Derive formula (3.6.3) for the probability of B.

†*6. A random graph with n vertices is generated by connecting pairs of vertices at random. Each of the $\binom{n}{2}$ possible edges is inserted with probability p, independently of other edges. Find the mean and variance of the number of triangles in the graph.

Hint: Define $\binom{n}{3}$ indicator variables.

*5.6 Conditional Expectation

Suppose that we wish to find the expected value of $h(X,Y)$, where X and Y are discrete variates with joint probability function f. By (5.1.4) and (4.5.6) we have

$$E\{h(X,Y)\} = \sum\sum h(x,y)f(x,y) = \sum\sum h(x,y)f_1(x)f_2(y|x)$$

where the sum is over all pairs of values (x,y). This double sum may be iterated to give two single sums,

$$E\{h(X,Y)\} = \sum_x \Big[\sum_{y|x} h(x,y)f_1(x)f_2(y|x)\Big];$$

that is, first we hold x fixed and sum over y, and then we sum over x. Now since $f_1(x)$ is constant within the first sum, we have

$$E\{h(X,Y)\} = \sum_x f_1(x)\Big[\sum_{y|x} h(x,y)f_2(y|x)\Big]. \qquad (5.6.1)$$

* This section may be omitted on first reading.

The sum in square brackets represents the expected value of $h(X,Y)$ taken with respect to the conditional distribution of Y given that $X = x$. We call this the <u>conditional expectation</u> of $h(X,Y)$ given that $X = x$, and write

$$E_{Y|X=x}\{h(X,Y)\} = \sum_{y|x} h(x,y)f_2(y|x). \tag{5.6.2}$$

This will be a function of x, say $k(x)$. We write $E_{Y|X}\{h(X,Y)\}$ to denote the variate $k(X)$.

Equation (5.6.1) says that we may obtain the expected value of $h(X,Y)$ in two stages. First we find its conditional expectation given X, and then we take the expectation of the result with respect to the marginal distribution of X; that is,

$$E\{h(X,Y)\} = E_X[E_{Y|X}\{h(X,Y)\}] = \sum k(x)f_1(x) \tag{5.6.3}$$

where the subscripts on E denote the distribution with respect to which expectation is being taken.

The conditional mean and variance

The mean of the conditional distribution of Y given that $X = x$ is a function of x:

$$m(x) = E_{Y|X=x}(Y) = \sum_{y|x} yf_2(y|x).$$

We take expectation with respect to X and use (5.5.3) to obtain

$$E_X\{m(X)\} = E_X\{E_{Y|X}(Y)\} = E(Y) = \mu_Y. \tag{5.6.4}$$

Hence <u>the unconditional mean of Y equals the expected value of the conditional mean</u>. The conditional mean of Y given X is sometimes called the <u>regression of Y on X</u>. (See Section 7.5.)

The variance of the conditional distribution of Y given that $X = x$ is also a function of x:

$$v(x) = E_{Y|X=x}\{[Y - m(x)]^2\} = \sum_{y|x} [y - m(x)]^2 f_2(y|x).$$

Note that in computing the conditional variance, we consider deviations from the condition mean $m(x)$ rather than from the unconditional mean μ_Y. Because of this, the expected value of the conditional variance is in general less than the unconditional variance. To prove this, we note that by (5.2.3)

$$v(x) = E_{Y|X=x}(Y^2) - [m(x)]^2.$$

Taking expectation with respect to X gives

$$E_X\{v(X)\} = E_X\{E_{Y|X}(Y^2)\} - E_X\{[m(X)]^2\}. \quad (5.6.5)$$

Now (5.6.3) and (5.2.3) give

$$E_X\{E_{Y|X}(Y^2)\} = E(Y^2) = \sigma_Y^2 + \mu_Y^2.$$

Also by (5.2.3) and (5.6.4) we have

$$E_X\{[m(X)]^2\} = \text{var}_X\{m(X)\} + [E_X\{m(X)\}]^2 = \text{var}_X\{m(X)\} + \mu_Y^2.$$

Upon substituting these results into (5.6.5) and solving for σ_Y^2, we obtain

$$\sigma_Y^2 = E_X\{v(X)\} + \text{var}_X\{m(X)\}. \quad (5.6.6)$$

Hence the unconditional variance equals the expected value of the conditional variance plus the variance of the conditional mean. The second term on the right of (5.6.6) is zero if and only if $m(X)$ is a constant, in which case the conditional and unconditional means are the same.

Results (5.6.4) and (5.6.6) are useful for determining μ_Y and σ_Y^2 in cases where the mean and variance of the conditional distribution are easy to derive.

Example 5.6.1. Use (5.6.4) and (5.6.6) to determine the mean and variance of Y in Example 4.5.4.

Solution. Given that $X = x$, Y has a binomial distribution with index x, and hence, from Section 5.3, the conditional mean and variance of Y are

$$m(x) = xp; \quad v(x) = xp(1-p).$$

Also, X has a Poisson distribution with mean μ, so that

$$E(X) = \text{var}(X) = \mu.$$

(See the note following Example 5.6.2.) By (5.1.5) and (5.2.4), we have

$$E\{m(X)\} = pE(X) = p\mu; \quad \text{var}\{m(X)\} = p^2\text{var}(X) = p^2\mu;$$

$$E\{v(X)\} = p(1-p)E(X) = p(1-p)\mu.$$

Now (5.6.2) and (5.6.4) give

$$\mu_Y = p\mu;$$
$$\sigma_Y^2 = p(1-p)\mu + p^2 = p\mu.$$

This argument does not require that we derive the distribution of Y, which is an advantage in more complicated situations. In the present case, these results could also be obtained by showing that Y has a Poisson distribution (Example 4.5.4) and using the results of Section 5.3.

Example 5.6.2. A machine is subject to periodic breakdowns. The number of breakdowns in a year follows a Poisson distribution with mean ν. When the machine breaks down, the time required to repair it is a variate with mean μ and variance σ^2. Find the mean and variance of the total repair time in one year.

Solution. Let X be the number of breakdowns in a year, and let Y be the total repair time. Given that x breakdowns occur, the total repair time may be written as a sum

$$Y \equiv Y_1 + Y_2 + \ldots + Y_x$$

where Y_i is the repair time for the ith breakdown. Assuming the Y_i's to be independent, we find that the conditional mean and variance of Y are

$$m(x) = E(Y_1) + E(Y_2) + \ldots + E(Y_x) = x\mu;$$

$$v(x) = \text{var}(Y_1) + \text{var}(Y_2) + \ldots + \text{var}(Y_x) = x\sigma^2.$$

But since X has a Poisson distribution, its mean and variance are equal to ν, and hence

$$E\{m(X)\} = \mu E(X) = \mu\nu; \quad \text{var}\{m(X)\} = \mu^2\text{var}(X) = \mu^2\nu;$$

$$E\{v(X)\} = \sigma^2 E(X) = \sigma^2\nu.$$

Now (5.6.4) and (5.6.6) give

$$\mu_Y = \mu\nu; \quad \sigma_Y^2 = \nu(\sigma^2 + \mu^2)$$

as required.

Note. The expected value of a function of X only can be computed from either the joint distribution or the marginal distribution:

$$E\{h(X)\} = \sum\sum h(x)f(x,y) = \sum h(x)f_1(x) = E_X\{h(X)\}.$$

It is therefore possible to simplify the notation by dropping the subscript X as we have done in Examples 5.6.1 and 5.6.2. In this simplified notation, (5.6.4) and (5.6.6) become

$$\mu_Y = E\{m(X)\}; \qquad \sigma_Y^2 = E\{v(X)\} + \text{var}\{m(X)\}.$$

Problems for Section 5.6

†1. In a rain-making experiment, each of n clouds is either seeded or not depending upon the toss of a balanced coin. If a cloud is seeded, it produces rain with probability p_1. If it is not seeded, it produces rain with probability p_2. Different clouds are independent. Find the mean and variance of the total number of rain producing clouds. (The experiment is intended to determine whether seeding increases the probability of rain. Can you see any defect in the proposed design?)

2. The number of copies of a magazine which are purchased in a month is a variate with mean μ and variance σ^2. The buyer of a copy reads it and then either destroys it (probability $1-p$), or else passes it on to someone who has not read it (probability p). Each person receiving the copy does this independently, and the population of possible readers may be assumed to be infinite. Find the mean and variance of the total number of readers of all copies purchased in a month.

†3. The eggs of a certain insect are found in clusters. The number of eggs per cluster is distributed according to a Poisson distribution with mean μ. The probability of finding y clusters in a field of specified area is given by

$$f(y) = \binom{-r}{y}p^r(1-p)^y \quad \text{for} \quad y = 0,1,2,\ldots,$$

where $r > 0$ and $0 < p < 1$. Find the mean and variance of the total number of eggs in a field.

4. Balls are distributed one at a time into n cells. Each ball has probability $1/n$ of going into each cell, independently of preceding balls. Let X_r be the number of empty cells after the rth ball has been placed. Show that, given X_{r-1},

$$E(X_r) = \frac{n-1}{n}X_{r-1}; \quad E\{X_r(X_r-1)\} = \frac{n-2}{n}X_{r-1}(X_{r-1}-1).$$

Hence determine the unconditional mean and variance of X_r.

Review Problems: Chapter 5

†1. Through hard experience, 18th century gamblers found that they made money betting on the occurrence of one or more sixes in four throws of a die but lost money betting on the occurrence of one or more double sixes in 24 throws of two dice. Show that this is to be expected, assuming the bets are even money wagers.

2. A salesman has probability 0.2 of selling his product in any given trial, trials being independent. If his commission is \$5 on a sale, and if each salespitch takes 10 minutes, how many hours a day should he work to make an average salary of \$50 per day?

3. A firm sends out 1000 salesmen. Each asks 10 persons whether they like a certain product. If the probability that a person likes the product is .4, find the probability that a salesman reports x persons who like the product. How many salesmen would you expect to report that 3 or fewer of their prospects like the product?

†4. The probability that a student knows the correct answer to a question on a multiple choice examination is p. If he doesn't know the correct answer, he chooses one of the k possible answers at random. Suppose that 1 mark is awarded for the correct answer, and m marks are deducted for an incorrect answer. What should m be in order that his expected mark for the question will be p?

5. In Problem 4.5.4, compute the means, variances, and covariance of X and Y.

6. Let X and Y denote the numbers obtained in two rolls of a balanced die, and define $U \equiv X + Y$, $V \equiv X - Y$.
   (a) Show that U and V are uncorrelated but not independent.
   (b) Find the joint probability function of U and V, and the conditional probability function of V given that $U = 7$.

†7. In Problem 5.1.6, let X be the number of people infected in a group of k people whose pooled blood test was positive. Find the probability function of X, and determine the mean and variance.

*8. In a certain plant species, spores are contained in sets of four, each set being arranged in a linear chain with three links. When a set of spores is ejected from the plant, some of the links may break, so that the set may be ejected as four single spores, two pairs of spores, etc.. It may be assumed that there is a constant chance θ that a link will break, independently of other links.

A large number of sets are ejected, and a count is then made showing that there are n_1 single spores, n_2 pairs of spores, n_3 sets of three, and n_4 sets of four. Show that the probabilities p_i corresponding to the relative frequencies $n_i/\sum n_i$ are

$$p_1 = 2\theta(1+\theta)/(1+3\theta) \qquad p_2 = \theta(1-\theta)(2+\theta)/(1+3\theta)$$
$$p_3 = 2\theta(1-\theta)^2/(1+3\theta) \qquad p_4 = (1-\theta)^3/(1+3\theta).$$

Why do the n_i's not have a multinomial distribution?

CHAPTER 6. CONTINUOUS VARIATES

There are many quantities such as time, weight, length, temperature, etc., which we naturally think of as continuous variables; that is, as variables capable of taking any real value in some range. These are represented in probability theory by continuous variates, which we define in Section 1 below. Generally speaking, continuous variates are handled mathematically in much the same way as discrete variates, with sums in the discrete case being replaced by integrals in the continuous case. However, there is a difference in change of variables problems, since the Jacobian of the transformation plays a role in the continuous case but not in the discrete case.

Section 2 introduces two important continuous distributions, the uniform distribution and the exponential distribution. Section 3 discusses transformations based on the cumulative distribution function. Section 4 considers models for lifetimes in situations where the risk of failure may increase or decrease with age. The distribution of the waiting time for random events to occur is derived in Section 5.

The important normal family of distributions is introduced in Section 6. Some of its properties are derived, and the use of normal distribution tables is described. Experience has shown that many measurements have approximately normal distributions, and the Central Limit Theorem (Section 7) helps to account for this phenomenon. In Section 8, normal approximations to the Poisson and binomial distributions are obtained from the Central Limit Theorem. The χ^2, F, and t distributions are defined in Section 9, and their tabulation and properties are described. These distributions have many important applications in Statistics.

6.1 Definitions and Notation

The cumulative distribution function F of a real-valued variate X is defined as follows:

$$F(t) = P(X \le t); \quad -\infty \le t \le \infty. \tag{6.1.1}$$

We noted in Section 4.1 that F is a non-decreasing function, with

$$F(-\infty) = 0; \quad F(\infty) = 1. \tag{6.1.2}$$

If X is discrete, then F is a step function with a discontinuity at each variate value x. In this case, a very small change in t may produce a large change in F(t).

When X represents a continuous variable such as time or weight, one would expect very small changes in t to produce correspondingly small changes in $P(X \le t)$; that is, F(t) should be a continuous function of t. Hence it is natural to develop a theory for variates with continuous cumulative distribution functions. This mathematical theory can be developed under quite general conditions. However, it is sufficient for practical purposes to consider c.d.f.'s which are not only continuous, but reasonably smooth as well. Hence we restrict our attention to variates whose c.d.f.'s are smooth, or have a finite number of corners. Such variates are called continuous.

Definition. A real-valued variate X with c.d.f. F is called a continuous variate if F(x) is a continuous function of x whose derivative

$$f(x) = \frac{d}{dx} F(x), \tag{6.1.3}$$

exists and is continuous except possibly at a finite number of points.

The function f defined by (6.1.3) is called the probability density function (p.d.f.) of X. Note that f need not be defined at finitely many points.

Since F is non-decreasing, it follows that $f(x) \ge 0$ for all x. Also, integrating (6.1.3) and taking $F(-\infty) = 0$ gives

$$F(x) = \int_{-\infty}^{x} f(t)dt. \tag{6.1.4}$$

Because of (6.1.4), the c.d.f. is sometimes called the probability integral of X. Since $F(\infty) = 1$, (6.1.4) gives

$$\int_{-\infty}^{\infty} f(t)dt = 1. \tag{6.1.5}$$

As a result, the probability density function f need be defined only up to a constant multiple. The constant can then be determined from (6.1.5).

An actual measurement of a time, weight, etc. will necessarily involve only finitely many decimals, and will therefore correspond to some small interval of real values. Hence, in the continuous case, we are concerned with assigning probabilities to intervals of X-values. If a and b are any real numbers with $a < b$, then

$$P(a < X \leq b) = P(X \leq b) - P(X \leq a) = F(b) - F(a).$$

Since f is the derivative of F, the fundamental theorem of calculus gives

$$P(a < X \leq b) = F(b) - F(a) = \int_a^b f(x)dx. \qquad (6.1.6)$$

Hence the probability that X lies between a and b is equal to the area under the curve $y = f(x)$ from $x = a$ to $x = b$. We noted in Section 4.1 that the probability histogram of a discrete variate has a similar area property. In fact, the height of a probability histogram satisfies all of the requirements for a p.d.f.. Thus a probability density function can be regarded as a generalized probability histogram.

If we let a approach b in (6.1.6), then since F is continuous, $F(a)$ approaches $F(b)$, and $P(a < X \leq b)$ tends to zero as $a \to b$. Hence the probability that a continuous variate assumes any particular real value is zero. Thus $f(b)$ does <u>not</u> give the probability that $X = b$, or indeed any probability. <u>A probability density function is useful only as an integrand</u>. It is because of this that f may be left undefined or defined arbitrarily at a finite number of points. Also, with continuous variates it is not necessary to distinguish between open and closed intervals, because the probability of the endpoints is zero.

Suppose that $b - a$ is small, and that $f(x)$ exists and is continuous for $a < x < b$. Then the area under the p.d.f. from $x = a$ to $x = b$ can be approximated by the area of a rectangle with base $b - a$ and height $f(t)$, where t is a point in (a,b):

$$P(a < X \leq b) = \int_a^b f(x)dx \approx (b - a)f(t). \qquad (6.1.7)$$

Usually t is taken to be the midpoint of (a,b).

<u>Example 6.1.1</u>. A manufacturing process produces fibres of varying lengths. The length X of a randomly chosen fibre has a continuous distribution with p.d.f.

$$f(x) = \begin{cases} kxe^{-x} & \text{for } x > 0; \\ 0 & \text{otherwise.} \end{cases}$$

Evaluate the constant k, derive the c.d.f. of X, and evaluate the following:

$$P(1 < X < 3); \quad P(X > 5); \quad P(3.9 < X < 4.1).$$

Solution. Since $f(x) = 0$ for $x < 0$, there is no probability on the negative axis, and $F(x) = 0$ for $x < 0$. For $x \geq 0$, we have

$$F(x) = \int_{-\infty}^{x} f(t)dt = \int_{0}^{x} kte^{-t}dt.$$

Integration by parts (or a table of integrals) gives

$$F(x) = k[-(t+1)e^{-t}]_{0}^{x} = k[1-(x+1)e^{-x}]$$

for $x > 0$. As $x \to \infty$, $(x+1)e^{-x} \to 0$, and $F(x) \to k$. Since the total probability must be 1, we have

$$k = F(\infty) = 1.$$

This result could also be obtained from (6.1.5).

Substituting $k = 1$ gives

$$f(x) = xe^{-x}; \quad F(x) = 1-(x+1)e^{-x} \quad \text{for} \quad x > 0.$$

These functions are graphed in Figures 6.1.1 and 6.1.2. The required probabilities may be obtained directly from the c.d.f. (Figure 6.1.2),

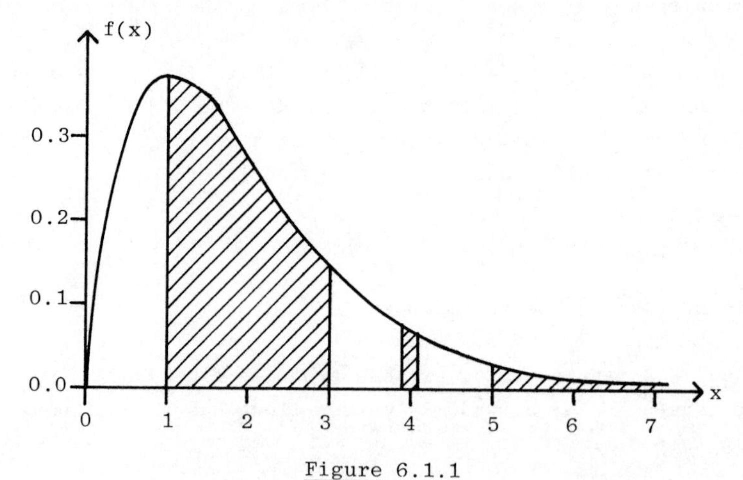

Figure 6.1.1

Probability Density Function for Fibre Length

or as integrals of the p.d.f. (Figure 6.1.1). Since $P(X = 3) = 0$ (the probability at any single point is zero), we have

$$P(1 < X < 3) = P(X < 3) - P(X \leq 1) = P(X \leq 3) - P(X \leq 1)$$
$$= F(3) - F(1) = 2e^{-1} - 4e^{-3} = 0.5366.$$

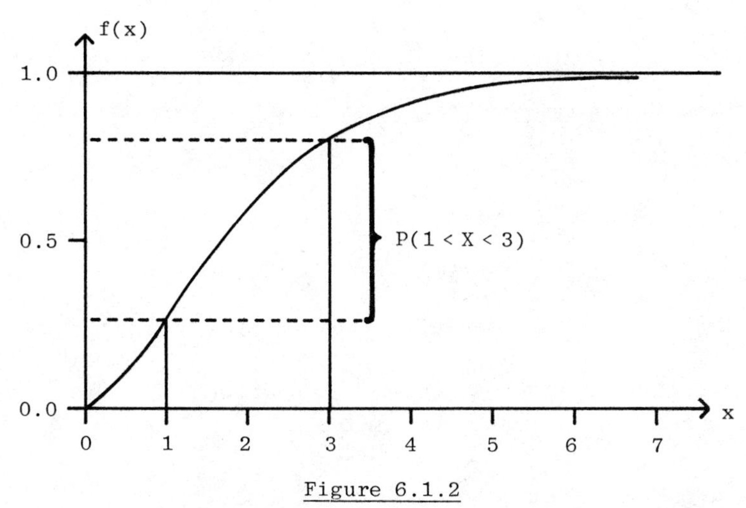

Figure 6.1.2

Cumulative Distribution Function for Fibre Length

Alternatively, this probability may be found as the area under the p.d.f. from $x = 1$ to $x = 3$:

$$P(1 < X < 3) = \int_1^3 f(x)dx = [-(x+1)e^{-x}]_1^3 = 0.5366.$$

Thus 53.66% of the fibres produced will have lengths between 1 unit and 3 units.

Similarly, $P(X > 5)$ can be found in two ways:

$$P(X > 5) = 1 - P(X \le 5) = 1 - F(5) = 6e^{-5} = 0.0404;$$
$$P(X > 5) = \int_5^\infty f(x)dx = [-(x+1)e^{-x}]_5^\infty .$$

The third probability is

$$P(3.9 < X < 4.1) = F(4.1) - F(3.9) = \int_{3.9}^{4.1} xe^{-x}dx$$
$$= 0.014665.$$

The approximation (6.1.7) with $t = 4$ gives

$$P(3.9 < X < 4.1) \approx (4.1 - 3.9)f(4) = 0.014653$$

which agrees closely with the exact result. The area under the p.d.f. from $x = 3.9$ to $x = 4.1$ is well approximated by the area of a rectangle with base 0.2 and height $f(4)$; see Figure 6.1.1.

Expectation

If X is a continuous variate with probability density function f, the mathematical expectation (mean value, expected value) of X is defined by

$$E(X) = \int_{-\infty}^{\infty} xf(x)dx \tag{6.1.8}$$

provided that the integral converges absolutely; that is,

$$\int_{-\infty}^{\infty} |xf(x)|dx < \infty.$$

Otherwise we say that X has no finite expectation. Absolute convergence is required in order that the limit of

$$\int_{a}^{b} xf(x)dx$$

as $a \to -\infty$ and $b \to \infty$ shall not depend upon the way in which the limit is taken (see Problem 6.1.9).

Because integrals have the same mathematical properties as sums, the definitions and results given for discrete distributions in Chapter 5 will also hold in the continuous case. In particular, (5.1.3) becomes

$$E\{h(X)\} = \int_{-\infty}^{\infty} h(x)f(x)dx \tag{6.1.9}$$

provided that the function h is reasonably well behaved, and the integral converges absolutely.

Example 6.1.2. Find the mean and standard deviation of the continuous distribution in Example 6.1.1.

Solution. Since $f(x) = xe^{-x}$ for $x > 0$, the rth moment is

$$m_r = E(X^r) = \int_{-\infty}^{\infty} x^r f(x)dx = \int_{0}^{\infty} x^{r+1}e^{-x}dx.$$

Now, by (2.1.13) and (2.1.15), we have

$$m_r = \Gamma(r+2) = (r+1)!$$

so that $m_1 = 2$ and $m_2 = 6$. The mean length of fibres produced is $\mu = m_1 = 2$ units. By (5.2.3), the variance is

$$\sigma^2 = E(X^2) - \mu^2 = m_2 - \mu^2 = 2.$$

The standard deviation is $\sigma = \sqrt{2}$ units.

The mean $\mu = 2$ is the "balance point" of the distribution - see Figure 6.1.1. If one made a cardboard cutout of the region between the p.d.f. and the x-axis and stood it vertically on the x-axis, it would balance if supported at the position $x = 2$.

Chebyshev's Inequality (5.2.7) implies that at least 8/9 of the probability should lie between $\mu - 3\sigma$ and $\mu + 3\sigma$. Here we have $\mu - 3\sigma = -2.24$ and $\mu + 3\sigma = 6.24$. Using the expression for F obtained earlier, we find that

$$P(-2.24 < X < 6.24) = P(0 < X < 6.24) = F(6.24) = 0.986.$$

Almost all of the probability is thus contained in the interval $\mu \pm 3\sigma$.

Quantiles

Let X be a continuous variate with cumulative distribution function F, and let α be a number between 0 and 1. The α-quantile of X (or of the distribution) is the variate value Q_α such that

$$P(X \leq Q_\alpha) = F(Q_\alpha) = \alpha. \tag{6.1.10}$$

The area under the p.d.f. to the left of Q_α is α, and the area to the right of Q_α is $1-\alpha$. The α-quantile is also called the α-fractile, or the 100αth percentile. The .5-quantile (or 50th percentile) is called the median, and will usually be denoted by m. The median is sometimes used in place of the mean to indicate the location of the "centre" of a continuous distribution. Half of the probability lies to the left of the median, and half to the right:

$$P(X < m) = P(X > m) = 0.5.$$

The median will be less than the mean for a distribution like that in Figure 6.1.1 which has a long tail to the right. A distribution with a long tail to the left will have $m > \mu$. If a distribution has a central point of symmetry, this point will be the median and also the mean (provided that the mean exists).

The quantiles (and median) of a discrete distribution are not well-defined, because then the c.d.f. is a step function (see Figure 4.1.1). Thus, there may not exist any variate value Q_α such that $F(Q_\alpha) = \alpha$.

Example 6.1.3. Find the median in Example 6.1.1.

Solution. Since $F(x) = 1 - (x+1)e^{-x}$ for $x > 0$, we have

$$0.5 = F(m) = 1 - (m+1)e^{-m}.$$

This equation must be solved numerically for m. Various procedures could be used; for instance, Newton's method, or the method of repeated bisection (see Section 9.2). Here we shall determine m by trial and error. Figure 6.1.1 suggests that $m < 2$, so we begin by trying $m = 1.6$. Since $F(1.6) = 0.475$ we see that $m > 1.6$, and next we try $m = 1.7$. We find that $F(1.7) = 0.507$, so $m < 1.7$. Continuing in this fashion, we find that $m = 1.678$, correct to three decimal places. Other quantiles can be found in a similar way. For instance, the .95-quantile satisfies the equation

$$.95 = 1 - (Q+1)e^{-Q}.$$

A trial and error solution gives $Q_{.95} = 4.744$, correct to three decimal places.

Change of Variables

Suppose that X is a continuous variate with probability density function f and cumulative distribution function F. We wish to find the probability distribution of $Y \equiv h(X)$, where h is a real-valued function. The recommended method is to find an expression for $G(y) = P(Y \leq y)$ in terms of F. From this expression for the cumulative distribution function G, one can then obtain g, the p.d.f. (or probability function) of Y.

Example 6.1.4. Find the p.d.f. of the variate $Y \equiv e^{-X}$ in Example 6.1.1.

Solution. Let G be the c.d.f. of Y. Then

$$\begin{aligned} G(y) &= P(Y \leq y) = P(e^{-X} \leq y) = P(-X \leq \log y) \\ &= P(X \geq -\log y) = 1 - P(X < -\log y) \\ &= 1 - F(-\log y) \end{aligned}$$

where F is the c.d.f. of X:

$$F(x) = 1 - (1+x)e^{-x} \quad \text{for} \quad x > 0.$$

Substituting $x = -\log y$ gives

$$G(y) = (1 - \log y)\, e^{\log y} = y(1 - \log y).$$

Since $0 < X < \infty$, the range of Y is $0 < Y < 1$. Since $G(y)$ is continuous and has a continuous derivative except at $y = 0$, Y is a continuous variate. The p.d.f. of Y is

$$g(y) = \frac{d}{dy} G(y) = \begin{cases} -\log y & \text{for } 0 < y < 1 \\ 0 & \text{otherwise.} \end{cases}$$

Note that $-\log y > 0$ for $0 < y < 1$, so $g(y) \geq 0$ as required. □

Monotonic transformations

The method described above can be used to establish a change of variables formula in an important special case. Suppose that $Y \equiv h(X)$ where h is either a strictly increasing or strictly decreasing function, and has a continuous derivative. Let y be a possible value of Y. Because the transformation is one-to-one, the equation $y = h(x)$ uniquely determines the corresponding X-value, $x = h^{-1}(y)$. If h is increasing, then $Y \leq y$ if and only if $X \leq x$, and hence

$$G(y) = P(Y \leq y) = P(X \leq x) = F(x).$$

If h is decreasing, then $Y \leq y$ if and only if $X \geq x$, and in this case

$$G(y) = P(Y \leq y) = P(X \geq x) = 1 - F(x).$$

We now differentiate using the chain rule to obtain

$$g(y) = \frac{dG(y)}{dy} = \pm \frac{dF(x)}{dy} = \pm \frac{dF(x)}{dx} \cdot \frac{dx}{dy} = \pm f(x) \cdot \frac{dx}{dy}.$$

Finally, we note that $\frac{dx}{dy}$ is positive when h is increasing and negative when h is decreasing. Therefore, the single formula

$$g(y) = f(x) \cdot \left|\frac{dx}{dy}\right| \tag{6.1.11}$$

holds for both increasing transformations and decreasing transformations.

Example 6.1.4 (continued). The transformation $y = e^{-x}$ is strictly decreasing, with inverse $x = -\log y$. Since $0 < y < 1$, we have

$$\left|\frac{dx}{dy}\right| = \left|-\frac{1}{y}\right| = \frac{1}{y}.$$

Since $f(x) = xe^{-x} = -\log y\, e^{\log y} = -y \log y$, (6.1.11) gives

$$g(y) = -\log y \quad \text{for } 0 < y < 1$$

which agrees with result obtained earlier. □

A probability density function is useful only as an integrand. Hence, when we change variables in the continuous case, we are effectively changing variables in an integral. It is for this reason that the derivative of the transformation comes into the formula. The derivative does not arise when we change variables in discrete distributions: see the end of Section 4.1.

Problems for Section 6.1.

†1. Let X be a continuous variate with probability density function

$$f(x) = kx(1-x) \quad \text{for} \quad 0 < x < 1.$$

(a) Evaluate the constant k, and find the c.d.f. of X.

(b) Determine the probability that X takes a value between 0.2 and 0.6, and show this probability on graphs of the p.d.f. and c.d.f.

(c) Evaluate $P(.39 < X < .41)$ using (6.1.7), and compare with the exact value.

(d) Find the median, the mean, and the variance of X.

2. A continuous variate X has probability density function

$$f(x) = \begin{cases} kx^{-3} & \text{for} \quad x \geq 1, \\ 0 & \text{otherwise}. \end{cases}$$

(a) Evaluate the constant k, derive the c.d.f., and evaluate the following:

$$P(2 \leq X \leq 3); \quad P(X > 4); \quad P(3.99 \leq X \leq 4.01).$$

(b) Derive the p.d.f. of Y, where $Y \equiv \log X$.

3. Let c, n, and k be positive constants. A continuous variate X with p.d.f.

$$f(x) = \begin{cases} kx^{-n-1} & \text{for} \quad x \geq c \\ 0 & \text{otherwise} \end{cases}$$

is said to have a Pareto distribution. Pareto distributions have been used as a model for income distribution, with c being the subsistence wage.

(a) Evaluate k (as a function of c and n), and derive the c.d.f.

(b) Find the median of the Pareto distribution.

(c) Find the mean and variance, indicating the conditions under which they fail to exist.

†4. Find the mean and variance of the continuous distribution with cumulative distribution function

$$F(x) = 1 - e^{-x^2/2} \quad \text{for} \quad x > 0.$$

5. A manufacturer must choose between two processes for producing components. The dimension X of the components (in cm.) is a continuous variate with p.d.f. as follows:

$$\text{Process 1:} \quad f(x) = 3/x^4 \quad \text{for} \quad 1 < x < \infty$$
$$\text{Process 2:} \quad f(x) = 4/x^5 \quad \text{for} \quad 1 < x < \infty.$$

Only components with dimensions between 1.1 and 2 cm. are acceptable.

(a) Which process produces the greater percentage of acceptable components?

(b) What is the average length of all components from process 1?

(c) What is the average length of all acceptable components from process 1?

6. Let X be a continuous variate with p.d.f.

$$f(x) = \begin{cases} 0 & \text{for} \quad x \le 0; \\ kx^3 & \text{for} \quad 0 < x < 1; \\ ke^{1-x} & \text{for} \quad x \ge 1. \end{cases}$$

(a) Evaluate k, and find the c.d.f. of X.

(b) Calculate $P(0.5 < X < 2)$ and $P(X > 2 | X > 1)$.

(c) Find the mean of the distribution.

†7. An isosceles triangle has two equal sides of unit length. The angle X between them is a continuous variate with p.d.f.

$$f(x) = kx(\pi - x) \quad \text{for} \quad 0 < x < \pi/2,$$

where k is a constant. Find the p.d.f. and the expected value of the triangle's area.

8. A <u>triangular distribution</u> has p.d.f.

$$f(x) = k(1 - |x|) \quad \text{for} \quad -1 < x < 1.$$

(a) Sketch the p.d.f., and evaluate the constant k.

(b) Find the mean, median, and variance of this distribution.

9. A <u>Cauchy distribution</u> has p.d.f.

$$f(x) = k/(1 + x^2) \quad \text{for} \quad -\infty < x < \infty.$$

(a) Find k, the c.d.f., and the median of the distribution.

(b) Show that the integral for $E(X)$ does not converge absolutely, so that the mean does not exist.

†10. A continuous distribution has c.d.f.

$$F(x) = kx^n/(1+x^n) \quad \text{for } x > 0$$

where n is a positive constant. Evaluate k, derive the p.d.f. of this distribution, and give an expression for the α-quantile.

11. Let X be a continuous variate with p.d.f.

$$f(x) = k(1-x^2) \quad \text{for } -1 < x < 1.$$

Find the p.d.f. of Y, where $Y \equiv \text{Sin}^{-1}X$.

12. A continuous variate X has p.d.f.

$$f(x) = kx^{n-1}(1+x)^{-2n} \quad \text{for } 0 < x < \infty.$$

Show that X^{-1} has the same distribution as X.

13. Let X be a continuous variate with p.d. and c.d. functions f and F. Define $Y \equiv aX + b$ where $a > 0$ and b are constants. Show that Y has p.d. and c.d. functions $\frac{1}{a} f(\frac{y-b}{a})$ and $F(\frac{y-b}{a})$.

†14. Show that the median m of the continuous distribution with p.d.f.

$$f(x) = k(x+1)^2 e^{-x} \quad \text{for } -1 < x < \infty$$

satisfies the equation

$$1 + (m+2)^2 - e^{m+1} = 0.$$

Use trial and error to evaluate m to three decimal places.

6.2 Uniform and Exponential Distributions

A continuous variate X whose p.d.f. is constant over the interval (a,b) and zero elsewhere,

$$f(x) = \begin{cases} k & \text{for } a < x < b \\ 0 & \text{elsewhere} \end{cases}$$

is said to have a uniform distribution (also called a rectangular distribution), and we write $X \sim U(a,b)$. Since the total area under the p.d.f. must be one, it follows that $k = 1/(b-a)$. Since all of the probability lies between a and b, $F(x) = P(X \le x) = 0$ for $x \le a$, and $F(x) = 1$ for $x \ge b$. For $a < x < b$ we have

$$F(x) = \int_{-\infty}^{x} f(t)dt = \int_{a}^{x} kdt = \frac{x-a}{b-a}.$$

The p.d.f. and c.d.f. are graphed in Figure 6.2.1. Note that F has

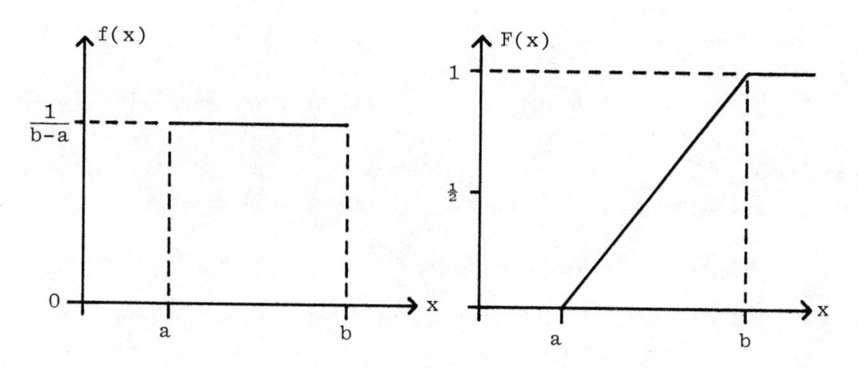

Figure 6.2.1
P.d.f. and c.d.f. for the uniform distribution U(a,b).

corners at $x = a$ and $x = b$, and f is discontinuous at these points. Note also that, if $b - a < 1$, then $f(x) > 1$ for $a < x < b$. It is possible for the probability density to exceed 1, so long as the total area under the p.d.f. equals 1.

An important special case is the uniform distribution on the interval (0,1). The p.d.f. and c.d.f. for U(0,1) are

$$f(x) = 1; \quad F(x) = x \quad \text{for} \quad 0 < x < 1. \tag{6.2.1}$$

By symmetry, the mean and median of this distribution are both equal to $\frac{1}{2}$. The second moment is

$$m_2 = E(X^2) = \int_{-\infty}^{\infty} x^2 f(x)dx = \int_0^1 x^2 dx = \frac{1}{3}$$

and hence the variance is

$$\sigma^2 = m_2 - \mu^2 = \frac{1}{3} - \frac{1}{4} = \frac{1}{12}.$$

Suppose that a balanced roulette wheel is spun, and let A be the angular displacement (in radians) of the wheel when it comes to rest. Then A has a uniform distribution on the interval $(0,2\pi)$, and $X \equiv A/2\pi$ has a uniform distribution on the interval (0,1). A set of n independent values $x_1, x_2, \ldots, x_n$ from U(0,1) could be generated by repeatedly spinning the wheel. These values may be transformed into independent values from any discrete or continuous distribution using the inverse probability integral transformation - see Section 6.3. The resulting values may then be used in simulation or Monte Carlo studies.

Mechanical devices such as roulette wheels are inconvenient to use when large numbers of values are needed. Much effort has been devoted to the development of computer methods for generating numbers which look like they come from $U(0,1)$. These numbers are called <u>random numbers</u>, or more accurately, <u>pseudo-random numbers</u>. In the most common method, integers are generated sequentially using

$$x_{n+1} \equiv ax_n (\text{mod } b),$$

where a and b are integers chosen so that a large number of distinct values x_n may be obtained before repeats occur. One can then divide by b to obtain numbers between 0 and 1.

<u>Exponential Distribution</u>

Let θ be any positive real number. A continuous variate X with probability density function

$$f(x) = \begin{cases} \frac{1}{\theta} e^{-x/\theta} & \text{for } x > 0 \\ 0 & \text{otherwise} \end{cases} \qquad (6.2.2)$$

is said to have an <u>exponential distribution with mean θ</u>. If $\theta = 1$, then $f(x) = e^{-x}$ for $x > 0$, and X is said to have a <u>unit exponential distribution</u>.

For $x > 0$, (6.1.4) gives

$$F(x) = \int_0^x \frac{1}{\theta} e^{-t/\theta} dt = [-e^{-t/\theta}]_0^x = 1 - e^{-x/\theta}, \qquad (6.2.3)$$

and $F(x) = 0$ for $x \le 0$. Note that $F(\infty) = 1$, so the total probability is 1 as it should be.

The α-quantile of X is obtained by solving the equation

$$\alpha = F(Q_\alpha) = 1 - \exp\{-Q_\alpha/\theta\}.$$

Upon rearranging and taking the natural logarithm, we obtain

$$Q_\alpha = -\theta \log(1-\alpha). \qquad (6.2.4)$$

In particular, the median (.5-quantile) is given by

$$m = -\theta \log(1 - .5) = \theta \log 2 = 0.69\theta.$$

The rth moment of the distribution is

$$m_r = E(X^r) = \int_{-\infty}^{\infty} x^r f(x) dx = \int_0^{\infty} \frac{1}{\theta} x^r e^{-x/\theta} dx.$$

Substituting $x = \theta u$, $dx = \theta du$ gives

$$m_r = \theta^r \int_0^\infty u^r e^{-u} du = \theta^r \Gamma(r+1) = r!\theta^r$$

by (2.1.13) and (2.1.15). Hence the mean is $\mu = m_1 = \theta$ as claimed. Note that the mean is greater than the median because of the long tail to the right. By (5.2.3), the variance is

$$\sigma^2 = m_2 - \mu^2 = 2\theta^2 - \theta^2 = \theta^2.$$

Hence the standard deviation of an exponential distribution is equal to the mean.

The p.d.f. and c.d.f. of an exponential distribution with mean θ are shown in Figure 6.2.2. Note that θ determines not only

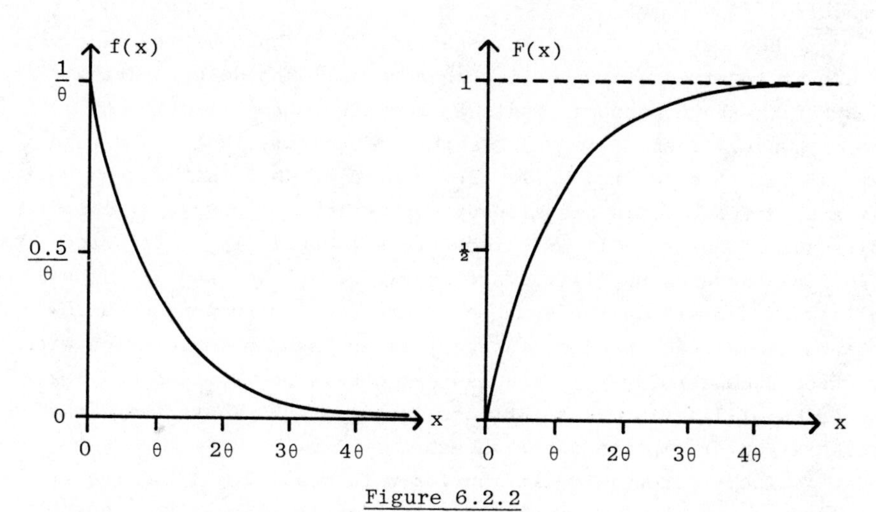

Figure 6.2.2

P.d.f. and c.d.f. for an exponential distribution with mean θ

the "centre" of the distribution, but also its spread. All exponential distributions have the same shape; they differ only in the measurement scale used. For suppose that X has an exponential distribution with mean θ, and define $T \equiv X/\theta$. Then

$$P(T \le t) = P(X \le t\theta) = F(t\theta) = 1 - e^{-t} \quad \text{for} \quad t > 0,$$

which is the c.d.f. of a unit exponential distribution. If we measure X in multiples of its mean value θ, the distribution becomes a unit exponential (mean 1).

An important characteristic of the exponential distribution

is its <u>memoryless property</u>: if X has an exponential distribution, then

$$P(X > b + c \mid X > b) = P(X > c) \tag{6.2.5}$$

for all non-negative b and c. To prove this result, we note that

$$P(X > x) = 1 - P(X \le x) = 1 - F(x) = e^{-x/\theta} \quad \text{for} \quad x \ge 0.$$

Now the definition of conditional probability (3.4.2) gives

$$P(X > b + c \mid X > b) = \frac{P(X > b + c, X > b)}{P(X > b)} = \frac{P(X > b + c)}{P(X > b)}$$

$$= \frac{e^{-(b+c)/\theta}}{e^{-b/\theta}} = e^{-c/\theta} = P(X > c)$$

as required.

Because of (6.2.5), exponential distributions are appropriate for use as lifetime distributions when there is no deterioration with age. For instance, suppose that X represents the lifetime of a radio tube in hours. Then $P(X > c)$ is the probability that a new tube lasts at least c hours, while $P(X > b + c \mid X > b)$ is the probability that a used tube, which has already worked for b hours, will last an additional c hours. If lifetimes are exponentially distributed, then by (6.2.5) these probabilities are equal for all b and c. The probability of failure in the next c hours does not depend upon how long the tube has already been operating. There is no deterioration with age; used tubes are just as good as new ones. See Section 6.4 for a discussion of lifetime distributions in situations where there may be deterioration or improvement with age.

The exponential distribution with mean $\theta = 1/\lambda$ also arises as the distribution of the waiting time between successive random events in a Poisson process of intensity λ; see Section 6.5. In Examples 1.3.4 and 1.4.2, the exponential distribution was considered as a possible model for waiting times between successive mining accidents. However, a comparison of observed and expected frequencies (Table 1.4.1) revealed rather a poor agreement. The explanation suggested was that the accident rate λ seemed to decrease with time (θ increased) for the accident data, whereas λ is assumed to be constant in a Poisson process.

A comparison of observed and expected frequencies, as in Example 1.4.2, is one way to investigate whether an assumed model is in good agreement with a set of data. With continuous distributions, this method involves an arbitrary grouping of the data into classes.

Another method, which does not require grouping, involves comparing the actual measurements with quantiles of the theoretical distribution. Quantile plots will be considered in Section 11.5.

Problems for Section 6.2

†1. For a certain type of electronic component, the lifetime X (in thousands of hours) has an exponential distribution with mean $\theta = 2$. What is the probability that a new component will last longer than 1000 hours? If a component has already lasted 1000 hours, what is the probability that it will last at least 1000 hours more?

2. The time from treatment to recurrence of a certain type of cancer is exponentially distributed with mean θ. Determine θ such that there is a 50% probability of a recurrence within 693 days.

3. Suppose that X, the lifetime (in days) of a certain type of electronic component, has an exponential distribution with mean θ. Let Y be the integer part of X, so that Y represents the number of completed days of life. Show that Y has a geometric distribution.

4. If X is uniformly distributed on $(-\frac{\pi}{2}, \frac{\pi}{2})$, what is the p.d.f. of $\tan X$?

†5. A continuous variate X is distributed in such a way that $1 - X^n$ has a uniform distribution on $(0,1)$. Find the p.d.f. of X.

6. The p.d.f. of the double exponential distribution is

$$f(x) = ke^{-|x|} \quad \text{for} \quad -\infty < x < \infty.$$

(a) Find k and the c.d.f. of the distribution.
(b) Show that odd moments are zero, and $m_{2r} = (2r)!$.

*6.3 Transformations Based on the Probability Integral

Let F be the probability integral (cumulative distribution function) of a continuous variate X. In general, F is non-decreasing and $0 \le F(x) \le 1$ for all x. In the following discussion, it is assumed that $F(x)$ is strictly increasing for all x such that $0 < F(x) < 1$.

The graph of a "typical" continuous c.d.f. is shown in Figure 6.3.1. Given any sequence of values $x_1, x_2, \ldots$ on the horizontal axis,

* This section may be omitted on first reading.

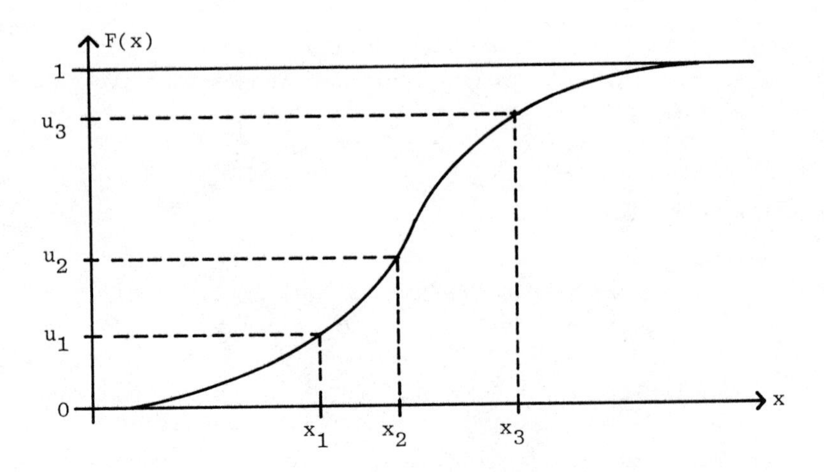

Figure 6.3.1

Probability Integral and Inverse P.I. Transformations

we may obtain the corresponding values $u_1, u_2, \ldots$ on the vertical axis by means of the probability integral transformation, $u_i = F(x_i)$. Conversely, given values $u_1, u_2, \ldots$ between 0 and 1, we may obtain the corresponding values $x_1, x_2, \ldots$ via the inverse probability integral transformation, $x_i = F^{-1}(u_i)$. We shall show that, if x_i is a value from the distribution F, then u_i is a value from the uniform distribution on $(0,1)$. Conversely if u_i is a value from $U(0,1)$, then x_i is a value from the distribution F.

First consider the probability integral transformation $U \equiv F(X)$, where F is the c.d.f. of X, and take $0 < u < 1$. Since F is strictly increasing, there is a unique x such that $u = F(x)$; furthermore, $U \le u$ if and only if $X \le x$. Hence U has c.d.f.

$$G(u) = P(U \le u) = P(X \le x) = F(x) = u \quad \text{for} \quad 0 < u < 1,$$

which, by (6.2.1), is the c.d.f. of $U(0,1)$.

Next suppose that U has a uniform distribution on $(0,1)$, and consider the inverse probability integral transformation $X \equiv F^{-1}(U)$, where F is strictly increasing. Then

$$P(X \le x) = P\{F^{-1}(U) \le x\} = P\{U \le F(x)\}.$$

But, since U has a uniform distribution on $(0,1)$, $P(U \le u) = u$, and hence

$$P(X \le x) = P\{U \le F(x)\} = F(x).$$

Therefore X has cumulative distribution function F.

Applications. Suppose that we have n observed values $x_1, x_2, \ldots, x_n$, and that we wish to know whether these could reasonably have come from a continuous distribution with a particular c.d.f. F. Using the probability integral transformation, we can obtain the corresponding probabilities

$$u_i = F(x_i), \quad i = 1,2,\ldots,n.$$

If the x_i's are independent observations from the distribution F, then the u_i's are independent observations from $U(0,1)$. Hence we can check the suitability of the proposed distribution F by determining whether the u_i's look like random numbers between 0 and 1.

The inverse p.i.t. makes it possible to generate a set of observations from a continuous distribution with any specified cumulative distribution function F. First we use a random number generator or tables of random numbers to obtain independent observations $u_1, u_2, \ldots, u_n$ from $U(0,1)$. Then we transform these via the inverse probability integral transformation to obtain the corresponding quantiles of F:

$$x_i = F^{-1}(u_i), \quad i = 1,2,\ldots,n.$$

Then $x_1, x_2, \ldots, x_n$ are independent observations from the distribution whose c.d.f. is F.

Such artificially generated values are useful in simulations. For instance, we might have a model of traffic flow which specifies that the time intervals between arrivals of successive cars at an intersection are like independent observations from a continuous distribution whose c.d.f. is F. The above method can be used to generate observations from this distribution, and hence simulate the arrival of traffic at the intersection. The likely effects of installing traffic signals, adjusting the length of the red light, introducing a left turn lane, etc., could then be investigated.

It is important to realize that any results obtainable through simulation could also be obtained without simulation via the rules of probability. The advantage of simulation is that it sometimes permits one to bypass difficult computational problems, and to obtain approximate results at a comparatively low cost in time and money. However with simulations, as with direct calculations, the results obtained will be reliable only if the assumed distribution F gives a reasonable approximation to the actual distribution.

Example 6.3.1. Generate a set of 10 values (a sample of size 10) from the exponential distribution with mean $\theta = \frac{1}{2}$.

Solution. From (6.2.3), the c.d.f. of an exponential distribution with mean θ is

$$F(x) = 1 - e^{-x/\theta} \quad \text{for} \quad x > 0.$$

Solving $u = F(x)$ for x in terms of u gives

$$x = F^{-1}(u) = -\theta \log(1-u). \tag{6.3.1}$$

If $U \sim U(0,1)$, then $-\theta \log(1-U)$ has an exponential distribution with mean θ.

The following ten 4-digit numbers were selected from a table of random numbers:

0.2214	0.8259	0.3403	0.2439	0.1343
0.9385	0.2584	0.7670	0.0007	0.6333

Using these as u-values in (6.3.1) with $\theta = \frac{1}{2}$ gives the following x's:

0.1251	0.8741	0.2080	0.1398	0.0721
1.3944	0.1495	0.7284	0.0004	0.5016

This is the required sample of size 10 from the exponential distribution with mean $\theta = \frac{1}{2}$.

Note that, whereas each u corresponds to a real interval of length 0.0001, the same is not true of the x's. Since $x = -\frac{1}{2}\log(1-u)$, we have $\frac{dx}{du} = \frac{1}{2(1-u)}$, and a small change in u will produce a large change in x if u is near 1. For example, changing u from 0.9995 to 0.9996 changes x from 3.800 to 3.912.

If we wish to obtain a sample of x's, none of which corresponds to a real interval of length greater than 0.001 say, then we must select the u's so that $\Delta x \le 0.001$. But

$$\Delta u \approx 2(1-u)\Delta x,$$

and hence we require

$$\Delta u \le 0.002(1-u). \tag{6.3.2}$$

For $u \le 0.95$, the right hand side of (6.3.2) is at least $(0.002)(0.05) = 0.0001$, and hence a 4-digit random number u will suffice. However for $u > 0.95$, the number of digits must be increased. This may be done by selecting a second 4-digit number from the random number tables. For instance, if the first number selected were

9873 and the second were 5238, we would take $u = 0.98735238$. Now for $u > 0.95$ we have $\Delta u = 10^{-8}$, and (6.3.2) will be satisfied unless $u > 0.999995$. In the latter case, a third number is drawn from the random number tables to reduce Δu to 10^{-12}, and so on. In this way, a table of random numbers, or a random number generator, may be used to generate x-values with any desired accuracy.

Simulation of discrete distributions

The inverse probability integral transformation can also be used to simulate a discrete distribution. In this case F has a discontinuity, or step, at each possible value of X, as in Figure 4.1.1. The random numbers $u_1, u_2, \ldots, u_n$ are represented by points on the vertical axis, and x_i is taken to be the X-value which corresponds to the step immediately above u_i. Thus in Figure 4.1.1, $x_i = 0$ for $0 \le u_i < \frac{1}{8}$; $x_i = 1$ for $\frac{1}{8} \le u_i < \frac{1}{2}$; $x_i = 2$ for $\frac{1}{2} \le u_i < \frac{7}{8}$; and $x_i = 3$ for $\frac{7}{8} \le u_i < 1$. The following are the X-values which correspond to the ten random numbers used in Example 6.3.1 above:

1	2	1	1	1
3	1	2	0	2

This is a sample of size ten from a binomial distribution with $n = 3$ and $p = \frac{1}{2}$.

Problems for Section 6.3.

1. Using a table of random numbers or a random number generator, obtain a random sample of size 10 from the uniform distribution on (0,1). Transform this into a sample from the exponential distribution with mean $\theta = 100$.
2. The following sample of size 100 was generated by the procedure described in problem 1:

134	305	31	111	46	28	1	101	140	2
33	28	43	262	46	12	130	83	27	74
144	129	102	179	72	11	10	280	10	68
68	114	1	71	132	240	5	261	69	96
128	9	64	77	6	299	27	26	19	208
414	37	14	46	31	32	0	12	146	118
105	67	53	17	89	17	131	88	62	76
125	172	187	56	22	341	63	70	5	29
59	12	47	17	183	155	34	204	240	129
581	88	362	34	6	143	171	115	12	43

Prepare an observed frequency histogram for the above data using an interval width of 50. Compare this with the expected frequency histogram for an exponential distribution with mean $\theta = 100$.

3. The number of persons arriving in a queue during a one minute interval has a Poisson distribution with mean $\mu = 1$. The number of minutes required to service a customer is Y, with geometric probability function

$$f(y) = \theta(1-\theta)^{y-1} \quad \text{for} \quad y = 1,2,\ldots ,$$

where $\theta = 0.8$. Service times for different customers are independent, and initially the queue is empty. Let Z be the length of the queue after 30 minutes time. The distribution of Z is difficult to work out analytically, but easy to simulate on the computer. First, generate 30 observations from the Poisson distribution to simulate arrivals in the queue. For each person arriving in the queue, generate a service time from the geometric distribution. With proper bookkeeping, the queue length after 30 minutes can now be determined. Repeat this procedure 100 times, and prepare a relative frequency histogram for Z. (This gives an estimate of the probability function of Z, whose accuracy can be improved by generating additional values of Z. Additional simulations could be undertaken to determine the effects of changes in μ and θ.)

*6.4 Lifetime Distributions

In the preceding section, the exponential distribution was shown to be appropriate for use as a lifetime distribution when there is a constant risk of failure which does not change with age. In most real-life situations, deterioration with age does (unfortunately) occur, and an item which has already operated for some time is likely to fail sooner than a new item. There are also cases in which the risk of failure decreases with age; for instance, the mortality rate for infants decreases during the first few months of life. In this section, we consider some lifetime distributions for which the risk of failure may change with age.

Let X be a non-negative continuous variate which represents the lifetime (survival time, failure time) of an item. Let f and F be the p.d.f. and c.d.f. of X. Two additional functions, the <u>survivor function</u> S and the <u>hazard function</u> h, are useful in discussing lifetime distributions, and these are defined as follows:

$$S(x) = P(X > x) = 1 - F(x); \tag{6.4.1}$$

* This section may be omitted on first reading.

$$h(x) = f(x) / S(x). \tag{6.4.2}$$

Note that $S(x)$ is the probability that an item survives beyong time x.

Given that an item survives beyond time x, the probability that it fails by time $x + \Delta x$ is

$$P(X \le x + \Delta x | X > x) = \frac{P(X > x, X \le x + \Delta x)}{P(X > x)} = \frac{P(x < X \le x + \Delta x)}{P(X > x)}$$

by the definition of conditional probability (3.4.2). Now if Δx is small, (6.1.7) gives

$$P(x < X \le x + \Delta x) \approx f(x)\Delta x,$$

and hence

$$P(X \le x + \Delta x | X > x) \approx \frac{f(x)}{S(x)} \Delta x = h(x)\Delta x.$$

Thus the hazard $h(x)$ represents the <u>instantaneous failure rate</u> at time x among all items which survive at time x. It is also called the <u>age-specific mortality rate</u>, or the <u>force of mortality</u>.

If $h(x)$ is an increasing function of x, the instantaneous failure rate increases with age: the older the item, the greater the probability of failure within a time interval of length Δx. The item is thus deteriorating with age, and we have <u>positive ageing</u>. If $h(x)$ is decreasing with x, the chance of failure within a time interval of length Δx decreases with age, and we have <u>negative ageing</u> (improvement with age). If $h(x)$ is constant, the failure rate does not change with age, and there is no ageing. These three cases are illustrated in Figure 6.4.1. Other types of hazard function are possible. For instance, the hazard function (mortality rate) for humans decreases through infancy and childhood, and then increases through middle and old age.

Since $f(x) = \frac{d}{dx} F(x) = -\frac{d}{dx} S(x)$, we have

$$h(x) = \frac{f(x)}{S(x)} = -\frac{1}{S(x)} \frac{dS(x)}{dx} = -\frac{d}{dx} \log S(x).$$

Integration with respect to x gives

$$\log S(x) = -\int h(x)dx = -\int_0^x h(t)dt + c$$

where c is a constant of integration. Since all items are assumed to be operational at time 0, we have

$$S(0) = P(X > 0) = 1$$

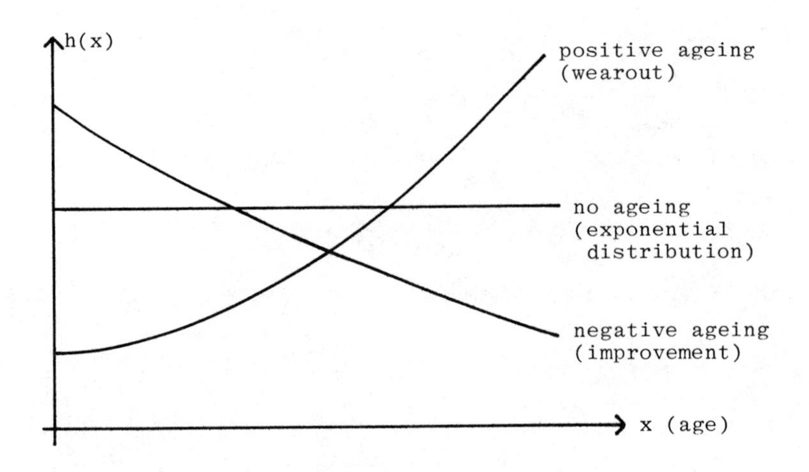

Figure 6.4.1
Monotonic hazard functions

and hence $c = 0$. It follows that

$$S(x) = \exp\{-\int_0^x h(t)dt\}. \tag{6.4.3}$$

Now (6.4.2) gives

$$f(x) = h(x) \cdot S(x) = h(x) \cdot \exp\{-\int_0^x h(t)dt\}. \tag{6.4.4}$$

Thus the hazard function uniquely determines the probability density function of the lifetime distribution.

Since X is assumed to be a continuous variate, it follows from (6.4.2) that h is a continuous function (except possibly at finitely many points), and that $h(x) \geq 0$ for $x \geq 0$. If we assume that all items must eventually fail, then $S(x) \to 0$ as $x \to \infty$, and (6.4.3) implies that

$$\lim_{x\to\infty} \int_0^x h(t)dt = \infty. \tag{6.4.5}$$

There are infinitely many ways to define a hazard function which satisfies these restrictions. We now consider the lifetime distributions which correspond to some simple choices of h.

Some lifetime distributions

(a) Suppose that the hazard function is constant:

$$h(x) = \lambda \quad \text{for} \quad x \geq 0.$$

Then (6.4.3) and (6.4.2) give

$$S(x) = \exp\{-\int_0^x \lambda dt\} = e^{-\lambda x} \quad \text{for} \quad x \geq 0;$$

$$f(x) = h(x) \cdot S(x) = \lambda e^{-\lambda x} \quad \text{for} \quad x \geq 0.$$

Thus X has an exponential distribution with mean $\theta = 1/\lambda$. If the instantaneous probability of failure does not depend upon the age of the item, then the lifetime distribution is exponential (see Section 6.2).

(b) Suppose that the risk of failure changes linearly with age:

$$h(x) = \alpha + \beta x \quad \text{for} \quad x \geq 0$$

where α and β are constants. Then by (6.4.4), the probability density function of the lifetime distribution is

$$f(x) = (\alpha + \beta x)\exp\{-\alpha x - \beta x^2/2\} \quad \text{for} \quad x \geq 0.$$

Since $h(x) \geq 0$ for all $x \geq 0$, we must have $\alpha \geq 0$ and $\beta \geq 0$. For $\beta = 0$, there is no ageing (exponential distribution), and for $\beta > 0$ there is positive ageing. This model cannot be used to represent negative ageing.

(c) Suppose that the logarithm of the hazard function changes linearly with age; that is,

$$h(x) = \exp\{\alpha + \beta x\} \quad \text{for} \quad x \geq 0.$$

For $\beta = 0$, the hazard function is constant, and the distribution is again exponential. For $\beta \neq 0$, we have

$$\int_0^x h(t)dt = e^{\alpha}(e^{\beta x} - 1)/\beta,$$

and by condition (6.4.5), β must be positive. Once again, the hazard function is increasing (positive ageing), and it is not possible to represent negative ageing by this model. By (6.4.4), the p.d.f. of the lifetime distribution is

$$f(x) = e^{\alpha+\beta x}\exp\{-e^{\alpha}(e^{\beta x} - 1)/\beta\} \quad \text{for} \quad x \geq 0,$$

where $-\infty < \alpha < \infty$ and $\beta \geq 0$. This is called a <u>Gompertz distribution</u>, and it has been used as a model for adult human mortality.

(d) Suppose that the risk of failure changes as a power of the age:

$$h(x) = cx^p \quad \text{for} \quad x \geq 0,$$

where c and p are constants. Then (6.4.5) and the condition $h(x) \geq 0$ imply that $c > 0$ and $p > -1$. (The case $p = -1$ is ruled out because then, by (6.4.3), $S(x) = 0$ for all $x > 0$.) Let $\beta = p + 1$ and $\lambda = c/\beta$. Then, by (6.4.3) and (6.4.4),

$$S(x) = \exp\{-\lambda x^\beta\} \quad \text{for} \quad x \geq 0; \tag{6.4.6}$$

$$f(x) = \lambda\beta x^{\beta-1}\exp\{-\lambda x^\beta\} \quad \text{for} \quad x \geq 0, \tag{6.4.7}$$

where $\beta > 0$ and $\lambda > 0$. This is called the Weibull distribution, and β is called its shape parameter. The hazard function is

$$h(x) = \lambda\beta x^{\beta-1} \quad \text{for} \quad x \geq 0.$$

For $\beta = 1$, the Weibull distribution simplifies to an exponential distribution (no ageing). There is positive ageing (deterioration) for $\beta > 1$, and negative ageing (improvement) for $0 < \beta < 1$. The Weibull distribution has been used as a model in reliability studies of ball bearings, gears, aircraft components, etc.

(e) In engineering applications, it is often assumed that $\log X$, the logarithm of the lifetime, has a normal distribution (see Section 6.6). Then X is said to have a lognormal distribution. The hazard function of the lognormal distribution is fairly complicated, but it can be shown that, as x increases, $h(x)$ first increases (positive ageing) and then decreases (negative ageing); see Problem 6.6.13.

(f) In some situations, it may be reasonable to suppose that certain events are occurring at random in time (Section 4.4), and that failure takes place as soon as a specified number of events has occurred. The lifetime will then have a gamma distribution - see Section 6.5.

Problems for Section 6.4

†1. Find the rth moment, the mean and the variance of the Weibull distribution (6.4.7). For what values of β is the mean greater than the median?

2. Let X have a Weibull distribution, and define $Y \equiv \log X$. Derive the p.d.f. of Y. (This is called a Type I extreme value distribution.)

3. The p.d.f. of an exponential distribution with guarantee time $c > 0$ is given by

$$f(x) = \lambda e^{-\lambda(x-c)} \quad \text{for} \quad x > c$$

where λ is a positive constant. Derive the survivor and hazard functions for this distribution.

†4. Let S be the survivor function of a continuous lifetime distribution with finite mean μ. The expected residual life at time x is defined as follows:

$$r(x) = E\{T - x | T \geq x\} \quad \text{for} \quad x \geq 0.$$

(a) Show that $xS(x) \to 0$ as $x \to \infty$, and hence that

$$r(x) = \frac{1}{S(x)} \int_x^\infty S(t)dt.$$

(b) Show that

$$S(x) = \frac{\mu}{r(x)} \exp\{-\int_0^x \frac{dt}{r(t)}\} \quad \text{for} \quad x \geq 0.$$

Note that the expected residual life function r uniquely determines S, and hence the p.d.f. of the lifetime distribution.

*6.5 Waiting Times in a Poisson Process

In Section 4.4, we imagined a system subject to instantaneous changes due to the occurrence of random events (e.g. radioactive decay). The following two assumptions were said to define a Poisson process with intensity parameter λ:

(1) The numbers of changes in disjoint time intervals are independent.

(2) If h is sufficiently small, the probability of one change in $(t, t+h)$ is λh where λ is constant with respect to t; the probability of more than one change in $(t, t+h)$ is negligible.

We consider two variates associated with a Poisson process:

$X_t \equiv$ number of events occurring in the fixed time interval $(0,t)$;
$T_x \equiv$ total waiting time for the xth event to occur.

In Section 4.4 we showed that X_t has a Poisson distribution with mean $\mu = \lambda t$, and probability function

$$f_t(x) = P(X_t = x) = (\lambda t)^x e^{-\lambda t}/x! \quad \text{for} \quad x = 0,1,2,\ldots .$$

* This section may be omitted on first reading.

Our aim in this section is to derive the distribution of T_x.

Similar problems were considered for Bernoulli trials in Sections 3.2 and 4.2. We found that the probability distribution of the number of events which occur in a fixed time (i.e. in a fixed number of trials) is binomial. The distribution of the waiting time for a specified number of successes to occur is negative binomial. We now extend these results from discrete time (number of trials) to continuous time.

Waiting Time for First Event. By definition, T_1 is the waiting time for the first event to occur in a Poisson process. Let G_1 be the c.d.f. of T_1, so that

$$G_1(t) = P(T_1 \le t).$$

The waiting time for the first event will be at most t if and only if at least one event occurs in $(0,t)$; that is, $T_1 \le t$ if and only if $X_t \ge 1$. Therefore

$$G_1(t) = P(T_1 \le t) = P(X_t \ge 1) = 1 - P(X_t = 0)$$
$$= 1 - f_t(0) = 1 - e^{-\lambda t} \quad \text{for} \quad t \ge 0.$$

This is a continuous function with a continuous derivative except at $t = 0$. Hence T_1 is a continuous variate; its p.d.f. is

$$g_1(t) = \frac{d}{dt} G_1(t) = \lambda e^{-\lambda t} \quad \text{for} \quad t > 0.$$

Upon comparing this with (6.2.2), we conclude that the waiting time for the first event has an exponential distribution with mean $\theta = 1/\lambda$. The exponential distribution is the continuous analogue of the geometric distribution in discrete time (Bernoulli trials).

If we begin waiting just after an event has occurred, then T_1 represents the waiting time between successive events. Therefore, the waiting time between consecutive events in a Poisson process follows an exponential distribution with mean $\theta = 1/\lambda$.

Waiting Time for Second Event. The waiting time for the second event will be at most t if and only if at least two events occur in the time interval $(0,t)$; that is, $T_2 \le t$ if and only if $X_t \ge 2$. Hence the c.d.f. of T_2 is

$$G_2(t) = P(T_2 \le t) = P(X_t \ge 2) = 1 - P(X_t \le 1)$$

$$= 1 - f_t(0) - f_t(1) = 1 - (1 + \lambda t)e^{-\lambda t} \quad \text{for} \quad t \geq 0.$$

The p.d.f. of T_2 is then

$$g_2(t) = \frac{d}{dt} G_2(t) = \lambda^2 t e^{-\lambda t} \quad \text{for} \quad t > 0.$$

Waiting Time for xth Event. The waiting time for the xth event will be at most t if and only if at least x events occur in $(0,t)$; that is, $T_x \leq t$ if and only if $X_t \geq x$. Therefore, T_x has c.d.f.

$$G_x(t) = P(T_x \leq t) = P(X_t \geq x) = 1 - P(X_t \leq x - 1)$$

$$= 1 - [1 + \lambda t + \frac{(\lambda t)^2}{2} + \ldots + \frac{(\lambda t)^{x-1}}{(x-1)!}]e^{-\lambda t} \quad \text{for} \quad t \geq 0.$$

Differentiating with respect to t and simplifying gives

$$g_x(t) = \frac{d}{dt} G_x(t) = \frac{\lambda(\lambda t)^{x-1} e^{-\lambda t}}{(x-1)!} \quad \text{for} \quad t > 0.$$

This is called a gamma distribution, and it is closely related to another important distribution, the χ^2 (chi-square) distribution, which is discussed in Section 6.9. If one defines $Y \equiv 2\lambda T_x$ and applies the change of variables formula (6.1.11), one finds that Y has a χ^2 distribution with $2x$ degrees of freedom. Thus

$$P(T_x \leq t) = P(2\lambda T_x \leq 2\lambda t) = P(\chi^2_{(2x)} \leq 2\lambda t). \tag{6.5.1}$$

The χ^2 distribution is tabulated, and because of (6.5.1), these tables can be used to compute probabilities for T_x; see Section 6.9. Alternatively, probabilities for T_x may be obtained by summing Poisson probabilities:

$$P(T_x \leq t) = P(X_t \geq x) = 1 - \sum_{i=0}^{x-1} f_t(x). \tag{6.5.2}$$

Formulas (6.5.1) and (6.5.2) together imply a connection between the Poisson and χ^2 distributions: if X has a Poisson distribution with mean μ, then

$$P(X \geq x) = P(\chi^2_{(2x)} \leq 2\mu). \tag{6.5.3}$$

Example 6.5.1. Telephone calls reach a switchboard at the average rate of 3 per minute (see Example 4.4.1). Find the probability that

(a) the time between successive calls exceeds one minute;

(b) the waiting time for the 3rd call is between two and three minutes;

(c) the waiting time for the 10th call is greater than five minutes.

Solution. (a) The waiting time between successive calls has an exponential distribution with mean $\theta = 1/\lambda = 1/3$, and c.d.f.

$$P(T_1 \le t) = 1 - e^{-3t} \quad \text{for} \quad t > 0.$$

The required probability is

$$P(T_1 > 1) = 1 - P(T_1 \le 1) = e^{-3} = 0.0498.$$

(b) The c.d.f. of T_3 is given by

$$G_3(t) = 1 - [1 + 3t + \tfrac{1}{2}(3t)^2]e^{-3t} \quad \text{for} \quad t > 0.$$

The required probability is

$$\begin{aligned} P(2 \le T_3 \le 3) &= G_3(3) - G_3(2) \\ &= 0.99377 - 0.93803 = 0.05574. \end{aligned}$$

(c) The waiting time for the 10th call exceeds 5 minutes if and only if fewer than 10 calls arrive in 5 minutes:

$$P(T_{10} > 5) = P(X_5 < 10) = P(X_5 \le 9).$$

Since X_5 has a Poisson distribution with mean $5\lambda = 15$, we have

$$P(T_{10} > 5) = \sum_{i=0}^{9} \frac{15^i e^{-15}}{i!} .$$

These Poisson probabilities may be computed recursively using (4.3.2) and then added to give $P(T_{10} > 5) = 0.06985$. Alternatively, (6.5.1) gives

$$P(T_{10} > 5) = 1 - P(T_{10} \le 5) = 1 - P(\chi^2_{(20)} \le 30)$$

which may be evaluated from tables of the χ^2 distribution (see Section 6.9).

Random Events in Space

Although the above discussion dealt only with random events in time, similar arguments may be applied for random events in the plane or in space.

Example 6.5.2. The distribution of flying bomb hits was considered in

Example 4.4.3, and we argued that the number of flying bomb hits in a region of area A should have a Poisson distribution with mean λA. Find the probability distribution of R, the distance from an arbitrarily chosen point in South London to the nearest bomb hit.

Solution. The distance to the nearest hit will be at most r if and only if there is at least one hit within a circle of radius r and area $A = \pi r^2$. The probability of x hits in a region of area A is

$$f(x) = (\lambda A)^x e^{-\lambda A}/x! \quad \text{for} \quad x = 0,1,2,\ldots .$$

The probability of at least one hit in a circle of area $A = \pi r^2$ is

$$1 - f(0) = 1 - \exp(-\lambda A) = 1 - \exp(-\lambda\pi r^2).$$

Hence the c.d.f. of R is

$$G(r) = P(R \le r) = 1 - \exp(-\lambda\pi r^2) \quad \text{for} \quad r > 0.$$

The p.d.f. of R is

$$g(r) = \frac{d}{dr} G(r) = 2\lambda\pi r \exp(-\lambda\pi r^2) \quad \text{for} \quad r > 0.$$

Note that if we take the centre of the circle to be a bomb hit, then R represents the distance from one bomb hit to its nearest neighbour.

Example 6.5.3. Supposing that stars are randomly distributed in space, find the probability distribution of the distance from an arbitrary point in space to the nearest star.

Solution. Under the assumption of a random distribution, the probability of finding x stars in a volume V of space is

$$f(x) = (\lambda V)^x e^{-\lambda V}/x! \quad \text{for} \quad x = 0,1,2,\ldots ,$$

where λ is the average number of stars per unit volume. The probability of at least one star in volume V is

$$1 - f(0) = 1 - e^{-\lambda V}.$$

Let R be the distance from an arbitrary point P to the nearest star. Then $R \le r$ if and only if there is at least one star in the sphere of radius r centred at P. Taking $V = \frac{4}{3}\pi r^3$, the volume of the sphere, we have

$$P(R \le r) = 1 - \exp(-\lambda \tfrac{4}{3}\pi r^3) \quad \text{for} \quad r > 0,$$

and differentiating with respect to r gives the p.d.f. of R. Note that if we take P to be the location of a star, R represents the distance to its nearest neighbour.

Problems for Section 6.5

1. In Example 6.5.2, show that the expected distance from a bomb hit to its nearest neighbour is $1/2\sqrt{\lambda}$.

2. In defining a Poisson process (Section 4.4), we required that the intensity parameter λ be constant over time. Suppose now that λ changes with time:

$$\lambda(t) = kt^{\beta-1} \quad \text{for} \quad 0 < t < \infty,$$

where k and β are positive constants. Show that $f_t(0)$, the probability of zero events in the time interval $(0,t)$, satisfies the differential equation

$$\frac{d}{dt} f_t(0) = -kt^{\beta-1} f_t(0)$$

and hence show that

$$f_t(0) = \exp\{-\frac{k}{\beta} t^{\beta}\}.$$

Find the p.d.f. of T_1, the waiting time to the first event. (T_1 has a Weibull distribution; see Section 6.4.)

6.6 The Normal Distribution

Let μ and σ be real numbers with $\sigma > 0$. A continuous variate X with probability density function

$$f(x) = \frac{1}{\sqrt{2\pi}\,\sigma} \exp\{-\frac{1}{2}(\frac{x-\mu}{\sigma})^2\}; \quad -\infty < x < \infty \tag{6.6.1}$$

is said to have a normal (or Gaussian) distribution with mean μ and variance σ^2. Experience has shown that many measurements have distributions that are approximately normal, and the Central Limit Theorem (Section 6.7) provides a partial explanation of this phenomenon.

The normal p.d.f. is a bell-shaped curve centred at μ, which is both the mean and median of the distribution; see Figure 6.6.1. The spread of the distribution depends upon the standard deviation σ, and 99.7% of the probability is contained in the interval $(\mu - 3\sigma,$ $\mu + 3\sigma)$. All normal distributions have the same shape; they differ only

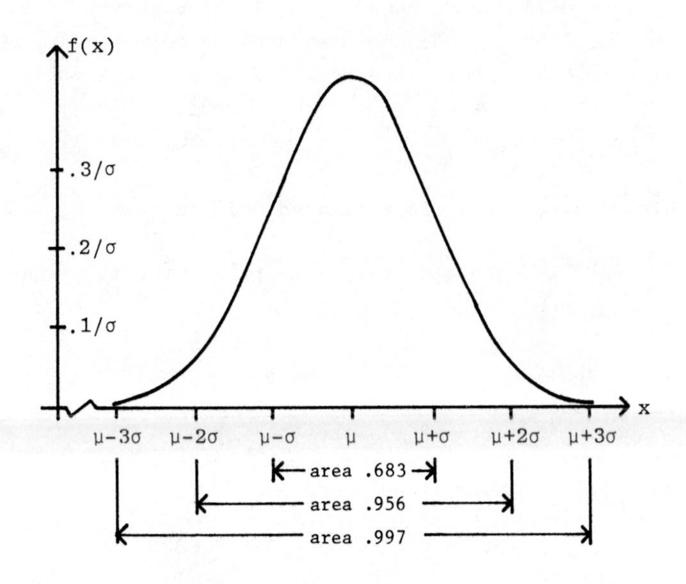

Figure 6.6.1
Probability Density Function of $N(\mu,\sigma^2)$

with respect to the location of the central value (μ), and the measurement scale (σ).

Standardized Normal Distribution N(0,1)

A variate X whose distribution is normal with mean $\mu = 0$ and variance $\sigma^2 = 1$ is said to have a standardized normal distribution, and we write $X \sim N(0,1)$. The p.d.f. of N(0,1) is

$$f(x) = \frac{1}{\sqrt{2\pi}} e^{-x^2/2} \quad \text{for} \quad -\infty < x < \infty; \tag{6.6.2}$$

the c.d.f. of N(0,1) is

$$F(x) = \int_{-\infty}^{x} f(u)du = \int_{-\infty}^{x} \frac{1}{\sqrt{2\pi}} e^{-u^2/2}du. \tag{6.6.3}$$

This integral cannot be evaluated algebraically. One must either perform the integration numerically, or else use tables - see Appendix B. Table B1 gives the value x such that $F(x) = p$ for $p = .50, .51, \ldots, .99$ and for selected larger values of p. Table B2 gives F(x) for $x = .00, .01, .02, \ldots, 3.09$. Since the standardized normal distribution is symmetrical about the origin, one has $P(X \le -x) = P(X \ge x)$, and therefore

$$F(-x) = 1 - F(x). \tag{6.6.4}$$

Because of this result, F(x) is tabulated only for $x \ge 0$.

Example 6.6.1. Let $X \sim N(0,1)$. Evaluate $P(-1 < X < 1)$, and determine x such that $P(-x \le X \le x) = 0.9$.

Solution. Since $F(-1) = 1 - F(1)$, we have

$$P(-1 < X < 1) = F(1) - F(-1) = 2F(1) - 1.$$

Table B2 gives $F(1) = 0.8413$, and hence

$$P(-1 < X < 1) = 2(0.8413) - 1 = 0.6826.$$

Similarly, (6.6.4) gives

$$P(-x \le X \le x) = F(x) - F(-x) = 2F(x) - 1.$$

We require $2F(x) - 1 = 0.9$, and this implies $F(x) = 0.95$. Now Table B1 gives $x = 1.645$.

Standard Form

Suppose that $X \sim N(\mu,\sigma^2)$, and let Z be the standard form of X: $Z \equiv \frac{X-\mu}{\sigma}$ (see Section 5.2). Since the transformation is monotonic, the p.d.f. of Z may be obtained from (6.1.11):

$$g(z) = f(x)\cdot\left|\frac{dx}{dz}\right| = \frac{1}{\sqrt{2\pi}\,\sigma}\exp\{-\frac{1}{2}(\frac{x-\mu}{\sigma})^2\}\cdot\sigma$$

$$= \frac{1}{\sqrt{2\pi}}e^{-z^2/2} \quad \text{for} \quad -\infty < z < \infty.$$

This is the p.d.f. of a standardized normal distribution. It follows that, if X has a normal distribution, the standard form of X has a standardized normal distribution:

$$X \sim N(\mu,\sigma^2) \Longrightarrow \frac{X-\mu}{\sigma} \sim N(0,1). \tag{6.6.5}$$

This result enables us to obtain probabilities for any normal distribution $N(\mu,\sigma^2)$ from tables for $N(0,1)$.

By a similar argument, one can show that

$$X \sim N(\mu,\sigma^2) \Longrightarrow aX+b \sim N(a\mu + b, a^2\sigma^2) \tag{6.6.6}$$

for any constants a,b.

Example 6.6.2. Suppose that IQ scores of first year mathematics students are normally distributed with mean 120 and standard deviation 10. What percentage of mathematics freshmen have IQ scores greater than 150? What percentage have scores which differ from the mean by 20 points or more?

Solution. Let X be the IQ score of a randomly chosen student. We are told that $X \sim N(120,10^2)$, and are asked to compute $P(X > 150)$ and $P\{|X-120| \ge 20\}$. We convert these into statements about the standardized variable $Z \equiv \frac{X-120}{10}$:

$$P(X > 150) = P(\frac{X-120}{10} > \frac{150-120}{10}) = P(Z > 3);$$

$$P(|X-120| \ge 20) = P(\left|\frac{X-120}{10}\right| \ge \frac{20}{10}) = P(|Z| \ge 2).$$

Since $Z \sim N(0,1)$, we can evaluate these probabilities from Table B2:

$$P(Z > 3) = 1 - P(Z \le 3) = 1 - F(3) = 0.00135;$$

$$P(|Z| \geq 2) = 1 - P(-2 < Z < 2) = 1 - [F(2) - F(-2)]$$
$$= 1 - [2F(2) - 1] = 2[1 - F(2)] = 0.0455.$$

Thus 0.135% have IQ scores greater than 150, and 4.55% have scores which differ from the mean by 20 points or more.

Linear Combinations

A very important property of the normal distribution is that any linear combination of independent normal variates has a normal distribution. If $X_1, X_2, \ldots, X_n$ are independent variates and $X_i \sim N(\mu_i, \sigma_i^2)$ for $i = 1, 2, \ldots, n$, then

$$\sum a_i X_i \sim N(\sum a_i \mu_i, \sum a_i^2 \sigma_i^2) \tag{6.6.7}$$

for all constants $a_1, a_2, \ldots, a_n$. This result may be proved by change of variables methods (Section 7.3), or by using moment generating functions (Example 8.4.3). (The definition of independence for continuous variates is similar to that given in Section 4.5 for discrete variates. See Section 7.1 for further discussion.)

Now suppose that $X_1, X_2, \ldots, X_n$ are independent variates having the same mean μ and variance σ^2, and let $S_n \equiv \sum X_i$ be their sum. The sample mean $\bar{X}$ is the arithmetic average of the X_i's:

$$\bar{X} \equiv \frac{1}{n} \sum X_i \equiv \frac{1}{n} S_n.$$

Since $E(S_n) = n\mu$ and $\text{var}(S_n) = n\sigma^2$ by (5.4.8) and (5.4.10), we have

$$E(\bar{X}) = \frac{1}{n} E(S_n) = \mu;$$

$$\text{var}(\bar{X}) = \frac{1}{n^2} \text{var}(S_n) = \frac{\sigma^2}{n}.$$

If the X_i's are normally distributed, then (6.6.7) implies that the linear combinations S_n and $\bar{X}$ are also normally distributed:

$$S_n \sim N(n\mu, n\sigma^2); \quad \bar{X} \sim N(\mu, \sigma^2/n). \tag{6.6.8}$$

These results are exact and hold for all values of n provided that the X_i's are independent normal variates. If the X_i's are not normally distributed, the Central Limit Theorem (Section 6.7) implies that (6.6.8) holds as an approximation for n sufficiently large.

Example 6.6.3. Bolts produced by a certain machine have diameters which are normally distributed with mean 10 mm. and standard deviation 1 mm. The inner diameters of washers produced by a second machine are also normally distributed, but with mean 11 mm. and standard deviation 0.5 mm. If a bolt and washer are selected at random, what is the probability that the bolt will fit inside the washer?

Solution. Let X denote the inner diameter of the washer, and Y the diameter of the bolt. If the washer and bolt are selected in separate random drawings, X and Y will be independent. We are given that

$$X \sim N(11,0.25); \quad Y \sim N(10,1)$$

and we wish to compute

$$P(X > Y) = P(X - Y > 0).$$

By (6.6.7), $X - Y$ has a normal distribution with mean and variance

$$E(X - Y) = E(X) - E(Y) = 1;$$

$$\text{var}(X - Y) = \text{var}(X) + (-1)^2\text{var}(Y) = 1.25 = (1.118)^2.$$

By (6.6.5), the standardized variable $Z \equiv \frac{(X - Y) - 1}{1.118}$ has a standardized normal distribution. Thus we obtain

$$P(X - Y > 0) = P\left(\frac{X - Y - 1}{1.118} > \frac{0 - 1}{1.118}\right) = P(Z > -0.89) = 1 - F(-0.89)$$

where F is the c.d.f. of $N(0,1)$. Now (6.6.4) and Table B2 give

$$P(X - Y > 0) = F(0.89) = 0.8133.$$

There is an 81% chance that the bolt will fit inside the washer.

Example 6.6.4. Suppose that the IQ's of freshmen mathematics students are normally distributed with mean 120 and variance 100. (a) If a class of 25 students is selected at random, what is the probability that the average IQ of the class exceeds 125? (b) If two classes of 25 students each are selected at random, what is the probability that the average IQ's for the two classes differ by more than 5?

Solution. (a) Denote the IQ's of the 25 students by $X_1, X_2, \ldots, X_{25}$. The X_i's are independent $N(120,100)$. The average IQ of the class is given by the sample mean, $\overline{X} \equiv \frac{1}{25} \sum X_i$, and we wish to find $P(\overline{X} > 125)$. Since $\overline{X} \sim N(120, \frac{100}{25})$ by (6.6.8), the standardized variate $Z \equiv \frac{\overline{X} - 120}{2}$ is distributed as $N(0,1)$. Therefore

$$P(\overline{X} > 125) = P(Z > \frac{125 - 120}{2}) = 1 - F(2.5) = 0.0062$$

from Table B2.

(b) The average IQ for the first class is $\overline{X} \sim N(120,4)$, and that for the second class is $\overline{Y} \sim N(120,4)$. Assuming that there are no students in common, $\overline{X}$ and $\overline{Y}$ will be independent. Hence, by (6.6.7), $\overline{X} - \overline{Y}$ has a normal distribution with mean and variance

$$E(\overline{X} - \overline{Y}) = E(\overline{X}) - E(\overline{Y}) = 0;$$

$$\text{var}(\overline{X} - \overline{Y}) = \text{var}(\overline{X}) + \text{var}(\overline{Y}) = 8 = (2.83)^2.$$

The variate $Z \equiv \frac{(\overline{X} - \overline{Y}) - 0}{2.83}$ is thus distributed as $N(0,1)$, and the required probability is

$$P(|\overline{X} - \overline{Y}| > 5) = P(|Z| > \frac{5}{2.83}) = 2[1 - F(1.77)] = 0.0767$$

from Table B2. There is a 7.67% chance that the average IQ's for the two classes differ by more than 5.

Example 6.6.5. Suppose that n independent measurements $X_1, X_2, \ldots, X_n$ are to be made of an unknown quantity μ, and the sample mean $\overline{X}$ is to be used as an estimate of μ. The measurement procedure is subject to errors, so that the X_i's are variates. If there is no bias in the procedure, then $E(X_i) = \mu$ for $i = 1,2,\ldots,n$. Suppose that the standard deviation of X_i is 1 unit. How large must n be in order that, with probability 0.95, the estimate will be in error by at most 0.1 units?

Solution. The error in the estimate is $|\overline{X} - \mu|$, and we wish to determine the value of n such that

$$P(|\overline{X} - \mu| \le 0.1) = 0.95.$$

Since the X_i's have mean μ and variance 1, (6.6.8) gives $\overline{X} \sim N(\mu, \frac{1}{n})$, so that $Z \equiv \sqrt{n}(\overline{X} - \mu)$ has a standardized normal distribution. This result holds exactly if the X_i's are normally distributed; otherwise it holds approximately for n sufficiently large. Since $P(|Z| \le 1.960) = 0.95$ from Table B1, we have

$$0.95 = P(\sqrt{n}|\overline{X} - \mu| \le 1.960) = P(|\overline{X} - \mu| \le \frac{1.960}{\sqrt{n}}).$$

It follows that $0.1 = 1.960/\sqrt{n}$, and hence that

$$n = 100(1.960)^2 = 384.2.$$

It would be necessary to take 385 measurements to be 95% sure of obtaining the desired precision in the estimate.

Similarly, if there is to be a 95% probability that the estimate will be in error by at most 0.05 units, we obtain $0.05 = 1.960/\sqrt{n}$, and this gives $n = 400(1.96)^2$. To double the precision, one must take four times as many observations. In general, the precision of the estimate increases only as the square root of the sample size n.

Moments

The rth moment of the standardized normal distribution is

$$m_r = \int_{-\infty}^{\infty} x^r \cdot \frac{1}{\sqrt{2\pi}} e^{-x^2/2} dx.$$

This integral converges absolutely for all $r \geq 0$, so all of the moments exist. For $r \geq 2$, integration by parts gives

$$m_r = \frac{1}{\sqrt{2\pi}} \left[-x^{r-1} e^{-x^2/2} \right]_{-\infty}^{\infty} + (r-1) \int_{-\infty}^{\infty} x^{r-2} \cdot \frac{1}{\sqrt{2\pi}} e^{-x^2/2} dx$$

$$= (r-1) m_{r-2}.$$

The first moment is

$$m_1 = \int_{-\infty}^{\infty} x \cdot \frac{1}{\sqrt{2\pi}} e^{-x^2/2} dx = \frac{1}{\sqrt{2\pi}} \left[-e^{-x^2/2} \right]_{-\infty}^{\infty} = 0$$

and hence all odd moments are zero. This can also be seen from the fact that the p.d.f. is symmetrical about the origin.

The evaluation of m_0 requires a trick. We first complicate the problem by setting up a double integral, and then evaluate this integral by transforming to polar coordinates. Define

$$I = \int_{-\infty}^{\infty} e^{-x^2/2} dx = \int_{-\infty}^{\infty} e^{-y^2/2} dy = \sqrt{2\pi}\, m_0.$$

Then we have

$$I^2 = \int_{-\infty}^{\infty} e^{-x^2/2} dx \cdot \int_{-\infty}^{\infty} e^{-y^2/2} dy = \int_{-\infty}^{\infty} \int_{-\infty}^{\infty} e^{-(x^2+y^2)/2}\, dxdy.$$

Now we substitute $x = r\cos\theta$ and $y = r\sin\theta$ where $0 < r < \infty$ and $0 \le \theta < 2\pi$. The Jacobian of the transformation is $\pm r$, and since $x^2 + y^2 = r^2$ we have

$$I^2 = \int_0^\infty \int_0^{2\pi} e^{-r^2/2}\, r\,dr\,d\theta$$

$$= \int_0^\infty re^{-r^2/2}dr \cdot \int_0^{2\pi} d\theta = 2\pi.$$

Since $I > 0$, it follows that $I = \sqrt{2\pi}$, and hence that $m_0 = 1$. The recursive formula derived above now gives $m_2 = 1$, $m_4 = 3$, and in general

$$m_{2r} = (2r-1)(2r-3)\ldots(3)(1).$$

In particular, we have shown that the total area under the N(0,1) density (6.6.2) is equal to 1, and that this distribution has mean $m_1 = 0$ and variance $m_2 - m_1^2 = 1$. From these results, one can easily show that the total area under (6.6.1) is also 1, and that the constants μ and σ which appear in (6.6.1) are in fact the mean and standard deviation of the distribution.

A useful result for the gamma function (2.1.13) can be obtained as a corollary to the evaluation of I. Since the integrand is symmetrical about $x = 0$, we have

$$I = \int_{-\infty}^{\infty} e^{-x^2/2}dx = 2\int_0^\infty e^{-x^2/2}dx.$$

Now substituting $x = \sqrt{2u}$ gives

$$I = 2\int_0^\infty e^{-u} \cdot \frac{du}{\sqrt{2u}} = \sqrt{2}\int_0^\infty u^{-\frac{1}{2}}e^{-u}du = \sqrt{2}\,\Gamma(\tfrac{1}{2}).$$

Since $I = \sqrt{2\pi}$, it follows that

$$\Gamma(\tfrac{1}{2}) = \sqrt{\pi}. \qquad (6.6.9)$$

Problems for Section 6.6

†1. The lengths (in inches) of nails produced in a factory are normally distributed with mean 4.0 and variance 0.0016. What fraction of the total output will be within the permissible limits of 3.9 to 4.1 inches?

2. Examination scores obtained by a very large number of students have

approximately a normal distribution with mean 65 and variance 100. Determine the fraction of students in each of the grades A(≥ 80), B(70-80), C(60-70), D(50-60) and F(<50).

3. A creamery puts up a large number of packages of butter. The weights (in pounds) of these packages are known from experience to be normally distributed with variance 0.0001. What should the mean package weight be in order that 90% of the packages will weigh at least one pound?

†4. The volume (in fluid ounces) placed in a bottle by a bottling machine follows a normal distribution with mean μ and variance σ^2. Over a long period of time, it is observed that 5% of the bottles contain less than 31.5 ounces, while 15% contain more than 32.3 ounces.
   (a) Find μ and σ.
   (b) What proportion of the bottles contain more than 32.2 ounces?
   (c) Give an expression for the probability that, out of 10 bottles purchased, exactly 3 will contain more than 32.2 ounces.

5. Suppose that the diameters (in mm.) of eggs laid by a flock of hens are normally distributed with mean 40 and variance 4. The selling price per dozen is 58¢ for eggs less than 37 mm. in diameter, 70¢ for eggs greater than 42 mm. in diameter, and 66¢ for the remainder. What is the average selling price per egg produced by the flock?

6. Let X and Y be independent N(2,16) and N(3,9) variates, respectively. Evaluate the following:

$$P(X+Y>0), \quad P(X<Y+2), \quad P(2X+3Y-20<0).$$

†7. An elevator has an allowance for 10 persons or a total weight of 1750 pounds. Assuming that only men ride the elevator and that their weights are normally distributed with a mean of 165 pounds and a standard deviation of 10 pounds, what is the probability that the weight limit will be exceeded in a randomly chosen group of 10 men?

8. The sample mean $\overline{X}$ is to be used to estimate the mean μ of a normal distribution with standard deviation 4 inches. How large a sample should be taken in order that, with 90% probability, the estimate will be in error by at most one-half inch?

9. (a) Suppose that the lifetimes of television tubes are normally distributed. A study of the output of one manufacturer shows that 15% of tubes fail before 2 years, while 5% last

longer than 6 years. Find the mean and variance of the lifetime distribution, and the probability that the total lifetime of two tubes exceeds 10 years.

(b) Tubes made by a second manufacturer have the same mean lifetime, but a 20% smaller standard deviation. What fraction of these tubes will fail before 2 years? last longer than 6 years?

†10. The lifetimes of car mufflers are approximately normally distributed with mean 2 years and standard deviation 6 months.

(a) Find the probability that the total lifetime of n mufflers exceeds 7 years for $n = 1,2,\ldots,5$.

(b) I buy a new car and keep it for 7 years. As soon as one muffler fails, I replace it with a new one. Find the probability function of X, the number of replacement mufflers that I will require.

11. Show that, if $X \sim N(\mu,\sigma^2)$, then $aX + b \sim N(a\mu + b, a^2\sigma^2)$ for any constants a,b.

12. X is said to have a lognormal distribution with median θ and shape parameter β if $\log X$ has a normal distribution with mean $\mu = \log\theta$ and standard deviation $\sigma = 1/\beta$.

(a) Derive the p.d.f. of X.

(b) Show that the rth moment of X is

$$m_r = \exp\{\mu r + \tfrac{1}{2} r^2\sigma^2\}; \quad r = 0,1,2,\ldots$$

and hence obtain the mean and variance of X.

†*13. Note. This problem depends upon material in Section 6.4.

(a) Show that the hazard function for the lognormal distribution (Problem 6.6.12) is given by

$$h(x) = \beta f(z)/xF(z) \quad \text{for} \quad x > 0$$

where $z = -\beta\log(x/\theta)$, f is the $N(0,1)$ p.d.f., and F is the $N(0,1)$ c.d.f.

(b) Investigate the shape of the hazard function in (a) by plotting it for $\theta = 1$ and $\beta = 0.5,\ 1.0,\ 2.0$. Discuss the type of ageing which occurs in specimens whose lifetimes are lognormally distributed.

6.7 The Central Limit Theorem

Let $S_n \equiv \sum X_i$ be the sum of n independent discrete or continuous variates with the same mean μ and variance σ^2. By (5.4.8) and (5.4.10), S_n has mean $n\mu$ and variance $n\sigma^2$. The standardized variate

$$S_n^* \equiv \frac{S_n - n\mu}{\sigma\sqrt{n}}$$

has mean 0 and variance 1 for all n. Let F_n denote the c.d.f. of S_n^*, so that $F_n(z) = P(S_n^* \leq z)$ for any real number z. The Central Limit Theorem asserts that, for all real z,

$$\lim_{n\to\infty} F_n(z) = \int_{-\infty}^{z} \frac{1}{\sqrt{2\pi}} e^{-t^2/2} dt.$$

Thus, for sufficiently large n, S_n^* has approximately a standardized normal distribution, whatever the distributions of $X_1, X_2, \ldots, X_n$. In other words, for n sufficiently large, S_n is distributed approximately as $N(n\mu, n\sigma^2)$.

The Central Limit Theorem also holds for many sequences of variates which do not satisfy the above conditions. For instance, the requirement that the X_i's all have the same mean and variance can be replaced by a much weaker condition. For details, see volume 2 of *An Introduction to Probability Theory and its Applications,* by W. Feller. A special case of the Central Limit Theorem is proved using generating functions in Section 8.4.

The Central Limit Theorem provides some theoretical justification of the fact that variates with nearly normal distributions are frequently encountered in nature. Quantities such as the heights and weights of individuals in a relatively homogeneous biological population can often be regarded as the sums of large numbers of genetic and environmental effects which are more or less unrelated, each contributing a small amount to the sum. The Central Limit Theorem implies that their distributions should be nearly normal in shape. However, it is certainly not possible to justify all applications of the normal distribution in this way. In many cases it is known from previous experience with similar measurements that a normal distribution should apply, but the underlying mechanism which produces it is not understood.

Although the limiting distribution of S_n^* does not depend upon the distributions of the X_i's, the rapidity of the approach of S_n^* to normality depends very much on the shapes of these distributions. If they are symmetrical, the normal distribution may give quite a good approximation for very small values of n. However, if the distributions of the X_i's are badly skewed, n must sometimes be very large indeed before a satisfactory approximation is obtained. This important point is illustrated by the examples below and in Section 6.8.

Example 6.7.1. Let $X_1, X_2, \ldots, X_n$ be independent $U(0,1)$ variates. From Section 6.2, the X_i's have mean $\mu = \frac{1}{2}$ and variance $\sigma^2 = \frac{1}{12}$, so that $S_n \equiv \sum X_i$ has mean $\frac{n}{2}$ and variance $\frac{n}{12}$. The standardized variate $S_n^* \equiv (S_n - \frac{n}{2}) \div \sqrt{\frac{n}{12}}$ has approximately a standardized normal distribution for n large by the Central Limit Theorem. The exact p.d.f. of S_n^* for n small can be obtained by change of variables methods. For instance, the p.d.f. of S_2^* is

$$f_2(z) = \begin{cases} (\sqrt{6} + z)/6 & \text{for } -\sqrt{6} < z < 0 \\ (\sqrt{6} - z)/6 & \text{for } 0 < z < \sqrt{6} \end{cases}$$

(see Example 7.2.3). Figure 6.7.1 compares the exact p.d.f. of S_n^* with the standardized normal p.d.f., for $n = 1,2,3$. Note that the agreement is good even for n as small as 3.

This result suggests a method for generating observations from a normal distribution on the computer. A random number generator can be used to obtain n values $x_1, x_2, \ldots, x_n$ from $U(0,1)$, and one can then compute $y = (\sum x_i - \frac{n}{2}) / \sqrt{\frac{n}{12}}$. By the Central Limit Theorem, y can be regarded as an observation from $N(0,1)$ if n is sufficiently large. Figure 6.7.1 suggests that sufficient accuracy should be obtained with only a few random numbers ($n = 5$ or 6, say). Another method for generating values from $N(0,1)$ is by means of the inverse probability integral transformation; see Section 6.3. □

Example 6.7.2. Yields of British Premium Savings Bonds.
Reference: R.L. Plackett, Stochastic Models of Capital Investments, *Journal of the Royal Statistical Society, Series B (1969)*, 1-28.

Instead of paying a fixed interest rate on a certain series of Savings Bonds, the British government ran a monthly lottery in which £100,000 in cash prizes was awarded by chance to bond holders. After an initial waiting period (which we ignore), each £1 bond was eligible for inclusion in the lottery each month, and each had one chance in 9600 of winning a prize. Each month, 2751 prizes were awarded as follows:

Amount (in £)	25	50	100	250	500	1000	5000
Number of prizes	2480	200	30	20	10	10	1

Let X be the amount earned by a £1 bond in one month. Then $X = 0$ with probability 9599/9600. The probability of winning a prize is 1/9600; given that a prize is won, the chance of winning £25 is 2480/2751. Hence

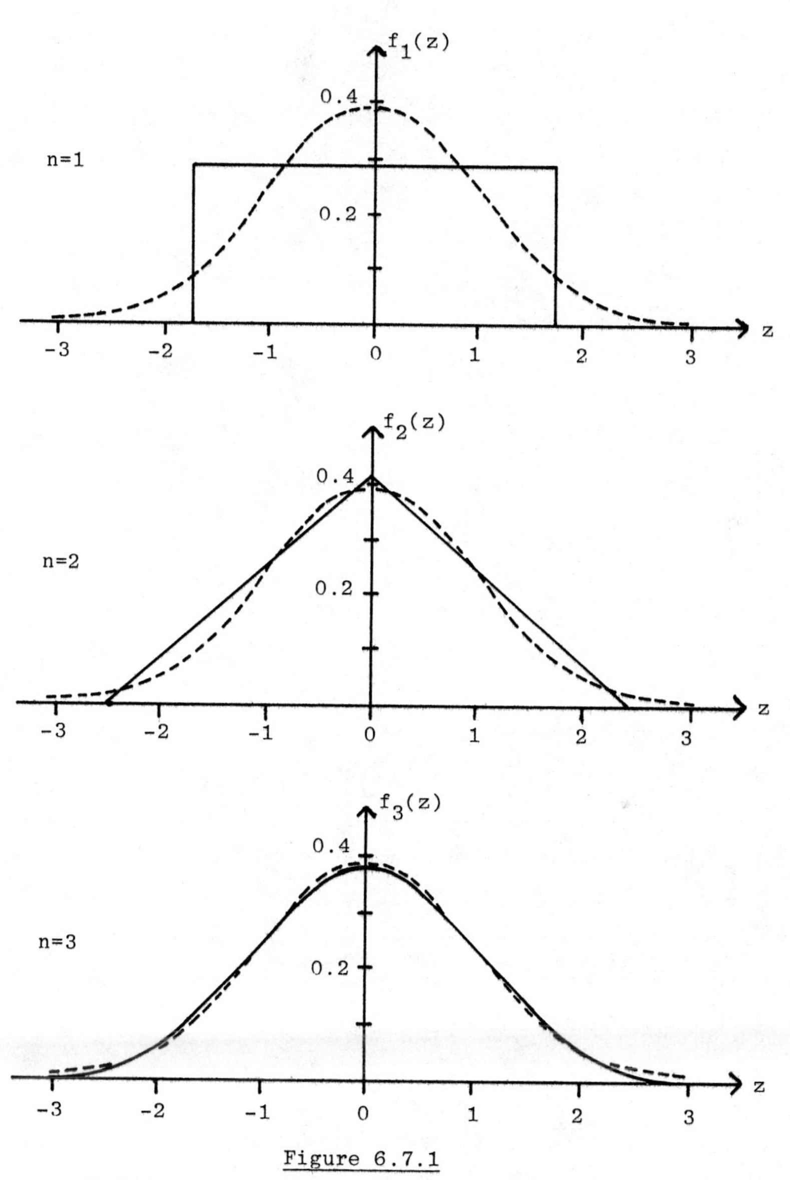

Figure 6.7.1

P.d.f. for the standard form of a sum of n independent uniform variates ($n = 1,2,3$). The standardized normal p.d.f. is shown with a broken line.

$$P(X = 25) = \frac{1}{9600} \cdot \frac{2480}{2751} = \frac{2480}{c}$$

where $c = 26,409,600$. The remaining probabilities in the following table may be obtained in a similar way.

x	0	25	50	100	250	500	1000	5000
$P(X=x)$	$\frac{9599}{9600}$	$\frac{2480}{c}$	$\frac{200}{c}$	$\frac{30}{c}$	$\frac{20}{c}$	$\frac{10}{c}$	$\frac{10}{c}$	$\frac{1}{c}$

By (5.1.1), the expected yield from a £1 bond is

$$E(X) = \sum xP(X = x) = \frac{100000}{c} = 0.003787$$

which corresponds to an annual interest rate of 4.54%. We also have

$$\text{var}(X) = \sum x^2 P(X = x) - [E(X)]^2 = 1.5562.$$

Note the extreme positive skewness of the distribution of X.

If an individual held n bonds, his total earnings for the month would be

$$S_n \equiv X_1 + X_2 + \ldots + X_n,$$

where X_i is the gain from the ith bond. The X_i's are negatively correlated, because a win on one bond would reduce the chance of winning on the others. However, any one individual was restricted by law to holding at most £1000 out of the £26,000,000 in the series. Because of this, the correlations among the X_i's will be extremely small and can safely be ignored. We therefore treat the X_i's as independent variates.

Suppose that $n = 1000$. Then (5.4.8) and (5.4.10) give

$$E(S_n) = 3.787; \quad \text{var}(S_n) = 1556.2 = (39.45)^2.$$

If the Central Limit Theorem applies, then

$$P(S_n < 0) = P(S_n^* < \frac{0 - 3.787}{39.45}) \approx F(-0.10) = 0.46$$

from tables of F, the standardized normal c.d.f.. Since negative values are impossible, it is clear that the Central Limit Theorem gives a very poor approximation when $n = 1000$.

In general, the Central Limit Theorem would give

$$P(S_n < 0) = P(S_n^* < \frac{0-0.003787n}{\sqrt{1.5562n}}) \approx F(-0.00304\sqrt{n}).$$

Even for $n = 500,000$, the approximation is not good; for

$$F(-0.00304\sqrt{500,000}) = F(-2.15) = 0.016$$

whereas the exact probability is zero. Because the distribution of the X_i's is so badly skewed, n must be very large indeed before the distribution of S_n becomes close to normal.

Problems for Section 6.7

†1. Suppose that n games of crown and anchor are played, and in each game a bet of \$6 is placed on "hearts" (Example 5.4.2). Find the approximate probability of making a profit at the game for $n = 10, 100$, and 1000.

2. The following table gives the probabilities of finding various numbers of typing errors on a page:

Number of errors	0	1	2	3	4	5	6	Total
Probability	.07	.20	.26	.23	.14	.07	.03	1

Find the mean and variance of the number of typing errors on a page, and compute the approximate probability that a 100-page manuscript contains at least 280 errors.

3. A die is weighted so that faces 1 and 6 are twice as probable as faces 2,3,4, and 5.

   (a) Find the mean and variance of the score in one roll of the die.

   (b) Find the approximate probability that the total score in 60 rolls of the die will lie between 190 and 230, inclusive.

   (c) Indicate, with reasons, whether the probability in (b) is larger or smaller than would be obtained for 60 rolls of a balanced die.

†4. Ice cream cones are available at the campus centre in three sizes: 15¢, 20¢, and 25¢. Of all cones sold, 30% are small, 20% are medium, and 50% are large. During one hectic lunch hour, 300 cones were sold. Find the approximate probability that the total cash taken in exceeded \$64.

5. An insurance company rounds 30,000 insurance premiums to the nearest dollar. Assuming that the fractional parts of the premiums are continuously and uniformly distributed between 0 and 1, what is the probability that the total amount owing will be altered by more than \$50? by more than \$100?

6. In the preceding problem, suppose that premiums are first rounded

to the nearest cent, and subsequently to the nearest dollar (with 50¢ being rounded upwards). Now what is the probability that the total amount owing will be altered by more than $50? by more than $100?

†7. Customers at Harvey's opt for 0,1,2, or 3 dill pickle slices on their hamburgers with probabilities .15, .50, .25, and .10 respectively, and 500 hamburgers are sold during lunch hour. A dill pickle makes 6,7, or 8 slices with probabilities .2, .6, and .2, respectively. What is the probability that 100 pickles will make enough slices for the lunch hour business?

*8. A point in the plane, initially at the origin O, is subjected to n independent random displacements, where n is large. Each displacement is of 2/n units to the north or to the east with equal probabilities. Let the final position of the point be P. Let R be the distance of P from the origin, and let Z be the angle between OP and an east-west line. Show that

$$P(R \le r) \approx 2F\left\{\sqrt{n(\tfrac{r^2}{2} - 1)}\right\} - 1 \quad \text{for} \quad \sqrt{2} < r < 2;$$

$$P(Z \le z) \approx 1 - F\left\{\sqrt{n}\ \frac{1 - \tan z}{1 + \tan z}\right\} \quad \text{for} \quad 0 < z < \frac{\pi}{2}$$

where F is the standardized normal c.d.f., and hence obtain the approximate p.d.f.'s of R and Z.

*9. The Central Limit Theorem can be verified empirically by simulating probability distributions on the computer (Section 6.3). Let X be a discrete variate taking values 1,2,...,6 with probabilities $p_1, p_2, \ldots, p_6$ ($\sum p_i = 1$). Choose any permissible set of values for the p_i's. Use the computer to generate ten observations from this distribution, and compute their total. Repeat 100 times and prepare a relative frequency histogram for the totals generated. Is the histogram well approximated by a normal distribution? Try this for several sets of p_i's. Can you detect any relationship between the shape of the distribution and the normality of the histogram?

6.8 Some Normal Approximations

In this section, the Central Limit Theorem is used to justify normal approximations to the Poisson, binomial, and hypergeometric distributions. We also discuss corrections for continuity which may be used to improve numerical accuracy when a discrete distribution is approximated by a continuous distribution.

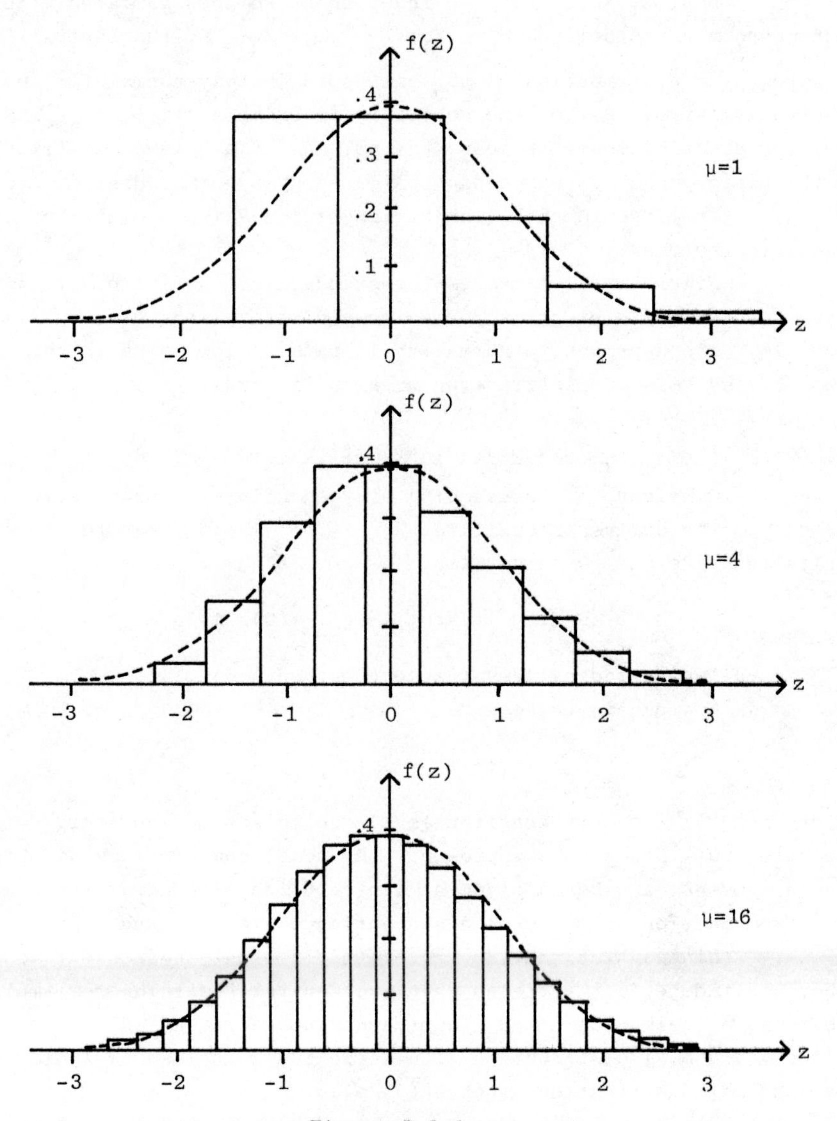

Figure 6.8.1

Histogram of $Z \equiv (X-\mu)/\sqrt{\mu}$, where X has a Poisson distribution with mean μ (μ=1,4,16). The standardized normal p.d.f. is shown with a broken line.

Approximation to the Poisson Distribution

Let $X_1, X_2, \ldots, X_n$ be independent Poisson variates with common mean m, and define $S_n \equiv X_1 + X_2 + \ldots + X_n$. By the Central Limit Theorem, the distribution of S_n is approximately normal for n sufficiently large. But by the Corollary to Example 4.5.5, S_n has a Poisson distribution with mean $\mu = nm$. Hence a Poisson distribution with a large mean μ can be approximated by a normal distribution $N(\mu,\mu)$. (From Section 5.3, the variance of a Poisson distribution is equal to the mean.)

Figure 6.8.1 compares the standardized normal p.d.f. with the standardized histograms for Poisson distributions with $\mu = 1, 4$, and 16. The approach to normality is rather slow owing to the skewness of the Poisson distribution when μ is small.

Approximation to the Binomial Distribution

Consider n tosses of a biassed coin with probability of heads p. We define the variate X_i to be 1 if the ith toss results in heads and 0 otherwise:

$$P(X_i = 1) = p; \quad P(X_i = 0) = 1 - p.$$

X_i is called an indicator variable (Section 5.6).

Now consider the sum

$$S_n \equiv X_1 + X_2 + \ldots + X_n.$$

Every head that occurs contributes 1 to this sum, and every tail contributes 0. Hence S_n represents the total number of heads obtained in n tosses, and has a binomial distribution with parameters n and p. From Section 5.3, its mean and variance are np and $np(1-p)$.

Since $X_1, X_2, \ldots, X_n$ refer to different independent tosses, they are independent variates, and the Central Limit Theorem implies that the distribution of S_n approaches normality as n increases. Hence a binomial distribution with large index n can be approximated by a normal distribution $N(np, np(1-p))$.

Figure 6.8.2 compares the standardized normal p.d.f. with the standardized histograms for binomial distributions with $(n,p) =$ (10,0.5), (20,0.5), and (20,0.1). When $p = 0.5$, the histogram is symmetrical about the origin, and converges very rapidly to the normal shape. However, the convergence is much slower when $p = 0.1$. For a given n, the normal approximation to the binomial distribution is much more accurate for p near 0.5 than for p near 0 or 1.

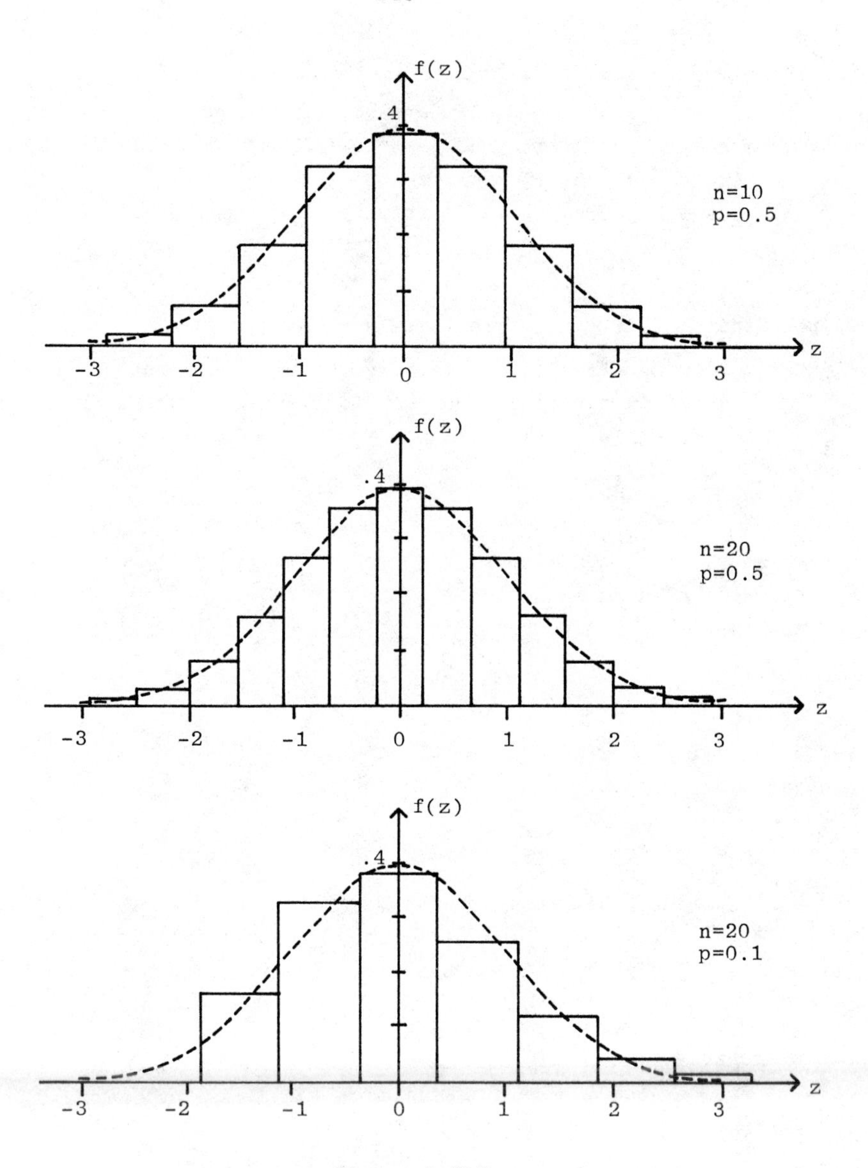

Figure 6.8.2

Histogram of $Z \equiv (X-np)/\sqrt{np(1-p)}$, where X has a binomial distribution with parameters n and p. The standardized normal p.d.f. is shown with a broken line.

Approximation to the Hypergeometric Distribution

Suppose that n balls are drawn at random without replacement from an urn containing a white balls and b black balls, and let X denote the number of white balls in the sample. From Section 2.3, X has a hypergeometric distribution, with probability function

$$f(x) = \binom{a}{x}\binom{b}{n-x}/\binom{a+b}{n} \quad \text{for} \quad x = 0,1,2,\ldots .$$

Suppose that n is large, and that $a + b \gg n$. Then

(i) from Section 2.5, the hypergeometric distribution is well approximated by a binomial distribution with index n and probability parameter $p = \frac{a}{a+b}$;

(ii) for n large, the binomial distribution is close to normal in shape.

Hence, under these conditions, the hypergeometric distribution can be approximated by a normal distribution. From Section 5.3, the mean and variance are

$$\mu = np; \quad \sigma^2 = np(1-p)\,\frac{a+b-n}{a+b-1}.$$

For given values of n and $a+b$, the approximation will be the most accurate for p near 0.5 $(\frac{a}{b} \approx 1)$.

Corrections for Continuity

Suppose that the probability histogram of a discrete variate X is being approximated by a continuous probability density function f (see Figure 6.8.3). The probability of a statement about X can

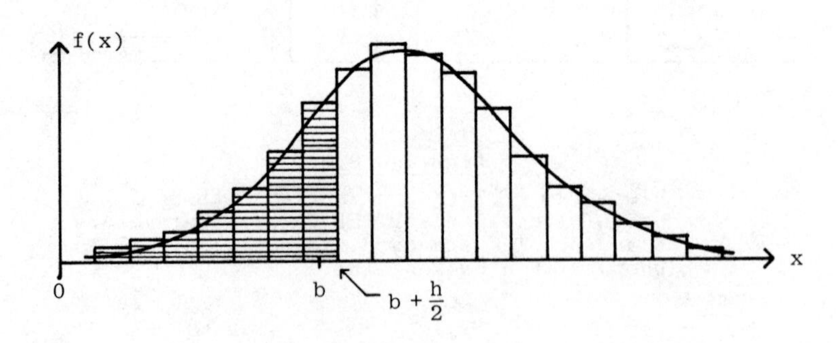

Figure 6.8.3.
Correcting for continuity

be interpreted as an area under the histogram, and then approximated by the corresponding area under f. For instance, suppose that we wish to calculate $P(X \le b)$, where b is a possible value of X. The histogram contains a rectangle corresponding to b. Suppose that this rectangle has base h and is centred on b, so that its base extends from $b - \frac{h}{2}$ to $b + \frac{h}{2}$. Then

$$\begin{aligned} P(X \le b) &= \text{Area under histogram to the left of } b + \tfrac{h}{2} \\ &\approx \text{Area under } f \text{ to the left of } b + \tfrac{h}{2} \\ &= \int_{-\infty}^{b+\frac{h}{2}} f(x)dx = F(b + \tfrac{h}{2}) \end{aligned}$$

where F is the c.d.f. of the continuous distribution. The term $\frac{h}{2}$ is called a correction for continuity. If the possible values of X are consecutive integers and b is an integer, then $h = 1$ and

$$P(X \le b) \approx F(b + 0.5).$$

In general, one can easily determine the appropriate continuity correction by first interpreting the desired probability as an area under the histogram, and then using the corresponding area under the approximating p.d.f. Of course, if the distribution being approximated is continuous, as in Example 6.7.1, no correction for continuity should be used. (If X is a continuous variate, then h is effectively zero.)

Example 6.8.1. Calculate the probability of at least 1200 and at most 1300 heads in 2500 tosses of a balanced coin.

Solution. Let X be the number of heads in 2500 tosses. Then X has a binomial distribution with parameters $n = 2500$ and $p = 0.5$. The mean and variance are $np = 1250$ and $np(1-p) = 625 = (25)^2$. Now the normal approximation to the binomial distribution gives

$$\begin{aligned} &P(1200 \le X \le 1300) \\ &\approx \text{area under } N(1250,625) \text{ from } 1199.5 \text{ to } 1300.5. \end{aligned}$$

We now standardize to obtain

$$P(1200 \le X \le 1300) \approx F(\frac{1300.5-1250}{25}) - F(\frac{1199.5-1250}{25})$$

$$= F(2.02) - F(-2.02) = 2F(2.02) - 1$$

where F is the standardized normal c.d.f., and Table B2 gives

$$P(1200 \le X \le 1300) \approx 2(0.9783) - 1 = 0.9566.$$

If the correction for continuity is ignored, we obtain

$$P(1200 \le X \le 1300) \approx 2F(2.00) - 1 = 0.9545.$$

The correction makes little difference in this case because the rectangle width $(h = 1)$ is small in comparison with the standard deviation $(\sigma = 25)$.

Example 6.8.2 (See Example 2.5.3). A candidate obtains 52% of the N votes in an election, where N is very large. What is the probability that he leads in a poll of 100 votes?

Solution. Let X be the number of votes he receives out of 100. Then X has a hypergeometric distribution with parameters $n = 100$, $a = pN$ and $b = (1-p)N$, where $p = 0.52$. The mean and variance are

$$\mu = np = 52; \quad \sigma^2 = np(1-p)\frac{a+b-n}{a+b-1} = 24.96\,\frac{N-100}{N-1}\,.$$

If $N \gg 100$, the variance is close to 24.96, and the distribution of X is approximately $N(52, 24.96)$. Now the probability that the candidate receives more than half the votes is

$$P(X > 50) \approx \text{area under } N(52,24.96) \text{ from } 50.5 \text{ to } \infty$$

$$= 1 - F\left(\frac{50.5 - 52}{5.00}\right) = F(0.3) = 0.6179$$

from Table B2. The correction for continuity makes a substantial difference in this case because the standard deviation is rather small.

Problems for Section 6.8

†1. Suppose that X has a binomial distribution with parameters $(10, \frac{1}{2})$. Sketch the probability histogram of X, and shade in the areas corresponding to the following probabilities:

$$P(X < 4), \quad P(X = 5), \quad P(7 < X \le 9).$$

Hence determine the appropriate continuity corrections, compute approximate probabilities, and compare with the exact values.

2. Let X be the number of aces in 180 rolls of a true die. Cal-

culate the approximate probability that $28 \le X \le 32$. If this experiment were performed by 20 people, how many of the 20 would you expect to report that their observed value of X lay outside this range?

3. Phone calls arrive at an exchange at the rate of 3 per minute on the average. Use a normal approximation to determine the approximate probability of
   (a) at least 200 calls in a one-hour period;
   (b) exactly 200 calls in a one-hour period.

†4. A coin is tossed 100 times and 43 heads are observed. If the coin were balanced, what would be the probability of obtaining such a large deviation from the expected number? Would you consider this to be very strong evidence of bias in the coin?

5. A high school has 600 male and 600 female students, of whom 36 come out to dancing class. If these 36 are a random sample from the entire student body, what is the (approximate) probability that at least 15 male-female couples can be formed?

6. In a club of n voting members, a block of s members votes unanimously on any issue. The other $t = n - s$ members vote at random $(p = 1/2)$. All n members always vote, and t is fairly large. How large must s be (in terms of t) such that there should be an 84.1% chance that the block of s voters carries any issue for which it votes?

†7. If the unemployment rate in Canada is 5%, what is the probability that a random sample of 10,000 will contain between 475 and 525 unemployed? between 450 and 550 unemployed? How large a sample would it be necessary to choose so that, with 95% probability, the percentage of unemployed in the sample would lie between 4.9% and 5.1%?

6.9 The χ^2, F and t Distributions

In this section, three continuous distributions are defined, and their tabulation and properties are briefly described. These distributions are closely related to the normal distribution, and they have important statistical applications.

Chi-Square (χ^2) Distribution

Let ν be a positive constant. A continuous variate X with p.d.f.

$$f(x) = k_\nu x^{\frac{\nu}{2}-1} e^{-\frac{x}{2}} \quad \text{for} \quad x > 0 \tag{6.9.1}$$

is said to have a chi-square distribution with ν degrees of freedom, and we write $X \sim \chi^2_{(\nu)}$. Although the distribution is defined for all real $\nu > 0$, in most applications ν will be a positive integer, and most sets of tables list only integer values of ν.

The rth moment of this distribution is

$$m_r = E(X^r) = k_\nu \int_0^\infty x^{\frac{\nu}{2}+r-1} e^{-\frac{x}{2}} dx.$$

Substituting $x = 2u$ and using (2.1.13) gives

$$m_r = k_\nu 2^{\frac{\nu}{2}+r} \int_0^\infty u^{\frac{\nu}{2}+r-1} e^{-u} du = k_\nu 2^{\frac{\nu}{2}+r} \Gamma(\tfrac{\nu}{2} + r).$$

In particular, since the total probability is $m_0 = 1$, we have $k_\nu = 1/2^{\frac{\nu}{2}}\Gamma(\frac{\nu}{2})$, and hence

$$m_r = 2^r\Gamma(\tfrac{\nu}{2} + r)/\Gamma(\tfrac{\nu}{2}) = 2^r(\tfrac{\nu}{2} + r - 1)(\tfrac{\nu}{2} + r - 2)\ldots(\tfrac{\nu}{2})$$

by repeated application of (2.1.14). Thus we have $m_1 = \nu$, $m_2 = (\nu + 2)\nu$, and $\sigma^2 = m_2 - m_1^2 = 2\nu$ by (5.2.3). Thus $\chi^2_{(\nu)}$ has mean ν and variance 2ν.

For small ν, the p.d.f. of $\chi^2_{(\nu)}$ has a long tail to the right. Figure 6.9.1 shows the density function for the cases $\nu = 2,4$ and 6. As ν increases, the centre of the distribution shifts to the right, and the density function becomes more symmetrical in shape. In fact, for large ν, $\chi^2_{(\nu)}$ may be approximated by $N(\nu, 2\nu)$.

The c.d.f. of $\chi^2_{(\nu)}$ is

$$F(x) = P(X \le x) = \int_0^x k_\nu u^{\frac{\nu}{2}-1} e^{-\frac{u}{2}} du \quad \text{for} \quad x > 0.$$

This integral must be evaluated numerically (except when ν is an even integer). Table B4 gives selected percentiles (quantiles) of $\chi^2_{(\nu)}$ for $\nu = 1,2,\ldots,30$. For instance, the entry in Table B4 for $\nu = 4$ and $F = 0.9$ is 7.779, while that for $\nu = 4$ and $F = 0.1$ is 1.064. Thus, if X has a χ^2 distribution with 4 degrees of

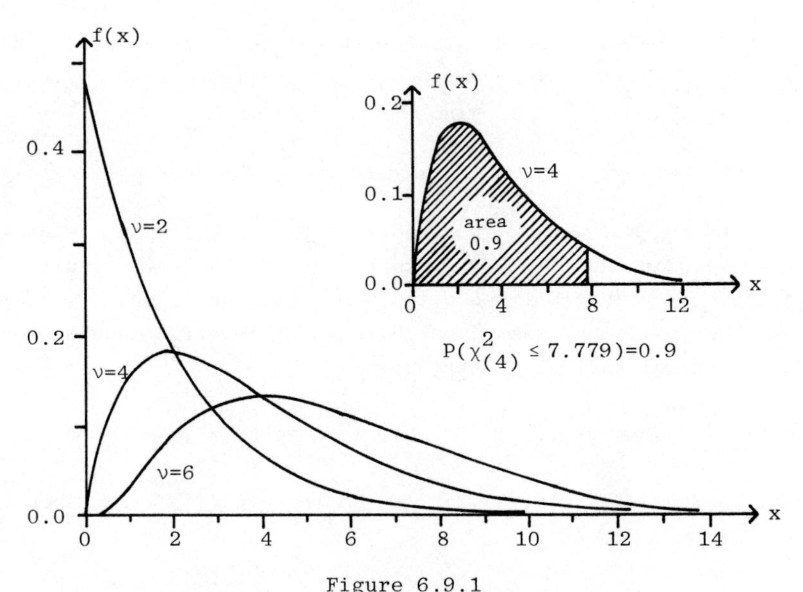

Figure 6.9.1

P.d.f.'s of χ^2 Distributions with 2,4 and 6 Degrees of Freedom

freedom, then $P(X \le 1.064) = 0.1$ and $P(X \le 7.779) = 0.9$.

If $X \sim \chi^2_{(\nu)}$ where ν is large, the following approximation may be used:

$$\sqrt{\frac{9\nu}{2}}\left\{\left(\frac{X}{\nu}\right)^{1/3} - 1 + \frac{2}{9\nu}\right\} \approx N(0,1). \tag{6.9.2}$$

For instance, to determine the 95th percentile of $\chi^2_{(\nu)}$, we note that the 95th percentile of $N(0,1)$ is 1.645 from Table B1. Now we solve the equation

$$\sqrt{\frac{9\nu}{2}}\left\{\left(\frac{x}{\nu}\right)^{1/3} - 1 + \frac{2}{9\nu}\right\} = 1.645$$

to obtain the approximate 95th percentile of $\chi^2_{(\nu)}$:

$$x = \nu\left[1 - \frac{2}{9\nu} + 1.645\sqrt{\frac{2}{9\nu}}\right]^3 .$$

For $\nu = 30$, we obtain $x = 43.77$, which agrees exactly with the value given in Table B4.

The χ^2 distribution has an important <u>additive property</u>. <u>A sum of independent χ^2 variates has a χ^2 distribution</u>; the degrees of freedom for the sum is equal to the sum of the degrees of freedom. Thus, if $X_1, X_2, \ldots, X_n$ are independent χ^2 variates with $\nu_1, \nu_2, \ldots, \nu_n$ degrees of freedom, then

$$X_1 + X_2 + \ldots + X_n \sim \chi^2_{(\nu)} \tag{6.9.3}$$

where $\nu = \nu_1 + \nu_2 + \ldots + \nu_n$. Two proofs of this result for the case $n = 2$ are given in Examples 7.2.2 and 8.4.2. The general result (6.9.3) may then be established by mathematical induction.

The following example establishes an important connection between the normal and χ^2 distributions.

Example 6.9.1. Show that, if $Z \sim N(0,1)$, then $Z^2 \sim \chi^2_{(1)}$.

Solution. Define $X \equiv Z^2$ and let G be the c.d.f. of X. Then

$$G(x) = P(X \le x) = P(Z^2 \le x) = P(-\sqrt{x} \le Z \le \sqrt{x})$$
$$= F(\sqrt{x}) - F(-\sqrt{x}) \quad \text{for} \quad x > 0$$

where F is the c.d.f. of $N(0,1)$. Now (6.6.4) gives

$$G(x) = 2F(\sqrt{x}) - 1.$$

The p.d.f. of X is obtained by differentiation:

$$g(x) = \frac{dG(x)}{dx} = 2\frac{dF(\sqrt{x})}{dx} = 2f(\sqrt{x})\cdot\frac{d\sqrt{x}}{dx}$$

where f is the p.d.f. of $N(0,1)$:

$$f(z) = \frac{dF(z)}{dz} = \frac{1}{\sqrt{2\pi}}e^{-z^2/2}.$$

Thus the p.d.f. of X is

$$g(x) = \frac{1}{\sqrt{2\pi}}x^{-\frac{1}{2}}e^{-\frac{x}{2}} \quad \text{for} \quad x > 0.$$

Since $\Gamma(\frac{1}{2}) = \sqrt{\pi}$ by (6.6.9), this agrees with (6.9.1) in the special case $\nu = 1$. It follows that X has a χ^2 distribution with one degree of freedom. □

Now let $Z_1, Z_2, \ldots, Z_n$ be independent $N(0,1)$ variates. Then $Z_1^2, Z_2^2, \ldots, Z_n^2$ are independent $\chi^2_{(1)}$ variates. Hence, by the additive property (6.9.3),

$$Z_1^2 + Z_2^2 + \dots + Z_n^2 \sim \chi^2_{(n)}. \qquad (6.9.4)$$

A sum of squares of n independent $N(0,1)$ variates has a χ^2 distribution with n degrees of freedom.

The χ^2 distribution is also closely related to the exponential, gamma, and Poisson distributions; see Section 6.5, and Problems 6.9.5 and 6.9.12.

F (variance-Ratio) Distribution

Let n and m be positive real numbers. A continuous variate X with p.d.f.

$$f(x) = k_{n,m} x^{\frac{n}{2}-1} \left(1 + \frac{n}{m}x\right)^{-\frac{n+m}{2}} \quad \text{for} \quad x > 0 \qquad (6.9.5)$$

is said to have an F distribution with n numerator and m denominator degrees of freedom, and we write $X \sim F_{n,m}$. This distribution is named in honour of R.A. Fisher, who first derived it. An important use of this distribution is in the comparison of two independent variance estimates, and for this reason it is also called the variance-ratio distribution. In most applications, n and m will be positive integers.

The constant $k_{n,m}$ is determined by the condition that the total probability must equal 1. It can be shown that

$$k_{n,m} = \left(\frac{n}{m}\right)^{\frac{n}{2}} \Gamma\left(\frac{n+m}{2}\right) / \Gamma\left(\frac{n}{2}\right)\Gamma\left(\frac{m}{2}\right).$$

See Example 7.2.2.

The probability density function of $F_{n,m}$ is graphed in Figure 6.9.2 for several pairs of values (n,m). The F distribution, like the χ^2 distribution, has a long tail to the right. Table B5 gives the 90th, 95th, 99th and 99.9th percentiles of $F_{n,m}$ for selected pairs of values (n,m). For instance, for $n=4$ and $m=8$, the entry in the table of 90th percentiles is 2.81. If X has an F distribution with 4 numerator and 8 denominator degrees of freedom, then $P(X \le 2.81) = 0.9$. Similarly, $P(X \le 3.84) = 0.95$, $P(X \le 7.01) = 0.99$, and $P(X \le 14.4) = 0.999$.

The reciprocal of an F-variate has an F-distribution with the degrees of freedom interchanged; that is,

$$X \sim F_{n,m} \Longrightarrow X^{-1} \sim F_{m,n}. \qquad (6.9.6)$$

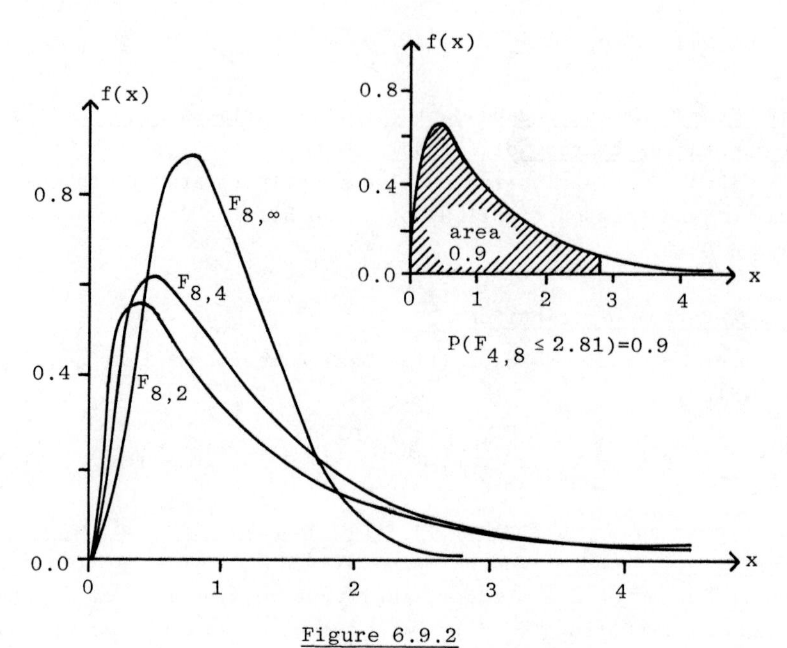

Figure 6.9.2

P.d.f. of the F Distribution $F_{n,m}$.

To prove this result, define $Y \equiv X^{-1}$. Since this is a monotonic transformation, the p.d.f. of Y can be obtained from (6.9.5) using the change of variables formula (6.1.11):

$$\begin{aligned} g(y) &= k_{n,m}\left(\frac{1}{y}\right)^{\frac{n}{2}-1}\left(1+\frac{n}{m}\,\frac{1}{y}\right)^{-\frac{n+m}{2}}\left(\frac{1}{y}\right)^{2} \\ &= k_{n,m}\left(\frac{n}{m}\right)^{-\frac{n+m}{2}} y^{\frac{m}{2}-1}\left(1+\frac{m}{n}y\right)^{-\frac{n+m}{2}} \\ &= k_{m,n}\, y^{\frac{m}{2}-1}\left(1+\frac{m}{n}y\right)^{-\frac{m+n}{2}} \qquad \text{for} \quad y > 0. \end{aligned}$$

This is the p.d.f. of an F distribution with m numerator and n denominator degrees of freedom.

It follows from (6.9.6) that

$$P(F_{n,m} \ge c) = P\left(\frac{1}{F_{n,m}} \le \frac{1}{c}\right) = P\left(F_{m,n} \le \frac{1}{c}\right). \tag{6.9.7}$$

Using this result, one can obtain the 10%, 5%, 1%, and 0.1% points of $F_{n,m}$ by taking reciprocals of the 90%, 95%, 99%, and 99.9%

points of $F_{m,n}$. For instance, Table B5 gives

$$P(F_{8,4} \le 3.95) = 0.9; \quad P(F_{8,4} \le 6.04) = 0.95.$$

Since $\frac{1}{3.95} = 0.253$ and $\frac{1}{6.04} = 0.166$, we have

$$P(F_{4,8} \ge 0.253) = 0.9; \quad P(F_{4,8} \ge 0.166) = 0.95.$$

Thus the 10% and 5% points of $F_{4,8}$ are 0.253 and 0.166.

Suppose that U and V are independent variates, with $U \sim \chi^2_{(n)}$ and $V \sim \chi^2_{(m)}$. A mean square is the ratio of a χ^2 variate to its degrees of freedom. Thus $U \div n$ and $V \div m$ are mean squares, and they are independent because U and V are independent. We shall prove in Example 7.2.2 that

$$\frac{U \div n}{V \div m} \sim F_{n,m}; \tag{6.9.8}$$

that is, a ratio of independent mean squares has an F distribution. Mean squares arise as variance estimates, and (6.9.8) says that a ratio of independent variance estimates has an F distribution.

t (Student's) Distribution

Let ν be a positive real number. A continuous variate X with probability density function

$$f(x) = k_\nu \left(1 + \frac{x^2}{\nu}\right)^{-\frac{\nu+1}{2}} \quad \text{for } -\infty < x < \infty \tag{6.9.9}$$

is said to have a t distribution with ν degrees of freedom, and we write $X \sim t_{(\nu)}$. This distribution is also called Student's distribution in honour of W.S. Gosset, who published under the pseudonym "Student". One important use of $t_{(\nu)}$ is in drawing statistical conclusions about the mean of a normal distribution when the variance is unknown.

The constant k_ν may be evaluated from the condition that the total area under (6.9.9) must equal 1, or by using the connection with the F distribution which is pointed out below. One finds that

$$k_\nu = \Gamma(\tfrac{\nu+1}{2})/\sqrt{\nu\pi}\ \Gamma(\tfrac{\nu}{2}).$$

The p.d.f. of Student's distribution is symmetrical about the origin, and is similar in shape to $N(0,1)$ but flatter. Figure 6.9.3 shows the p.d.f.'s of $t_{(1)}, t_{(5)}$, and $N(0,1)$. It is easy to show that (6.9.9) approaches the standardized normal p.d.f. as $\nu \to \infty$. For large ν, tables of $N(0,1)$ may be used to obtain approximate probabilities for $t_{(\nu)}$.

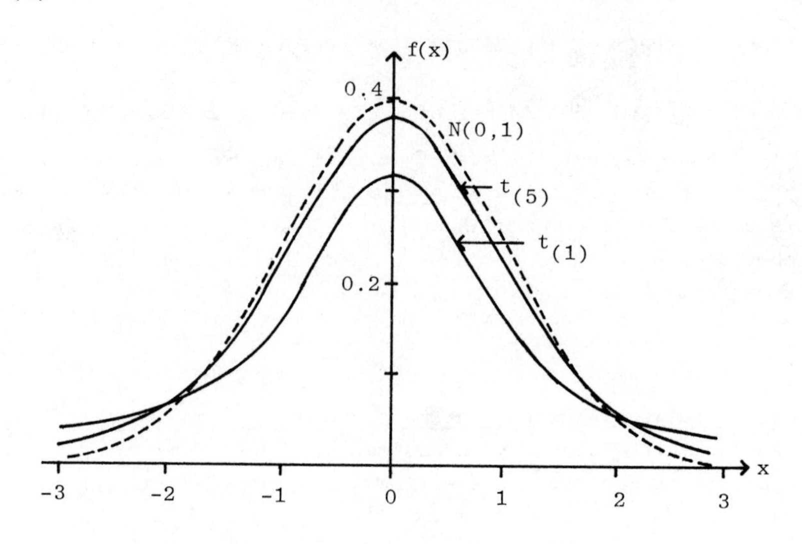

Figure 6.9.3
Comparison of the t Distribution with N(0,1)

Table B3 gives selected percentiles of $t_{(\nu)}$ for $\nu = 1,2,\ldots,30$, and for $\nu = 40,60,120$. The values given for $\nu = \infty$ in the last row of Table B3 also appear in Table B1 as percentiles of $N(0,1)$. Lower percentage points of $t_{(\nu)}$ are obtained by using the symmetry of the distribution about the origin, so that

$$F(-x) = 1 - F(x).$$

For instance, Table B3 gives $P(t_{(11)} \le 2.201) = 0.975$, and hence $P(t_{(11)} \le -2.201) = 0.025$. It follows that

$$P(-2.201 \le t_{(11)} \le 2.201) = 0.95.$$

It can be shown by the method used in Example 6.9.1 that

$$X \sim t_{(\nu)} \Longrightarrow X^2 \sim F_{1,\nu}. \tag{6.9.10}$$

Because of this result, probabilities for $t_{(\nu)}$ can also be obtained from tables of the F distribution. For instance, Table B4 gives

$P(F_{1,11} \le 4.84) = 0.95$. Hence

$$.95 = P(t^2_{(11)} \le 4.84) = P(-\sqrt{4.84} \le t_{(11)} \le \sqrt{4.84})$$
$$= P(-2.20 \le t_{(11)} \le 2.20),$$

which agrees with the result obtained from Table B3.

The following result can be proved by change of variables methods (see Problem 7.2.7). It underlies many of the statistical applications of the t distribution. Suppose that Z and U are independent variates, with $Z \sim N(0,1)$ and $U \sim \chi^2_{(\nu)}$. Then

$$Z \div \sqrt{U/\nu} \sim t_{(\nu)}; \tag{6.9.11}$$

that is, <u>the ratio of a standardized normal variate to the square root of an independent mean square has a t distribution</u>. Note that the degrees of freedom for the t distribution is equal to the degrees of freedom for the χ^2 distribution.

The t distribution with $\nu = 1$ degree of freedom is also called the <u>Cauchy distribution</u>. Its p.d.f. is

$$f(x) = \frac{1}{\pi(1+x^2)} \quad \text{for} \quad -\infty < x < \infty.$$

This distribution has the interesting property that although the p.d.f. is clearly symmetrical about the origin, the mean does not exist; see Problem 6.1.9.

Problems for Section 6.9

†1. Suppose that X has a χ^2 distribution with 5 degrees of freedom. Using tables, determine the values x_1, x_2 for which

$$P(X < x_1) = P(X > x_2) = 0.025.$$

Show these values on a sketch of the probability density function.

2. When $\nu = 2$, the χ^2 distribution simplifies to an exponential distribution with mean $\theta = 2$. Find the c.d.f., and verify the values given in Table B4 for $\nu = 2$.

3. If $X \sim \chi^2_{(1)}$, the argument in Example 6.9.1 gives

$$P(X \le x) = 2F(\sqrt{x}) - 1$$

where F is the c.d.f. of $N(0,1)$. Use this result to find the 90th, 95th, and 99th percentiles of $\chi^2_{(1)}$, and verify that these agree with the values given in Table B4.

4. Use (6.9.2) to obtain approximate 90th, 95th, and 99th percentiles of the χ^2 distribution with 20 degrees of freedom, and compare with the values given in Table B4.

†5. (a) Show that, if X has an exponential distribution with mean θ, then $2X/\theta$ has a χ^2 distribution with 2 degrees of freedom.

(b) Define $T \equiv X_1 + X_2 + \ldots + X_n$, where the X_i's are independent exponential variates with the same mean θ. Show that $2T/\theta$ has a χ^2 distribution with $2n$ degrees of freedom.

(c) The lifetimes of electron tubes are exponentially distributed with mean 100 hours. Fifteen tubes are purchased, and one of them is put into use. As soon as one tube fails, it is immediately replaced by another until the entire supply is used up. What is the probability that the supply will last longer than 2000 hours?

6. (a) Show that, if U has a uniform distribution on $(0,1)$, then $-2\log U$ has a χ^2 distribution with two degrees of freedom.

(b) Define $V \equiv U_1 U_2 \ldots U_n$ where the U_i's are independent $U(0,1)$ variates. Show that $-2\log V$ has a χ^2 distribution with $2n$ degrees of freedom, and find the p.d.f. of V.

7. Show that the coefficient of skewness for a χ^2 distribution is always positive, but decreases to zero as $\nu \to \infty$.

†8. Using tables, find the 5th and 95th percentiles of an F distribution with 6 numerator and 12 denominator degrees of freedom.

†9. Let $X \sim t_{(30)}$. Use tables to determine the values x_1, x_2 such that

$$P(-x_1 \le X \le x_1) = 0.90; \quad P(X > x_2) = 0.995.$$

Compare with the corresponding values from standardized normal distribution tables.

10. Show that the p.d.f. of Student's distribution with ν degrees of freedom tends to the standardized normal p.d.f. as $\nu \to \infty$.

11. Show that, if $X \sim t_{(\nu)}$, then $X^2 \sim F_{1,\nu}$.

12. A continuous variate T with p.d.f.

$$f(t) = \lambda^p t^{p-1} e^{-\lambda t}/\Gamma(p) \quad \text{for} \quad t > 0 \tag{6.9.12}$$

is said to have a <u>gamma distribution</u> with scale parameter λ and shape parameter p, where λ and p are positive constants.

(a) Find an expression for the rth moment of (6.9.12), and hence obtain the mean and variance.

(b) Show that $2\lambda T$ has a χ^2 distribution with $2p$ degrees of freedom. (Note: this establishes (6.5.1).)

†*13. Note. This problem depends upon material in Section 6.4.
Plot the hazard function for the gamma distribution (6.9.12) when $\lambda = 1$ and $p = 0.5, 1.0, 2.0$. Discuss the type of ageing which occurs in specimens whose lifetimes have a gamma distribution. What would be the effect on the graph of changing λ?

14. The beta function is defined by

$$\beta(p,q) = \int_0^1 u^{p-1}(1-u)^{q-1}\,du. \tag{6.9.13}$$

The integral converges for all positive real values of p and q. It can be shown (Problem 7.2.8) that

$$\beta(p,q) = \Gamma(p)\Gamma(q)/\Gamma(p+q). \tag{6.9.14}$$

A continuous variate X with p.d.f.

$$f(x) = x^{p-1}(1-x)^{q-1}/\beta(p,q) \quad \text{for} \quad 0 < x < 1 \tag{6.9.15}$$

is said to have a beta distribution with parameters p and q.

(a) Show that the rth moment of (6.9.15) is

$$m_r = (p+r-1)^{(r)}/(p+q+r-1)^{(r)},$$

and hence find the mean and variance.

(b) Show that $\frac{p}{q}(X^{-1}-1)$ has an F distribution with $2q$ numerator and $2p$ denominator degrees of freedom.

(c) Ceramic parts are manufactured in large batches. Because of production hazards, the fraction X of a batch which is saleable is a random variable. It is known, from past experience, that X has approximately a beta distribution with parameters $p = 6$ and $q = 3$. Using part (b), find the 5th and 95th percentiles of this distribution.

15. Show that the rth moment of the t distribution with ν degrees of freedom exists for $r < \nu$; that all odd moments which exist are zero; and that

$$m_{2r} = \nu^r \frac{(2r-1)(2r-3)\ldots(3)(1)}{(\nu-2)(\nu-4)\ldots(\nu-2r)} \quad \text{for} \quad \nu > 2r.$$

Hence show that X has variance $\nu/(\nu-2)$ for $\nu > 2$, and coefficient of kurtosis $\gamma_2 = 6/(\nu-4)$ for $\nu > 4$.

*16. Show that the F distribution (6.9.5) has rth moment

$$m_r = \Gamma(\tfrac{n}{2} + r)\Gamma(\tfrac{m}{2} - r)(\tfrac{m}{n})^r/\Gamma(\tfrac{n}{2})\Gamma(\tfrac{m}{2})$$

provided that $m > 2r$. Hence show that the mean and variance are

$$\mu = \frac{m}{m-2} \quad \text{for} \quad m > 2; \qquad \sigma^2 = \frac{2m^2(n+m-2)}{n(m-2)^2(m-4)} \quad \text{for} \quad m > 4.$$

Review Problems: Chapter 6

†1. My neighbour and I have identical floodlamps whose lifetimes are exponentially distributed with mean $\theta = 300$ hours. Each of us burns his floodlamp for six hours each night. Find the probability that (a) my floodlamp lasts longer than 60 nights;
(b) both floodlamps last longer than 60 nights;
(c) both lamps burn out on the same night.

2. Electrical parts are delivered by the manufacturer in large batches. The fraction X of good parts in a batch may be assumed to be a continuous variate with p.d.f.

$$f(x) = kx^{19}(1-x)^2 \quad \text{for} \quad 0 < x < 1.$$

(a) Find the c.d.f. of X and evaluate the constant k.
(b) Find the p.d.f. of the fraction of bad parts in a batch.
(c) Compute the probability that a batch contains between 5% and 20% defectives.
(d) Show that the median of X is approximately 0.88.

3. According to the Maxwell-Boltzman Law, the velocity V of a gas molecule is a continuous variate with p.d.f.

$$f(v) = kv^2\exp(-\beta v^2) \quad \text{for} \quad v > 0,$$

where β and k are positive constants. The kinetic energy of a molecule of mass m is $Y \equiv \frac{1}{2}mV^2$.
(a) Find the p.d.f. and expected value of Y.
(b) Show that $k = 4\beta^{3/2}/\Gamma(1/2)$.

†4. Evaluate k and find the p.d.f. for the continuous distribution with c.d.f.
(a) $F(x) = kx^2(1+x)^{-2}$ for $x > 0$.
(b) $F(x) = k[1 - e^{-\theta\sin x}]$ for $0 < x < \frac{\pi}{2}$, where $\theta > 0$.

5. Let X be a continuous variate with p.d.f.

$$f(x) = k \exp\{-|x|^{\beta}\} \quad \text{for} \quad -\infty < x < \infty.$$

Evaluate the constant k and find the moments of this distribution.

6. Suppose that heights (in inches) of adult Caucasian males are normally distributed with mean 69 and variance 9.
   (a) What fraction of adult males will have heights in excess of 72 inches? 76 inches?
   (b) What is the probability that, out of 1000 randomly selected males, at least two will be more than 76 inches tall?

†7. Find the approximate probability that, after 20 rounds of the game described in Problem 4.1.8, neither player has lost more than five dollars.

8. If 48% of births result in daughters, what is the approximate probability that there will be more daughters than sons in 1000 births?

9. A market research organization wants to test the claim that 60 percent of all housewives in a certain area prefer Brand A cleanser to all competing brands. It is decided to select at random 18 housewives and to reject the claim if fewer than 9 of them prefer brand A. What is the probability that the market research organization will thus reject the claim even though it is true?

†10. A manufacturer wants to sell boxes of screws such that there is a probability of at least .8 that 100 screws in the box are not defective. If the manufacturing process produces a defective screw with probability .015, how many screws should the manufacturer put in each box?

11. In repeated throws with a perfect coin, what is the probability of getting
   (a) at least 55% heads in the first 20 throws?
   (b) at least 55% heads in the first 100 throws?
   (c) at least 55% heads in the first 500 throws?

   Give an approximate formula for the number of heads which would be exceeded with probability .025 in n tosses.

†12. The diameters of apples grown in an orchard are normally distributed with mean 3.4 and standard deviation 0.4 inches.
   (a) What proportion of the apples have diameters greater than 4.0 inches? What proportion have diameters less than 3.0 inches?
   (b) If two apples are selected at random, what is the probability that the difference in their diameters exceeds 0.5 inches?
   (c) If four apples are selected at random, what is the probability

that their average diameter exceeds 4.0 inches?

(d) If ten apples are selected at random, what is the probability that exactly two have diameters less than 3 inches and exactly one has diameter greater than 4 inches?

13. A baseball player has a 20% chance of hitting the ball at each trip to the plate, with successive times at bat being independent.

(a) What is the probability that he gets three hits in ten times at bat?

(b) What is the probability that he gets his third hit on his tenth time at bat?

(c) What is the probability that he gets 35 or more hits in 144 times at bat?

14. Of the 100,000 adults in the Kitchener-Waterloo area, only 64% are eligible for jury duty. Each year prospective jurors are selected at random from the community and contacted to determine their eligibility. What is the probability that, out of 1600 people contacted, at least 1000 will be eligible? Justify any approximations you use.

†15. (a) In a particular county, a very large number of people are eligible for jury duty, and half of these are women. The judge is supposed to prepare a jury list by randomly selecting individuals from all those eligible. In a recent murder trial, the jury list of 82 contained 58 men and 24 women. If the jury list were properly selected, what would be the probability of obtaining such an extreme imbalance in the sexes?

(b) One percent of those eligible for jury duty belong to a racial minority group. What is the probability that there will be no representative of this minority group on a randomly chosen jury list of 82? What is the probability that there will be no representative on any of ten jury lists prepared over a period of several years?

16. Let D denote the diameter of a tree cut in a lumbering operation, and let A be the cross-sectional area. Given that $\log D$ has a normal distribution with mean μ and variance σ^2, find the p.d.f. of A.

17. (a) The United Fruit company has found by experience that the weight of bananas per crate is normally distributed with standard deviation 5 kilograms. Government regulations stipulate that at most 1% of crates marked at 100 kg. are permitted to contain less than 100 kg. At what weight should the mean be set in order to just satisfy the regulations?

(b) A zoo keeper wishes to purchase 320 kilogrammes of bananas.

If he buys three of the crates from (a), what is the probability that he will get 320 kg. or more?

18. Define $R \equiv X/Y$, where $X \sim N(\mu, a^2)$ and $Y \sim N(\nu, b^2)$, independently of X. Show that, if ν is large in comparison with b, then

$$P(R \leq r) \approx P(X \leq rY) = F\left(\frac{r\nu - \mu}{\sqrt{a^2 + r^2b^2}}\right)$$

where F is the c.d.f. of the standardized normal distribution. If $\mu = 0$, $\nu = a = 1$, and $b = 0.1$, estimate the probability that R will exceed 1, and indicate the accuracy of your estimate.

19. Show that $E\{(X - c)^2\}$ is a minimum for $c = E(X)$, and that $E\{|X - c|\}$ is a minimum when c is the median.

CHAPTER 7. BIVARIATE CONTINUOUS DISTRIBUTIONS

Bivariate continuous distributions are defined in Section 1, and change of variables problems are considered in Section 2. In Section 3, we prove some results which will be needed in deriving statistical methods for analysing normally distributed measurements. Sections 4 and 5 deal with properties and applications of the bivariate normal distribution. The discussion and examples are mostly confined to the two-variable case, but the extension to multivariate distributions is straightforward.

7.1 Definitions and Notation

The joint cumulative distribution function F of two real-valued variates X and Y was defined in Section 4.5. F is a function of two variables:

$$F(x,y) = P(X \le x, \ Y \le y) \quad \text{for all real } x,y. \tag{7.1.1}$$

Suppose that $F(x,y)$ is continuous, and that the derivative

$$f(x,y) = \frac{\partial^2}{\partial x \partial y} F(x,y) \tag{7.1.2}$$

exists and is continuous (except possibly along a finite number of curves). Furthermore, suppose that

$$\int_{-\infty}^{\infty}\int_{-\infty}^{\infty} f(x,y)dxdy = 1. \tag{7.1.3}$$

Then X and Y are said to have a bivariate continuous distribution, and f is called their joint probability density function.

Condition (7.1.3) is needed to rule out the possibility of a concentration of probability in a one-dimensional subspace of the (x,y) plane. For instance, if half the probability is spread uniformly over the unit square and the other half is spread uniformly over its main diagonal $x = y$, conditions (7.1.1) and (7.1.2) are satisfied, but (7.1.3) is not; see Problem 7.1.8. In the univariate case, the continuity of F is enough to ensure that there is not a concentration of probability at any single point (0-dimensional subspace).

The joint p.d.f. is a non-negative function. The probability that (X,Y) belongs to a region R in the (x,y) plane is given by the volume under the surface $z = f(x,y)$ above the region R; that is,

$$P\{(X,Y) \in R\} = \iint_R f(x,y)dxdy. \tag{7.1.4}$$

In particular, we have

$$F(x,y) = P(X \le x, Y \le y) = \int_{-\infty}^{x} ds \int_{-\infty}^{y} f(s,t)dt. \tag{7.1.5}$$

The marginal c.d.f. of X is the c.d.f. of X alone:

$$F_1(x) = P(X \le x) = P(X \le x, Y \le \infty) = \int_{-\infty}^{x} ds \int_{-\infty}^{\infty} f(s,y)dy. \tag{7.1.6}$$

Differentiating with respect to x gives the marginal p.d.f. of X:

$$f_1(x) = \frac{d}{dx} F_1(x) = \int_{-\infty}^{\infty} f(x,y)dy. \tag{7.1.7}$$

The marginal p.d.f. of X can thus be found by integrating the unwanted variable Y out of the joint p.d.f. The marginal c.d.f. and p.d.f. of Y may be defined similarly.

Example 7.1.1. Let X and Y be continuous variates with joint p.d.f.

$$f(x,y) = \begin{cases} k(x^2 + 2xy) & \text{for } 0 < x < 1 \text{ and } 0 < y < 1 \\ 0 & \text{otherwise.} \end{cases}$$

(a) Evaluate the normalizing constant k.
(b) Find the joint c.d.f. of X and Y and the marginal p.d.f. of X.
(c) Compute the probability of the event "$X \le Y$".

Solution. (a) The total volume under the p.d.f. must be 1, and hence

$$1 = \int_{-\infty}^{\infty} dx \int_{-\infty}^{\infty} f(x,y)dy = k\int_0^1 dx \int_0^1 (x^2 + 2xy)dy$$

$$= k\int_0^1 dx \left[x^2 y + xy^2\right]_{y=0}^{y=1} = k\int_0^1 (x^2 + x)dx$$

$$= k\left[\frac{x^3}{3} + \frac{x^2}{2}\right]_0^1 = \frac{5k}{6}.$$

Therefore $k = \frac{6}{5}$.

(b) All of the probability is located inside the unit square. By (7.1.1),

$$F(x,y) = 0 \quad \text{for} \quad x \le 0 \quad \text{or} \quad y \le 0;$$
$$F(x,y) = 1 \quad \text{for} \quad x \ge 1 \quad \text{and} \quad y \ge 1.$$

Also, for $y > 1$ and $0 \le x \le 1$ we have

$$F(x,y) = P(X \le x,\ Y \le y) = P(X \le x,\ Y \le 1) = F(x,1).$$

Similarly, for $x > 1$ and $0 \le y \le 1$ we find that

$$F(x,y) = F(1,y).$$

Therefore, it is necessary to determine $F(x,y)$ only for $0 \le x \le 1$ and $0 \le y \le 1$. We then have

$$F(x,y) = \int_{-\infty}^{x} ds \int_{-\infty}^{y} f(s,t)dt = k\int_{0}^{x} ds \int_{0}^{y} (s^2 + 2st)dt$$

$$= k\int_{0}^{x} ds \left[s^2 t + st^2\right]_{t=0}^{t=y} = \int_{0}^{x} (s^2 y + sy^2)ds$$

$$= k\left[\frac{s^3 y}{3} + \frac{s^2 y^2}{2}\right]_{s=0}^{s=x} = \frac{6}{5}\left[\frac{x^3 y}{3} + \frac{x^2 y^2}{2}\right]$$

$$= \frac{1}{5}(2x^3 y + 3x^2 y^2).$$

As a check, we differentiate to obtain the joint p.d.f.:

$$\frac{\partial^2 F(x,y)}{\partial x \partial y} = \frac{1}{5}\frac{\partial}{\partial x}(2x^3 + 6x^2 y) = \frac{1}{5}(6x^2 + 12xy) = f(x,y).$$

The marginal p.d.f. of X may be obtained via (7.1.7). For $0 < x < 1$ we have

$$f_1(x) = \int_{-\infty}^{\infty} f(x,y)dy = k\int_{0}^{1} (x^2 + 2xy)dy$$

$$= k\left[x^2 y + xy^2\right]_{y=0}^{y=1} = \frac{6}{5}(x^2 + x),$$

and $f_1(x) = 0$ otherwise. Alternatively, we may find the marginal c.d.f. of X and then differentiate:

$$F_1(x) = P(X \le x) = P(X \le x,\ Y \le \infty) = P(X \le x,\ Y \le 1)$$

$$= F(x,1) = \tfrac{1}{5}(2x^3 + 3x^2) \quad \text{for} \quad 0 \le x \le 1;$$

$$f_1(x) = \frac{d}{dx}F_1(x) = \tfrac{1}{5}(6x^2 + 6x) \quad \text{for} \quad 0 < x < 1.$$

(c) We wish to determine the probability that (X,Y) lies in the region R, where R is the half of the unit square lying above the main diagonal:

$$R = \{(x,y);\quad 0 < y < 1 \quad \text{and} \quad 0 \le x \le y\}.$$

Now, by (7.1.4),

$$P(X \le Y) = \int_R\int f(x,y)\,dxdy = k\int_R\int (x^2 + 2xy)\,dxdy$$

$$= k\int_0^1 dy \int_0^y (x^2 + 2xy)\,dx = \frac{6}{5}\int_0^1 dy\left[\frac{x^3}{3} + x^2y\right]_{x=0}^{x=y}$$

$$= \frac{2}{5}\int_0^1 4y^3 dy = \frac{2}{5}\left[y^4\right]_0^1 = \frac{2}{5}.$$

Alternatively, we could write

$$R = \{(x,y);\quad 0 < x < 1 \quad \text{and} \quad x \le y \le 1\}.$$

We would then obtain

$$P(X \le Y) = k\int_0^1 dx \int_x^1 (x^2 + 2xy)\,dy = k\int_0^1 dx\left[x^2y + xy^2\right]_{y=x}^{y=1}$$

which leads to the same result as before. □

Independence. Variates X and Y are said to be independent if and only if their joint c.d.f. factors,

$$F(x,y) = F_1(x)F_2(y) \tag{7.1.8}$$

for all x and y. By (7.1.2) and (7.1.5), an equivalent condition for X and Y to be independent is that their joint p.d.f. factors,

$$f(x,y) = f_1(x)f_2(y) \tag{7.1.9}$$

for all x and y (except possibly along a finite number of curves).

In many applications, the independence of variates X and Y

is a result of their dependence on different independent experiments. However, in other situations the following test for independent may be useful.

Factorization Criterion. Suppose that the joint p.d.f. of X and Y factors,

$$f(x,y) = g(x)h(y), \tag{7.1.10}$$

for all x and y (except possibly along finitely many curves). Then X and Y are independent variates with marginal p.d.f.'s proportional to g and h, respectively.

Proof. Suppose that (7.1.10) holds for some functions g and h. Then, by (7.1.7),

$$f_1(x) = \int_{-\infty}^{\infty} f(x,y)dy = g(x)\int_{-\infty}^{\infty} h(y)dy = k_1 g(x),$$

$$f_2(y) = \int_{-\infty}^{\infty} f(x,y)dx = h(y)\int_{-\infty}^{\infty} g(x)dx = k_2 h(y),$$

where k_1 and k_2 are constants. Furthermore, by (7.1.3) we have

$$1 = \int_{-\infty}^{\infty}\int_{-\infty}^{\infty} f(x,y)dxdy = \int_{-\infty}^{\infty} g(x)dx\int_{-\infty}^{\infty} h(y)dy = k_1 k_2.$$

Now, for all x and y, we have

$$f_1(x)f_2(x) = k_1 k_2 g(x)h(y) = f(x,y),$$

and therefore X and Y are independent variates.

Note. To obtain a factorization criterion for the case when X and Y are discrete, merely replace "p.d.f." by "probability function" and integrals by sums. The factorization criterion is a consequence of the fact that a p.d.f. or probability function need be defined only up to a multiplicative constant, which may then be determined from the condition that the total probability is 1. Use of the factorization criterion will be illustrated in Section 7.2.

Conditional Distributions

Suppose that X and Y are continuous variates with joint and marginal probability density functions f, f_1, and f_2. The conditional p.d.f. of Y given that $X = x$ is defined as follows:

$$f_2(y|x) = \frac{f(x,y)}{f_1(x)} \quad \text{for} \quad -\infty < y < \infty, \tag{7.1.11}$$

provided that $f_1(x) \neq 0$.

Some caution is necessary in dealing with conditional distributions in the continuous case. As we noted in Section 6.1, the probability that a continuous variate X takes on any particular real value is zero. Hence the event "$X = x$" has probability zero, and the definition of conditional probability (3.4.2) does not apply in this case. In order to give (7.1.11) an interpretation in terms of conditional probability, it is necessary to consider the event "$X = x$" as the limit of a sequence of events with non-zero probabilities.

Let h be a positive constant, and suppose that f_1 is positive over the interval $[x, x+h]$. Then $P(x \le X \le x+h) > 0$, and the definition of conditional probability gives

$$P(Y \le y | x \le X \le x+h) = \frac{P(x \le X \le x+h, Y \le y)}{P(x \le X \le x+h)} = \frac{F(x+h,y) - F(x,y)}{F_1(x+h) - F_1(x)}.$$

It is now not difficult to show that

$$\lim_{h \to 0} P(Y \le y | x \le X \le x+h) = \int_{-\infty}^{y} \frac{f(x,t)}{f_1(x)} dt = \int_{-\infty}^{y} f_2(t|x) dt.$$

Hence when "$X = x$" is considered to be the limit approached by the sequence of intervals $[x, x+h]$, conditional probabilities for Y can be obtained by integrating (7.1.11). However, as the following example shows, this will not generally be true if "$X = x$" is regarded as the limit of some other sequence of events. If X is a continuous variate, conditional probabilities given that $X = x$ are not uniquely determined until we specify the way in which the condition $X = x$ arises.

Example 7.1.2. Let X and Y be continuous variates with joint p.d.f.

$$f(x,y) = e^{-(x+y)} \quad \text{for} \quad 0 \le x < \infty \quad \text{and} \quad 0 \le y < \infty.$$

Taking $h > 0$, we find that

$$P(Y \le 2, 0 \le X \le h) = \int_0^2 dy \int_0^h e^{-(x+y)} dx = (1 - e^{-2})(1 - e^{-h});$$

$$P(Y \le 2, 0 \le X \le hY) = \int_0^2 dy \int_0^{hy} e^{-(x+y)} dx = \frac{h}{1+h} - \frac{1 + h - e^{-2h}}{e^2(1+h)};$$

$$P(0 \le X \le h) = \int_0^\infty dy \int_0^h e^{-(x+y)} dx = 1 - e^{-h};$$

$$P(0 \le X \le hY) = \int_0^\infty dy \int_0^{hy} e^{-(x+y)} dx = \frac{h}{1+h}.$$

For $h > 0$, the latter two probabilities are nonzero, and we may apply the definition of conditional probability (3.4.2) to obtain

$$p_1(h) = P(Y \le 2 | 0 \le X \le h) = 1 - e^{-2};$$

$$p_2(h) = P(Y \le 2 | 0 \le X \le hY) = 1 - \frac{1 + h - e^{-2h}}{e^2 h}.$$

Letting $h \to 0$ we obtain

$$p_1 = \lim_{h \to 0} p_1(h) = 1 - e^{-2} = 0.865;$$

$$p_2 = \lim_{h \to 0} p_2(h) = 1 - 3e^{-2} = 0.594.$$

Both p_1 and p_2 may be interpreted as conditional probabilities of $Y \le 2$ given that $X = 0$. In the first case, the condition $X = 0$ arises as the limit of a sequence of intervals $[0,h]$, and the definition (7.1.11) will yield the correct value. In the second case, $X = 0$ arises as the limit of a sequence of intervals $[0,hY]$, and then (7.1.11) is not applicable.

Problems for Section 7.1

†1. Let X and Y be continuous variates with joint c.d.f.

$$F(x,y) = kxy(x + y) \quad \text{for} \quad 0 < x < 1 \quad \text{and} \quad 0 < y < 1.$$

(a) Evaluate the constant k, and find the joint p.d.f. of X and Y.

(b) Find the marginal and conditional p.d.f.'s of X.

(c) Evaluate the following:

$P(X < 0.5,\ Y < 0.5)$, $P(X < 0.5)$, $P(X < 0.5 | Y < 0.5)$.

2. Let X and Y be continuous variates with joint p.d.f.

$$f(x,y) = ke^{-2x-3y} \quad \text{for} \quad 0 < x < \infty \quad \text{and} \quad 0 < y < \infty.$$

(a) Evaluate k and find the joint c.d.f. of X and Y.

(b) Find the marginal p.d.f. and c.d.f. of X.

(c) Obtain an expression for $P(2X + 3Y < t)$, and hence find the p.d.f. of T, where $T \equiv 2X + 3Y$.

3. X and Y are continuous variates. The conditional p.d.f. of X given $Y = y$ is

$$f_1(x|y) = \frac{x+y}{1+y} e^{-x} \quad \text{for} \quad 0 < x < \infty \quad \text{and} \quad 0 < y < \infty.$$

The marginal p.d.f. of Y is

$$f_2(y) = \tfrac{1}{2}(1+y)e^{-y} \quad \text{for} \quad 0 < y < \infty.$$

Find the marginal p.d.f. of X.

†4. Suppose that X and Y have a continuous uniform distribution over a triangle:

$$f(x,y) = \begin{cases} 2 & \text{for } x+y \le 1,\ x \ge 0, \text{ and } y \ge 0; \\ 0 & \text{otherwise.} \end{cases}$$

Evaluate the following:

$P(X < 0.5)$, $P(X < Y)$, $P(X < 0.5 | Y < 0.5)$, $P(X + Y < 0.5)$.

5. Suppose that X and Y are uniformly distributed over the unit circle:

$$f(x,y) = \begin{cases} 1/\pi & \text{for } 0 \le x^2 + y^2 < 1 \\ 0 & \text{otherwise.} \end{cases}$$

Find the marginal and conditional probability functions of X. Show that X and Y are uncorrelated but not independent.

6. Let X and Y be continuous variates with joint c.d.f.

$$F(x,y) = (1-e^{-x})(1-e^{-y})(1+ke^{-x-y}); \quad 0 < x < \infty,\ 0 < y < \infty,$$

where k is a constant.

(a) Find the marginal c.d.f. and p.d.f. of X.

(b) Show that $-1 \le k \le 1$ is a necessary condition for $F(x,y)$ to be a valid cumulative distribution function.

†7. Let X and Y be continuous variates with joint p.d.f.

$$f(x,y) = kx(x+y) \quad \text{for} \quad 0 < x < 1 \quad \text{and} \quad 0 < y < 1.$$

(a) Evaluate k, and find the marginal p.d.f.'s of X and Y.

(b) Find the means, variances, and correlation coefficient of X and Y.

8. A balanced coin is tossed. If heads occurs, two random numbers X and Y are generated. If tails occurs, only one random number U is generated, and we take $X = Y = U$. Find the joint c.d.f. of X and Y. Show that (7.1.3) does not hold, but that otherwise the conditions for a bivariate continuous distribution are satisfied.

9. Let Y be the number of good parts in a sample of ten parts selected at random from a batch in Review Problem 2 of Chapter 6. Find

(a) the conditional distribution of Y given that $X = x$;

(b) the marginal distribution of Y.

10. Let X and Y be continuous variates with joint p.d.f.

$$f(x,y) = \frac{1}{2\pi\sqrt{1-\rho^2}} \exp\{-\frac{1}{2(1-\rho^2)}(x^2 - 2\rho xy + y^2)\}$$

for $-\infty < x < \infty$ and $-\infty < y < \infty$, where ρ is a constant and $-1 < \rho < 1$. This is a <u>bivariate normal distribution</u>; see Section 7.4.

(a) Show that the correlation coefficient of X and Y is equal to ρ.

(b) Show that $\rho = 0$ is a necessary and sufficient condition for X and Y to be independent.

†11. A conveyor belt moves at constant speed past a stationary server. Items are distributed along the belt at random, and approach the server at the average rate of λ per hour. The server requires T hours to service an item after removing it from the belt, and T has an exponential distribution with mean θ. Any items passing him during the service period are missed. Let X be the number of items missed while one is serviced.

(a) Find the conditional distribution of X given that $T = t$.

(b) Show that the marginal distribution of X is geometric.

12. A flat table top is ruled with parallel lines one unit apart. A needle of length $b < 1$ is twirled and tossed onto the table, so that the direction of motion is perpendicular to the lines. Show that the probability that the needle will cut one of the parallel lines when it comes to rest is $2b/\pi$.

*13. A chord of a circle of radius r can be selected "at random" in the following three ways:

(i) Point A is selected on the circumference, and the chord is drawn at an angle θ to the tangent at A, where θ is uniform on $(0,\pi)$.

(ii) Point A is selected on the circumference, and the centre of the chord is located on the diameter through A at distance X from A, where X is uniform on $(0,2r)$.

(iii) The centre of the chord is located at (X,Y) where (X,Y) is uniformly distributed over the interior of the circle.

(a) Show that the probability of a chord greater than r in length is $2/3$ in (i), $\sqrt{3}/2$ in (ii), and $3/4$ in (iii).

(b) Find the p.d.f., the mean, and the variance of the chord length in each of the three cases.

7.2 Change of Variables

Suppose that X and Y have a bivariate continuous distribution with joint c.d. and p.d. functions F and f. In this section we discuss methods for finding the probability distribution of one or more functions of X and Y.

First consider a <u>one-to-one transformation</u>

$$U \equiv h(X,Y); \qquad V \equiv k(X,Y).$$

Then corresponding to any pair of values (u,v), there will be a unique pair of values (x,y) such that $u = h(x,y)$ and $v = k(x,y)$. The <u>Jacobian</u> of the transformation is the two-by-two determinant of partial derivatives,

$$\frac{\partial(x,y)}{\partial(u,v)} = \begin{vmatrix} \frac{\partial x}{\partial u} & \frac{\partial x}{\partial v} \\ \frac{\partial y}{\partial u} & \frac{\partial y}{\partial v} \end{vmatrix}. \tag{7.2.1}$$

Suppose that these derivatives are continuous and that the Jacobian is nonzero over the region where $f(x,y) > 0$. Then to obtain the joint p.d.f. of U and V, we multiply the joint p.d.f. of X and Y by the absolute value of the Jacobian,

$$g(u,v) = f(x,y) \cdot \left|\frac{\partial(x,y)}{\partial(u,v)}\right|, \tag{7.2.2}$$

and then substitute for x and y in terms of u and v. This result follows from the rules for changing variables in double integrals.

It may be extended in an obvious way to one-to-one transformations of three or more variates.

It is sometimes easier to work out the partial derivatives of the new variables u and v with respect to x and y to obtain

$$\frac{\partial(u,v)}{\partial(x,y)} = \begin{vmatrix} \frac{\partial u}{\partial x} & \frac{\partial u}{\partial y} \\ \frac{\partial v}{\partial x} & \frac{\partial v}{\partial y} \end{vmatrix}.$$

The reciprocal of this will be the required Jacobian:

$$\frac{\partial(x,y)}{\partial(u,v)} = 1/\frac{\partial(u,v)}{\partial(x,y)} . \qquad (7.2.3)$$

Example 7.2.1. Let X and Y be independent variates, each having a unit exponential distribution. Find the joint distribution of their sum $X + Y$ and their ratio X/Y.

Solution. The marginal p.d.f.'s of X and Y are

$$f_1(x) = e^{-x} \quad \text{for} \quad x > 0; \quad f_2(y) = e^{-y} \quad \text{for} \quad y > 0.$$

Since they are independent, their joint p.d.f. is

$$f(x,y) = f_1(x)f_2(y) = e^{-(x+y)} \quad \text{for} \quad x > 0 \quad \text{and} \quad y > 0.$$

Define $U \equiv X + Y$ and $V \equiv X/Y$. Then

$$\frac{\partial(u,v)}{\partial(x,y)} = \begin{vmatrix} \frac{\partial u}{\partial x} & \frac{\partial u}{\partial y} \\ \frac{\partial v}{\partial x} & \frac{\partial v}{\partial y} \end{vmatrix} = \begin{vmatrix} 1 & 1 \\ \frac{1}{y} & -\frac{x}{y^2} \end{vmatrix} = -\frac{x+y}{y^2}.$$

The transformation is one-to-one, and solving $u = x + y$, $v = x/y$ gives

$$x = \frac{uv}{v+1}; \quad y = \frac{u}{v+1}.$$

Hence, by (7.2.3) and (7.2.2), the joint p.d.f. of U and V is

$$g(u,v) = e^{-(x+y)} \cdot \frac{y^2}{x+y} = e^{-u} \cdot \frac{u}{(v+1)^2}.$$

Since (x,y) ranges over the positive quadrant, we find that (u,v)

does too, so this expression holds for all $u > 0$ and $v > 0$.

Note that, for all (u,v), $g(u,v)$ factors into a function of u times a function of v. Hence, by the factorization criterion, U and V are independent variates. Their marginal p.d.f.'s are

$$g_1(u) = k_1 ue^{-u} \quad \text{for} \quad u > 0; \quad g_2(v) = k_2(v+1)^{-2} \quad \text{for} \quad v > 0,$$

where k_1 and k_2 are normalizing constants and $k_1k_2 = 1$. Since

$$1 = \int_0^\infty g_2(v)dv = k_2\int_0^\infty (v+1)^{-2}dv = k_2\left[-\frac{1}{v+1}\right]_0^\infty = k_2,$$

it follows that $k_2 = 1$ and hence that $k_1 = 1$. □

<u>Example 7.2.2</u>. Suppose that X and Y are independent variates, where X has a χ^2 distribution with n degrees of freedom, and Y has a χ^2 distribution with m degrees of freedom. Define

$$U \equiv X + Y; \qquad V \equiv \frac{X \div n}{Y \div m}.$$

Show that U has a χ^2 distribution with $n+m$ degrees of freedom, and that V has an F distribution with n numerator and m denominator degrees of freedom.

<u>Solution</u>. By (6.9.1), the joint p.d.f. of X and Y is

$$f(x,y) = f_1(x)f_2(y) = k_n x^{\frac{n}{2}-1} e^{-\frac{x}{2}} \cdot k_m y^{\frac{m}{2}-1} e^{-\frac{y}{2}}$$

for $x > 0$ and $y > 0$. As in Example 7.2.1, we obtain

$$\frac{\partial(u,v)}{\partial(x,y)} = \begin{vmatrix} 1 & 1 \\ m/ny & -mx/ny^2 \end{vmatrix} = -\frac{m(x+y)}{ny^2},$$

and hence the joint p.d.f. of U and V is

$$g(u,v) = k_n k_m x^{\frac{n}{2}-1} y^{\frac{m}{2}-1} e^{-\frac{x+y}{2}} \cdot \frac{ny^2}{m(x+y)}.$$

Solving $u = x + y$, $v = mx/ny$ for x and y gives

$$y = \frac{mu}{nv+m} = u(1 + \frac{n}{m}v)^{-1}; \quad x = \frac{n}{m}uv(1 + \frac{n}{m}v)^{-1}$$

Upon substituting for x and y, we obtain

$$g(u,v) = k_n k_m (\frac{n}{m})^{\frac{n}{2}} u^{\frac{n+m}{2}-1} e^{-\frac{u}{2}} v^{\frac{n}{2}-1} (1 + \frac{n}{m} v)^{-\frac{n+m}{2}} .$$

Since (x,y) ranges over the positive quadrant, so does (u,v), and so this expression holds for $u > 0$ and $v > 0$.

As in Example 7.2.1, the joint p.d.f. of U and V factors, and thus U and V are independent. Their marginal p.d.f.'s are

$$g_1(u) = c_1 u^{\frac{n+m}{2}-1} e^{-\frac{u}{2}} \quad \text{for } u > 0;$$

$$g_2(v) = c_2 v^{\frac{n}{2}-1} (1 + \frac{n}{m} v)^{-\frac{n+m}{2}} \quad \text{for } v > 0,$$

where

$$c_1 c_2 = k_n k_m (\frac{n}{m})^{\frac{n}{2}} .$$

Upon comparing these expressions with (6.9.1) and (6.9.5), we conclude that $U \sim \chi^2_{(n+m)}$ and $V \sim F_{n,m}$ as required.

Since $c_1 = k_{n+m}$, the normalizing constant for a $\chi^2_{(n+m)}$ distribution, it follows that

$$c_2 = k_n k_m (\frac{n}{m})^{\frac{n}{2}} / k_{n+m} .$$

Now, using the expression derived for k_ν in Section 6.9, we obtain

$$c_2 = (\frac{n}{m})^{\frac{n}{2}} \Gamma(\frac{n+m}{2}) / \Gamma(\frac{n}{2}) \Gamma(\frac{m}{2}) .$$

Thus, as a byproduct of the change of variables, we have evaluated the normalizing constant for the F-distribution (6.9.5).

More General Transformations

Two methods are commonly used in change of variables problems involving transformations that are not one-to-one.

Method 1. Define one or more additional variables to make the transformation one-to-one, change variables via (7.2.2) or its generalization, and then integrate out the unwanted variables.

Method 2. Obtain the c.d.f. for the variate(s) of interest as a multiple integral, then differentiate to get the p.d.f.

These two methods are illustrated in the following examples. Neither method can be applied routinely in all cases; in general, change of variables problems require ingenuity. With both methods, one must be careful to obtain the correct region of integration.

Example 7.2.3. Find the p.d.f. of $X+Y$, where X and Y are independent $U(0,1)$ variates.

Solution 1. The joint p.d.f. of X and Y is

$$f(x,y) = f_1(x)f_2(y) = 1 \quad \text{for} \quad 0<x<1 \quad \text{and} \quad 0<y<1.$$

Define $U \equiv X+Y$, and take $V \equiv X-Y$ to make the transformation one-to-one. Solving $u = x+y$ and $v = x-y$ gives

$$x = \frac{u+v}{2}, \qquad y = \frac{u-v}{2},$$

and the Jacobian of the transformation is

$$\frac{\partial(x,y)}{\partial(u,v)} = \begin{vmatrix} 1/2 & 1/2 \\ 1/2 & -1/2 \end{vmatrix} = -1/2.$$

By (7.2.2), the joint p.d.f. of U and V is

$$g(u,v) = f(x,y) \cdot \left|\frac{\partial(x,y)}{\partial(u,v)}\right| = \frac{1}{2}$$

$$\text{for} \quad 0 < \frac{u+v}{2} < 1 \quad \text{and} \quad 0 < \frac{u-v}{2} < 1.$$

This is a uniform distribution over the square bounded by the four lines $u+v=0$, $u-v=0$, $u+v=2$, and $u-v=2$ (see Figure 7.2.1). Integrating $g(u,v)$ over v now gives the marginal p.d.f. of U:

$$g_1(u) = \int_{-\infty}^{\infty} g(u,v)dv.$$

It is apparent from Figure 7.2.1 that there are two cases to be considered. For $u \le 1$, v ranges from $-u$ to $+u$, and hence

$$g_1(u) = \int_{-u}^{u} \frac{1}{2}\, dv = u \quad \text{for} \quad 0 < u \le 1.$$

However, for $u > 1$, v ranges from $u-2$ to $2-u$, and thus

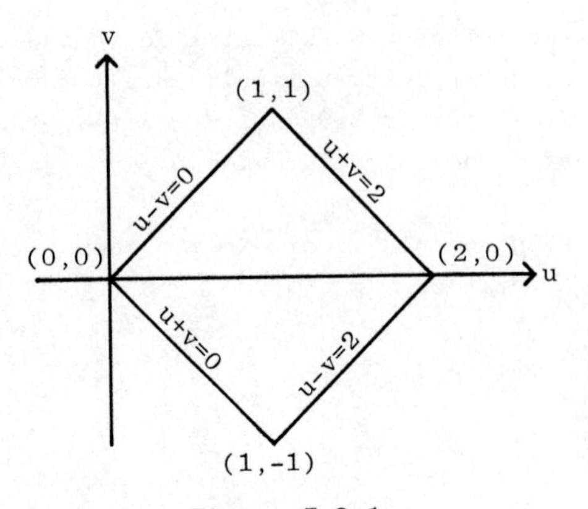

Figure 7.2.1
Range of U and V in Example 7.2.3

$$g_1(u) = \int_{u-2}^{2-u} \frac{1}{2}\, dv = 2 - u \quad \text{for} \quad 1 < u < 2.$$

Finally, we have $g_1(u) = 0$ for $u < 0$ or $u > 2$.

Solution 2. Define $U \equiv X + Y$, and denote the c.d.f. and p.d.f. of U by G and g, respectively. Then

$$G(u) = P(U \le u) = P(X + Y \le u)$$

$$= \iint_{x+y \le u} f(x,y)\,dxdy = \int_R\!\!\int dxdy$$

$$= \text{area of region } R$$

where R is the region inside the unit square within which $x + y \le u$ (see Figure 7.2.2). Clearly, $G(u) = 0$ for $u \le 0$, and $G(u) = 1$ for $u \ge 2$. For $0 < u \le 1$, R is a triangle with base and height u, so that

$$G(u) = \frac{1}{2} u^2 \quad \text{for} \quad 0 < u \le 1.$$

For $1 < u \le 2$, R consists of the entire unit square with the exception of a triangle of base and height $2 - u$; so that

$$G(u) = 1 - \tfrac{1}{2}(2 - u)^2 \quad \text{for} \quad 1 < u < 2.$$

The p.d.f. of U is now obtained by differentiating:

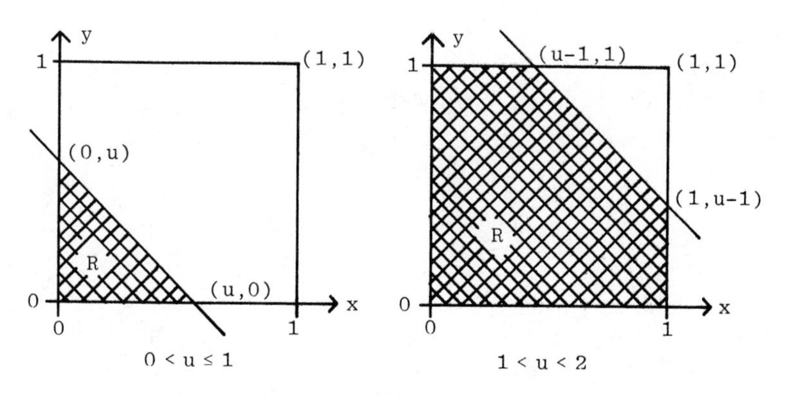

Figure 7.2.2

Region of Integration in Example 7.2.3

$$g(u) = \frac{d}{du} G(u) = \begin{cases} u & \text{for} \quad 0 < u < 1 \\ 2-u & \text{for} \quad 1 < u < 2. \end{cases}$$

Note. From Section 6.2, the mean and variance of $U(0,1)$ are $\frac{1}{2}$ and $\frac{1}{12}$, respectively, and hence

$$E(U) = \frac{1}{2} + \frac{1}{2} = 1; \quad \text{var}(U) = \frac{1}{12} + \frac{1}{12} = \frac{1}{6}.$$

The standard form of U is thus $\sqrt{6}(U-1) \equiv Z$, say, so that $U \equiv 1 + Z/\sqrt{6}$. By (6.1.11), the p.d.f. of Z is

$$g(u) \cdot \left|\frac{du}{dz}\right| = \begin{cases} (\sqrt{6} + z)/6 & \text{for} \quad -\sqrt{6} < z < 0 \\ (\sqrt{6} - z)/6 & \text{for} \quad 0 < z < \sqrt{6}. \end{cases}$$

This result was quoted in Example 6.7.1, and the p.d.f. is graphed in Figure 6.7.1.

Example 7.2.4. Distribution of $X_{(r)}$.

A system is made up of n identical components which function simultaneously, but independently of one another. The system ceases to function as soon as the rth component fails. Find the probability distribution of the system's lifetime.

Solution. Because the components are identical, their lifetimes $X_1, X_2, \ldots, X_n$ will have the same distribution. Let F and f denote their common c.d.f. and p.d.f.. Then

$$P(X_i \le x) = F(x); \quad P(X_i > x) = 1 - F(x).$$

We assume that $X_1, X_2, \ldots, X_n$ are independent.

The system fails as soon as the rth component fails. Hence the lifetime of the system is $X_{(r)}$, the rth smallest of $X_1, X_2, \ldots, X_n$. We denote the c.d.f. and p.d.f. of $X_{(r)}$ by G_r and g_r.

If $r = 1$, the system fails as soon as one of the components has failed. An example of such a system would be a set of n electronic components connected in series. The lifetime of the system is $X_{(1)}$, the smallest of the X_i's. The smallest of the X_i's exceeds x if and only if all of the X_i's are greater than x:

$$P\{X_{(1)} > x\} = P\{X_1 > x, X_2 > x, \ldots, X_n > x\}.$$

Because the X_i's are independent, this probability may be written as a product:

$$\begin{aligned} P\{X_{(1)} > x\} &= P(X_1 > x)\cdot P(X_2 > x)\cdot \ldots \cdot P(X_n > x) \\ &= [1 - F(x)]^n. \end{aligned}$$

It follows that

$$\begin{aligned} G_1(x) &= P\{X_{(1)} \le x\} = 1 - [1 - F(x)]^n; \\ g_1(x) &= \frac{d}{dx} G_1(x) = n[1 - F(x)]^{n-1} \frac{dF(x)}{dx} \\ &= n[1 - F(x)]^{n-1} f(x). \end{aligned}$$

If $r = n$, the system does not fail until all of its components have failed (e.g. electronic components connected in parallel). The system's lifetime is then $X_{(n)}$, the largest of the X_i's. The largest of the X_i's will be at most x if and only if every X_i is at most x. Therefore,

$$\begin{aligned} G_n(x) &= P\{X_{(n)} \le x\} = P\{X_1 \le x, X_2 \le x, \ldots, X_n \le x\} \\ &= P(X_1 \le x)\cdot P(X_2 \le x)\cdot \ldots \cdot P(X_n \le x) \\ &= [F(x)]^n; \end{aligned}$$

$$g_n(x) = \frac{d}{dx} G_n(x) = n[F(x)]^{n-1} f(x).$$

In general, the probability that exactly s components have failed by time x is equal to

$$\binom{n}{s}p^s(1-p)^{n-s}$$

where $p = F(x)$. Now the rth smallest of the X_i's will be at most x if and only if at least r components fail before time x; that is,

$$G_r(x) = P\{X_{(r)} \le x\} = \sum_{s=r}^{n} \binom{n}{s}p^s(1-p)^{n-s}.$$

Differentiating with respect to x gives the p.d.f. of $X_{(r)}$:

$$g_r(x) = \sum_{s=r}^{n} \binom{n}{s}[sp^{s-1}(1-p)^{n-s} - (n-s)p^s(1-p)^{n-s-1}]\frac{dp}{dx}\ .$$

Since $p = F(x)$, $\frac{dp}{dx} = f(x)$. The above expression may be simplifeid by using (2.1.6), and then noting that all terms cancel except one:

$$g_r(x) = n[\sum_{s=r}^{n} \binom{n-1}{s-1}p^{s-1}(1-p)^{n-s} - \sum_{s=r}^{n} \binom{n-1}{s}p^s(1-p)^{n-s-1}]f(x)$$

$$= n\binom{n-1}{r-1}p^{r-1}(1-p)^{n-r}f(x)$$

where $p = F(x)$. This agrees with the results obtained above in the special cases $r = 1$ and $r = n$.

In the above, we have assumed that F is the c.d.f. of a continuous distribution. The same arguments apply in the discrete case, except that then the probability function of $X_{(r)}$ is obtained as a difference rather than as a derivative. The special case $r = n = 2$ was considered in Example 4.1.2.

Problems for Section 7.2

1. (a) Find the mean of the distribution with p.d.f.

$$f(x) = \cos x \quad \text{for} \quad 0 < x < \pi/2.$$

   (b) Two independent observations are taken from this distribution. Find the probability that the interval which they form contains the mean.

†2. Point A is taken X units along a horizontal axis, and point B is taken Y units along a vertical axis, where X and Y are independent continuous uniform variates defined on $(0,1)$. Find the probability density function for the area of the triangle formed by

A,B, and the origin.

3. Let X and Y be independent variates with p.d.f.'s

$$f_1(x) = e^{-x} \quad \text{for } 0 < x < \infty;$$
$$f_2(y) = ye^{-y} \quad \text{for } 0 < y < \infty.$$

Find the probability density function of the ratio $R \equiv X/Y$.

4. Let X and Y be independent standardized normal variates, and define

$$U \equiv (X+Y)/\sqrt{2}; \qquad V \equiv (X-Y)/\sqrt{2}.$$

Show that U and V are independent and have standardized normal distributions.

5. Let X and Y be independent standardized normal variates. Show, by direct integration, that

$$P(X^2 + Y^2 < r) = 1 - e^{-r/2},$$

and hence deduce the p.d.f. of $X^2 + Y^2$.

†6. Let X_1 and X_2 be independent and identically distributed variates with p.d.f.

$$f(x) = xe^{-x^2/2} \quad \text{for } 0 < x < \infty.$$

Find the p.d.f. of $R \equiv \sqrt{X_1^2 + X_2^2}$.

7. Suppose that Y and Z are independent, where Z has a standardized normal distribution and Y has a χ^2 distribution with ν degrees of freedom. Define $X \equiv Z \div \sqrt{Y/\nu}$. Show that X has Student's distribution with ν degrees of freedom.

8. Let X and Y be independent variates having gamma distributions with the same scale parameter $\lambda = 1$, but with possibly different shape parameters p and q (see Problem 6.9.12). Define

$$U \equiv X + Y; \qquad V \equiv \frac{X}{X+Y}.$$

(a) Find the joint p.d.f. of U and V. Show that they are independent, that U has a gamma distribution, and V has a beta distribution (see Problem 6.9.14).

(b) The beta function was defined in Problem 6.9.14. Show, as a corollary to the change of variables in (a), that

$$\beta(p,q) = \Gamma(p)\Gamma(q)/\Gamma(p+q)$$

for all positive real numbers p and q.

†9. An electronic system is made up of n components connected in

series. It operates only so long as all n components are working. The lifetime of the ith component is exponentially distributed with mean $1/\lambda_i$ $(i=1,2,\ldots,n)$, and lifetimes are independent. Show that the lifetime of the system is exponentially distributed with mean $1/\sum\lambda_i$. If there are $n=5$ identical components with mean lifetime 5000 hours, what is the probability that the system functions for more than 2000 hours?

10. Let $X_1, X_2, \ldots, X_n$ be independent $U(0,1)$ variates. Show that the rth smallest of the X_i's has a beta distribution (see Problem 6.9.14), and its expected value is

$$E\{X_{(r)}\} = \frac{r}{n+1}.$$

Find the expectation of the range $R \equiv X_{(n)} - X_{(1)}$.

†11. Let $X_1, X_2, \ldots, X_n$ be independent and identically distributed continuous variates with p.d. and c.d. functions f and F. Let Y and Z denote the smallest and largest of the X_i's. Find the joint p.d.f. of Y and Z.
Hint: Find expressions for $P(Z \le z)$ and $P(Y > y, Z \le z)$. Their difference is the joint c.d.f. of Y and Z.

*12. (Continuation) Assuming that the X_i's are independent $U(0,1)$, show that the sample range $R \equiv Z - Y$ has p.d.f.

$$f(r) = n(n-1)r^{n-2}(1-r) \quad \text{for } 0 < r < 1.$$

13. Let X and Y be independent $U(0,1)$ variates. Find the c.d.f. and p.d.f. of their absolute difference, $D \equiv |X-Y|$.

†*14. The centre of a circle of radius $2/3$ is randomly located within a square of side 1. Find the probability distribution of N, the number of corners of the square which lie inside the circle.

7.3 Transformations of Normal Variates

The main result to be proved in this section is that, if one applies an orthogonal transformation to a set of independent $N(0,1)$ variates, one obtains a new set of independent $N(0,1)$ variates. From this, one can derive several results which have important applications in the statistical analysis of measurements which are assumed to be independent normal.

Orthogonal Matrices

An $n \times n$ matrix $C = (c_{ij})$ is said to be *orthogonal* if

$$C^tC = CC^t = I$$

where $C^t = (c_{ji})$ is the transpose of C, and I is the $n \times n$ identity matrix. Each row vector of an orthogonal matrix C has length 1, and is orthogonal to every other row vector of C. A similar property holds for the columns of C. Let $\det(C)$ denote the determinant of C. Since $\det(I) = 1$ and $\det(C^t) = \det(C)$, (7.3.1) gives

$$1 = \det(C^tC) = \det(C^t)\det(C) = [\det(C)]^2,$$

and therefore $\det(C) = \pm 1$.

Suppose that we have a set of r mutually orthogonal vectors $V_1, V_2, \ldots, V_r$ of unit length $(1 \le r < n)$. Then there exists an orthogonal matrix C having $V_1, V_2, \ldots, V_r$ as its first r rows. Such a matrix C could be constructed by adding $n - r$ row vectors to form a set of n linearly independent vectors, and then applying the Gram-Schmidt orthogonalization method. However, we shall not need to do this. For the applications to be considered, all that we require is the existence of such an orthogonal matrix.

Orthogonal Transformations

Let $Z = (Z_i)$ and $U = (U_i)$ be n-dimensional column vectors, and consider the linear transformation $Z = CU$; that is,

$$\begin{aligned}
Z_1 &= c_{11}U_1 + c_{12}U_2 + \cdots + c_{1n}U_n \\
Z_2 &= c_{21}U_1 + c_{22}U_2 + \cdots + c_{2n}U_n \\
&\cdots\cdots\cdots\cdots \\
Z_n &= c_{n1}U_1 + c_{n2}U_2 + \cdots + c_{nn}U_n.
\end{aligned}$$

If C is an orthogonal matrix, the transformation is called an _orthogonal transformation_. Orthogonal transformations correspond to rotations and reflections in n-space, and such transformations do not alter lengths or volumes. Since $C^tC = I$, we have

$$\begin{aligned}
\sum Z_i^2 &= Z^tZ = (CU)^t(CU) = U^tC^tCU \\
&= U^tIU = U^tU = \sum U_i^2.
\end{aligned}$$

Thus the transformed vector Z has the same length as the original vector. Also, since $\partial Z_i/\partial U_j = c_{ij}$, the matrix of partial derivatives is just the matrix C, and the Jacobian of the transformation is equal to the determinant of C:

$$\frac{\partial(Z_1,Z_2,\ldots,Z_n)}{\partial(U_1,U_2,\ldots,U_n)} = \det(C) = \pm 1.$$

We are now ready to prove the main result.

Theorem 7.3.1. Let $U_1,U_2,\ldots,U_n$ be independent $N(0,1)$ variates, and define n new variates $Z_1,Z_2,\ldots,Z_n$ via an orthogonal transformation $Z \equiv CU$. Then $Z_1,Z_2,\ldots,Z_n$ are independent $N(0,1)$ variates.

Proof. Since the U_i's are independent $N(0,1)$, their joint p.d.f. is

$$f(u_1,u_2,\ldots,u_n) = \prod_{i=1}^{n} \frac{1}{\sqrt{2\pi}} \exp(-\tfrac{1}{2}u_i^2) = (2\pi)^{-n/2}\exp\{-\tfrac{1}{2}\sum u_i^2\}.$$

We have shown that the Jacobian of an orthogonal transformation is ± 1, so by (7.2.2), the joint p.d.f. of $Z_1,Z_2,\ldots,Z_n$ is

$$g(z_1,z_2,\ldots,z_n) = f(u_1,u_2,\ldots,u_n).$$

The range of $(z_1,z_2,\ldots,z_n)$, like that of $(u_1,u_2,\ldots,u_n)$, is the whole of real n-space. Since $\sum u_i^2 = \sum z_i^2$, we have

$$g(z_1,z_2,\ldots,z_n) = (2\pi)^{-n/2}\exp\{-\tfrac{1}{2}\sum z_i^2\} = \prod_{i=1}^{n} \frac{1}{\sqrt{2\pi}} e^{-z_i^2/2},$$

which is a product of $N(0,1)$ densities. It follows that $Z_1,Z_2,\ldots,Z_n$ are independent $N(0,1)$ variates. □

We noted above that, given r mutually orthogonal vectors of unit length, there exists an orthogonal matrix with these vectors as its first r rows. We use this result for $r = 1$ and $r = 2$ to obtain the following two theorems.

Theorem 7.3.2. Let $U_1,U_2,\ldots,U_n$ be independent $N(0,1)$ variates, and let $a_1,a_2,\ldots,a_n$ be constants with $\sum a_i^2 = 1$. Define

$$Z_1 \equiv \sum a_i U_i; \quad V \equiv \sum U_i^2 - (\sum a_i U_i)^2.$$

Then Z_1 and V are independent variates, with

$$Z_1 \sim N(0,1); \quad V \sim \chi^2_{(n-1)}.$$

Proof. Let C be an orthogonal matrix whose first row is $(a_1, a_2, \ldots, a_n)$, and define $Z \equiv CU$. By Theorem 7.3.1, $Z_1, Z_2, \ldots, Z_n$ are independent $N(0,1)$ variates. In particular, $Z_1 \equiv \sum a_i U_i \sim N(0,1)$. Also, since $\sum U_i^2 \equiv \sum Z_i^2$, we have

$$V \equiv \sum U_i^2 - (\sum a_i U_i)^2 \equiv \sum Z_i^2 - Z_1^2$$

$$\equiv Z_2^2 + Z_3^2 + \ldots + Z_n^2.$$

Since V equals a sum of squares of $n-1$ independent $N(0,1)$ variates, (6.9.4) implies that V has a χ^2 distribution with $n-1$ degrees of freedom. Since $Z_2, Z_3, \ldots, Z_n$ are distributed independently of Z_1, V is also independent of Z_1.

Theorem 7.3.3. Let $U_1, U_2, \ldots, U_n$ be independent $N(0,1)$ variates. Let $(a_1, a_2, \ldots, a_n)$ and $(b_1, b_2, \ldots, b_n)$ be orthogonal vectors of length 1, so that $\sum a_i b_i = 0$ and $\sum a_i^2 = \sum b_i^2 = 1$. Define

$$Z_1 \equiv \sum a_i U_i; \quad Z_2 \equiv \sum b_i U_i; \quad V \equiv \sum U_i^2 - (\sum a_i U_i)^2 - (\sum b_i U_i)^2.$$

Then Z_1, Z_2, and V are independent variates, with

$$Z_1 \sim N(0,1); \quad Z_2 \sim N(0,1); \quad V \sim \chi^2_{(n-2)}.$$

Proof. Take $Z \equiv CU$, where C is an orthogonal matrix with first row $(a_1, a_2, \ldots, a_n)$ and second row $(b_1, b_2, \ldots, b_n)$. Since $\sum U_i^2 \equiv \sum Z_i^2$, we have

$$V \equiv \sum Z_i^2 - Z_1^2 - Z_2^2 \equiv Z_3^2 + Z_4^2 + \ldots + Z_n^2.$$

The result now follows from Theorem 7.3.1 and (6.9.4).

Applications

1. Let $X_1, X_2, \ldots, X_n$ be independent variates with $X_i \sim N(\mu_i, \sigma_i^2)$ for $i = 1, 2, \ldots, n$, and consider the linear combination $\sum a_i X_i$ where the a_i's are constants. Define $U_i \equiv (X_i - \mu_i)/\sigma_i$, so that the U_i's are independent $N(0,1)$. Since $X_i \equiv \sigma_i U_i + \mu_i$, we have

$$\sum a_i X_i \equiv \sum a_i \sigma_i U_i + \sum a_i \mu_i.$$

Take $a_i' = a_i \sigma_i / \sqrt{\sum a_i^2 \sigma_i^2}$, so that $\sum (a_i')^2 = 1$. Then

$$\sum a_i X_i \equiv A \sum a_i' U_i + B$$

where $A = \sqrt{\sum a_i^2 \sigma_i^2}$ and $B = \sum a_i \mu_i$. Now Theorem 7.3.2 implies that $\sum a_i' U_i \sim N(0,1)$, and (6.6.6) implies that $\sum a_i X_i$ is normally distributed with mean B and variance A^2; that is

$$\sum a_i X_i \sim N(\sum a_i \mu_i, \sum a_i^2 \sigma_i^2).$$

This result was stated previously as (6.6.7).

2. Take $a_1 = a_2 = \ldots = a_n = 1/\sqrt{n}$ in Theorem 7.3.2, so that

$$Z_1 \equiv \sum \frac{1}{\sqrt{n}} U_i \equiv \frac{1}{\sqrt{n}} \sum U_i \equiv \sqrt{n}\,\bar{U}$$

where $\bar{U} \equiv \frac{1}{n} \sum U_i$ is the sample mean. Then

$$V \equiv \sum U_i^2 - n\bar{U}^2 \equiv \sum (U_i - \bar{U})^2.$$

Suppose that $Y_1, Y_2, \ldots, Y_n$ are independent $N(\mu, \sigma^2)$. Define $U_i \equiv (Y_i - \mu)/\sigma$, so that $U_1, U_2, \ldots, U_n$ are independent $N(0,1)$. Substituting for the U_i's gives

$$Z_1 \equiv \sqrt{n}\left(\frac{\bar{Y} - \mu}{\sigma}\right); \qquad V \equiv \sum (Y_i - \bar{Y})^2/\sigma^2.$$

Theorem 7.3.2 implies that Z_1 and V are independent, with $Z_1 \sim N(0,1)$ and $V \sim \chi^2_{(n-1)}$. These results will be used to set up tests of significance for μ and σ^2 based on n observations which are assumed to arise from a normal distribution; see Section 14.2.

3. Similarly, Theorem 7.3.3 can be used to derive the necessary distributional results for fitting a straight line to data, where deviations from the line are assumed to be independent normal variates. See Chapter 15 for details.

Problems for Section 7.3

1. Let $U \equiv aX + bY$ and $V \equiv cX + dY$ where X and Y are independent $N(0,1)$ variates and $ad \neq bc$. Derive the joint p.d.f. of U and V. Show that U and V are independent if and only if $ac + bd = 0$.
2. Let U be a column vector of n independent $N(0,1)$ variates. Consider a linear transformation $Y \equiv AU$, where A is an $n \times n$ non-singular matrix, and define $V = AA^t$.

(a) Show that the diagonal elements of V are the variances of the Y_i's, and the off-diagonal elements of V are the covariances. (V is called the <u>variance-covariance matrix</u> of $Y_1, Y_2, \ldots, Y_n$.)

(b) Show that the joint p.d.f. of the Y_i's is

$$g(y_1, y_2, \ldots, y_n) = (2\pi)^{-n/2}[\det(V)]^{-1/2}\exp\{-\tfrac{1}{2} Y^t V^{-1} Y\}.$$

(This is the p.d.f. of a multivariate normal distribution.)

*7.4 The Bivariate Normal Distribution

In Section 7.3 we proved that, if $X_1, X_2, \ldots, X_n$ are independent normal variates and if $a_1, a_2, \ldots, a_n$ are any constants, then $a_1X_1 + a_2X_2 + \ldots + a_nX_n$ is normally distributed. In many applications, one encounters variates which are not necessarily independent, but which nevertheless have the property that any linear combination of them is normally distributed. Then $X_1, X_2, \ldots, X_n$ are said to have a <u>multivariate normal distribution</u>. A satisfactory discussion of the general case requires results from the theory of symmetric matrices and quadratic forms. Only the bivariate case $(n = 2)$ will be considered here.

<u>Definition</u>. The pair of variates (X_1, X_2) has a <u>bivariate normal distribution</u> if and only if $aX_1 + bX_2$ is normally distributed for all constants a and b.

Suppose that X_1 and X_2 have a bivariate normal distribution. Upon taking $b = 0$ in the definition we see that X_1 has a normal distribution, say $X_1 \sim N(\mu_1, \sigma_1^2)$. Similarly, $X_2 \sim N(\mu_2, \sigma_2^2)$. By (5.4.11), (5.4.12) and (5.4.6) we have

$$E(aX_1 + bX_2) = a\mu_1 + b\mu_2;$$

$$\begin{aligned} \text{var}(aX_1 + bX_2) &= a^2\sigma_1^2 + b^2\sigma_2^2 + 2ab\,\text{cov}(X_1, X_2) \\ &= a^2\sigma_1^2 + b^2\sigma_2^2 + 2ab\sigma_1\sigma_2\rho \end{aligned}$$

where ρ is the correlation coefficient of X_1 and X_2. The joint distribution of X_1 and X_2 will clearly depend upon the five parameters $(\mu_1, \mu_2, \sigma_1, \sigma_2, \rho)$.

---

* This section may be omitted on first reading.

In Section 8.5 we shall show that the above definition uniquely determines the distribution of X_1 and X_2, and that it depends only on these five parameters. Furthermore, we shall show that $\underline{\rho = 0}$ in a bivariate normal distribution implies independence of X_1 and X_2. This is a special feature of the bivariate normal distribution; it is not generally true that uncorrelated variates are independent (see Section 5.4).

Suppose that (X_1,X_2) has a bivariate normal distribution, and define

$$Y_1 \equiv a_1X_1 + b_1X_2; \quad Y_2 \equiv a_2X_1 + b_2X_2$$

where a_1,a_2,b_1,b_2 are constants. Then any linear combination of Y_1 and Y_2 may be written as a linear combination of X_1 and X_2, and consequently has a normal distribution. It follows that (Y_1,Y_2) has a bivariate normal distribution. Hence if we begin with a bivariate normal distribution and take any two linear combinations, they also have a bivariate normal distribution.

Joint Probability Density Function

We now derive the joint probability density function for a bivariate normal distribution in the case $-1 < \rho < 1$. If $\rho = \pm 1$ the distribution is called singular, and it is not possible to write down its joint p.d.f. We showed in Section 5.4 that if $\rho = \pm 1$, then one variate is a linear function of the other:

$$X_1 \equiv \mu_1 \pm \frac{\sigma_1}{\sigma_2}(X_2 - \mu_2).$$

Hence any statement about X_1 and X_2 jointly can be converted into a statement about X_2 alone, and its probability can be obtained as an integral of the univariate normal p.d.f. of X_2.

Now taking $\rho^2 < 1$, we define three variates as follows:

$$Z_1 \equiv \frac{X_1-\mu_1}{\sigma_1}; \quad Z_2 \equiv \frac{X_2-\mu_2}{\sigma_2}; \quad Z_3 \equiv \frac{Z_2-\rho Z_1}{\sqrt{1-\rho^2}}. \qquad (7.4.1)$$

Since Z_1 and Z_2 are the standard forms of X_1 and X_2, we have

$$E(Z_i) = 0; \quad \text{var}(Z_i) = E(Z_i^2) = 1 \quad \text{for} \quad i = 1,2;$$
$$\rho = \text{cov}(Z_1,Z_2) = E(Z_1Z_2).$$

Using these results, it is easy to show that

$$E(Z_3) = 0; \quad var(Z_3) = 1; \quad cov(Z_1, Z_3) = 0.$$

Since Z_1 and Z_3 are linear combinations of X_1 and X_2, they have a bivariate normal distribution. Furthermore, since $cov(Z_1, Z_3) = 0$, they are independent (Section 8.5). It follows that Z_1 and Z_3 are independent $N(0,1)$, and hence their joint p.d.f. is

$$\frac{1}{2\pi} \exp\{-\tfrac{1}{2}(z_1^2 + z_3^2)\}.$$

We now transform back to X_1 and X_2. The Jacobian is

$$\frac{\partial(z_1, z_3)}{\partial(x_1, x_2)} = \begin{vmatrix} 1/\sigma_1 & 0 \\ \dfrac{-\rho}{\sigma_1\sqrt{1-\rho^2}} & \dfrac{1}{\sigma_2\sqrt{1-\rho^2}} \end{vmatrix} = \frac{1}{\sigma_1\sigma_2\sqrt{1-\rho^2}}.$$

Hence by (7.2.2), the joint p.d.f. of X_1 and X_2 is

$$f(x_1, x_2) = \exp\{-\tfrac{1}{2} Q\}/k \tag{7.4.2}$$

where $k = 2\pi\sigma_1\sigma_2\sqrt{1-\rho^2}$, and the exponent Q is given by

$$Q = z_1^2 + z_3^2 = \frac{z_1^2 - 2\rho z_1 z_2 + z_2^2}{1-\rho^2}$$

$$= \frac{1}{1-\rho^2}\left[\left(\frac{x_1-\mu_1}{\sigma_1}\right)^2 - 2\rho\left(\frac{x_1-\mu_1}{\sigma_1}\right)\left(\frac{x_2-\mu_2}{\sigma_2}\right) + \left(\frac{x_2-\mu_2}{\sigma_2}\right)^2\right]. \tag{7.4.3}$$

Note that the exponent Q has been expressed as the sum of squares of two independent normal variates Z_1, Z_3. Hence, by (6.9.4), the exponent in the joint p.d.f. of a nonsingular bivariate normal distribution has a χ^2 distribution with two degrees of freedom. When $\nu = 2$, (6.9.1) simplifies to the p.d.f. of an exponential distribution with mean 2, and hence

$$P(Q \le q) = 1 - e^{-q/2} \quad \text{for} \quad q > 0. \tag{7.4.4}$$

Form of the bivariate normal p.d.f.

Figure 7.4.1 shows the surface $y = f(x_1, x_2)$, where f is the bivariate normal p.d.f. (7.4.2). The probability that (X_1, X_2)

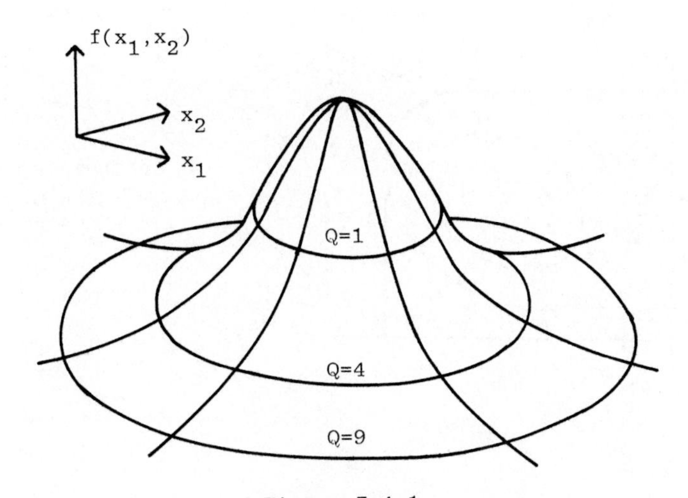

Figure 7.4.1

Bivariate normal p.d.f., showing contours of constant probability density.

falls in some region is equal to the volume above that region and under the surface. The surface has constant height above the curve $Q = q$, where q is any constant, and this curve is called a contour of constant probability density. For the bivariate normal distribution these contours are ellipses concentric at (μ_1, μ_2). The probability density is greater inside an ellipse than outside it. The contour $Q = 0$ consists of the single point (μ_1, μ_2) where the probability density is greatest. The contours $Q = 1$, $Q = 4$, and $Q = 9$ are shown in Figure 7.4.1.

By (7.4.4), the total probability contained inside the ellipse $Q = 1$ is

$$P(Q \le 1) = 1 - e^{-1/2} = 0.393.$$

Similarly, $P(Q \le 4) = 0.865$, and $P(Q \le 9) = 0.989$.

Figure 7.4.2 shows contours of constant probability density for three bivariate normal distributions. The outer contour shown is $Q = 9$. From (7.4.3), the range of values of X_1 within the contour $Q = 9$ is from $\mu_1 - 3\sigma_1$ to $\mu_1 + 3\sigma_1$. If $\rho = 0$, the major and minor axes of the ellipse are parallel to the coordinate axes as in Figure 7.4.2(ii). Figures (i) and (iii) illustrate the cases $\rho > 0$ and $\rho < 0$. If $\sigma_1 > \sigma_2$, the ellipses are more elongated horizontally than

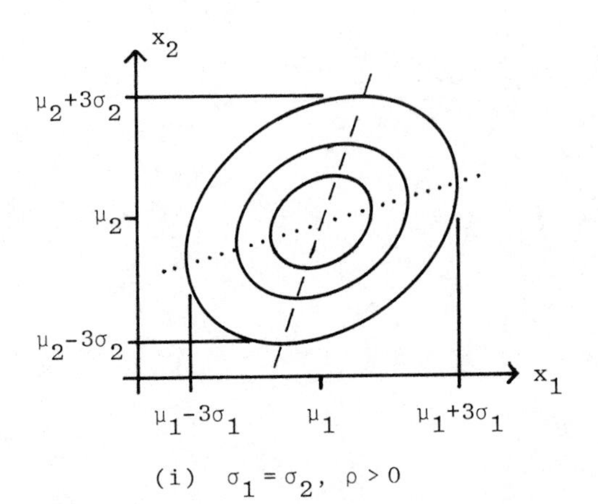

(i) $\sigma_1 = \sigma_2$, $\rho > 0$

The outer contour shown is $Q = 9$.
The regression of X_1 on X_2 is shown as - - - - -
The regression of X_2 on X_1 is shown as · · · · ·

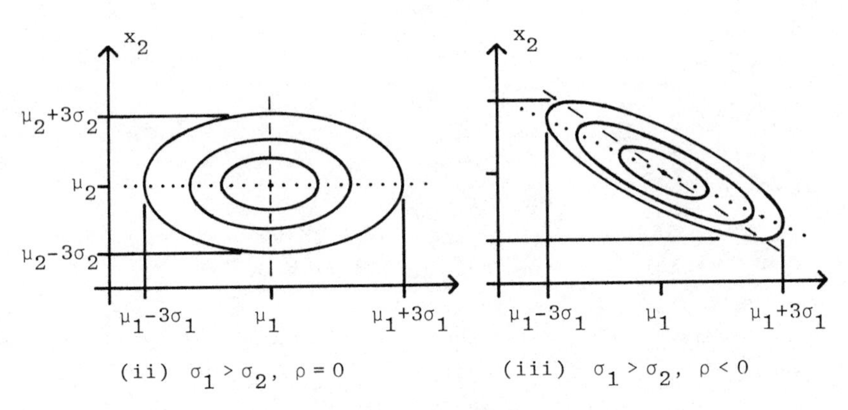

(ii) $\sigma_1 > \sigma_2$, $\rho = 0$ (iii) $\sigma_1 > \sigma_2$, $\rho < 0$

Figure 7.4.2

Elliptical contours of constant probability density in the bivariate normal distribution.

vertically, as in Figures (ii) and (iii). The dotted and broken lines in Figure 7.4.2 are regression lines; these will be discussed in the next section.

*7.5 Conditional Distributions and Regression

In the last section we saw that, if X_1 and X_2 have a bivariate normal distribution, their marginal distributions are normal. We shall now show that the conditional distributions are also normal.

---

* This section may be omitted on first reading.

Theorem 7.5.1. Let X_1 and X_2 have a non-singular bivariate normal distribution. Then the conditional distribution of X_2 given that $X_1 = x_1$ is normal with mean and variance

$$m_2 = \mu_2 + \rho\frac{\sigma_2}{\sigma_1}(x_1 - \mu_1); \qquad v_2^2 = \sigma_2^2(1-\rho^2). \tag{7.5.1}$$

Proof. From Section 7.4, the joint p.d.f. of X_1 and X_2 is

$$f(x_1,x_2) = \exp\{-\tfrac{1}{2}(z_1^2 + z_3^2)\}/k$$

where $k = 2\pi\sigma_1\sigma_2\sqrt{1-\rho^2}$, and the z_i's are defined in (7.4.1). Also, since $X_1 \sim N(\mu_1,\sigma_1^2)$, the marginal p.d.f. of X_1 is

$$f_1(x_1) = \frac{1}{\sqrt{2\pi}\sigma_1}\exp\{-\tfrac{1}{2}z_1^2\}.$$

Hence, by (7.1.11), the conditional p.d.f. of X_2 given that $X_1 = x_1$ is

$$f_2(x_2|x_1) = \frac{f(x_1,x_2)}{f_1(x_1)} = \exp\{-\tfrac{1}{2}z_3^2\}/c,$$

where $c = \sqrt{2\pi}\sigma_2\sqrt{1-\rho^2} = \sqrt{2\pi}v_2$. Now (7.4.1) gives

$$(1-\rho^2)z_3^2 = (z_2 - \rho z_1)^2 = \left(\frac{x_2-\mu_2}{\sigma_2} - \rho\frac{x_1-\mu_1}{\sigma_1}\right)^2$$

$$= [x_2 - \mu_2 - \rho\frac{\sigma_2}{\sigma_1}(x_1-\mu_1)]^2/\sigma_2^2.$$

Hence the conditional p.d.f. of X_2 is

$$f_2(x_2|x_1) = \frac{1}{\sqrt{2\pi}\,v_2}\exp\{-\frac{1}{2}\left(\frac{x_2-m_2}{v_2}\right)^2\},$$

and this is the p.d.f. of a normal distribution $N(m_2,v_2^2)$. □

Note that the variance of the conditional distribution does not depend up x_1, and is always smaller than σ_2^2 when $\rho \neq 0$. If X_1 and X_2 are correlated, the value of X_1 provides some information about X_2, and this shows up in the smaller variance of the conditional distribution.

The mean of the conditional distribution of X_2 given X_1 is called the regression of X_2 on X_1, and the reason for this terminology is explained below. In the bivariate normal distribution, the conditional mean m_2 is a linear function of x_1, and so we speak of the linear regression of X_2 on X_1. If we plot m_2 against x_1, we obtain a straight line through the point (μ_1,μ_2) with slope $\rho\sigma_2/\sigma_1$. This is shown by a dotted line in Figure 7.4.2. The broken line shows the regression of X_1 on X_2, which is also a straight line through (μ_1,μ_2), but with slope $\sigma_2/(\rho\sigma_1)$.

If X_1 and X_2 do not have a bivariate normal distribution, the conditional variance need not be constant, and the regression of one variate on the other need not be linear. In fact, we shall show that constancy of variance and linearity of regression, together with normality of the marginal and conditional distributions, are sufficient to imply that the joint distribution is bivariate normal.

Theorem 7.5.2. Suppose that X_1 has a normal distribution $N(\mu_1,\sigma_1^2)$. Suppose also that the conditional distribution of X_2 given that $X_1 = x_1$ is $N(\alpha+\beta x_1,\sigma^2)$ for all x_1, where α,β, and σ^2 are constants with respect to x_1. Then X_1 and X_2 jointly have a bivariate normal distribution, with parameters

$$\mu_1, \mu_2 = \alpha+\beta\mu_1; \quad \sigma_1^2, \sigma_2^2 = \sigma^2+\beta^2\sigma_1^2; \quad \rho = \beta\sigma_1/\sigma_2.$$

Proof. By (7.1.11), the joint p.d.f. of X_1 and X_2 is

$$f(x_1,x_2) = f_2(x_2|x_1)f_1(x_1)$$

$$= \frac{1}{\sqrt{2\pi}\,\sigma}\exp\{-\frac{1}{2}\left(\frac{x_2-\alpha-\beta x_1}{\sigma}\right)^2\}\cdot\frac{1}{\sqrt{2\pi}\,\sigma_1}\exp\{-\frac{1}{2}\left(\frac{x_1-\mu_1}{\sigma_1}\right)^2\}.$$

Upon rearranging the terms in the exponent, we find that this is a bivariate normal p.d.f. (7.4.3), with μ_2,σ_2^2 and ρ as defined above. □

Applications

The bivariate normal distribution frequently provides a satisfactory model for measurements of two different quantities taken on the same individual. For instance, one would expect the foot length X_1 and foot width X_2 of an adult male to have approximately a bivariate normal distribution. A shoe manufacturer might use this distribution to determine how many shoes of various lengths and widths he should produce. From actual foot measurements on 100 or so suitably chosen individuals, he could obtain reasonably precise estimates of $\mu_1, \mu_2, \sigma_1, \sigma_2$, and ρ. The proportion of adults who wear shoes of a specific size could then be obtained as an integral of the bivariate normal p.d.f. The bivariate and multivariate normal distributions can be used in a similar way to determine the size distributions for other articles of clothing.

Measurements of a quantity such as height, weight, or intelligence on two related individuals would also be expected to have a bivariate normal distribution. Sir Francis Galton considered the relationship between the heights of parent and child in his book Natural Inheritance, published in 1889. From data on 205 sets of parents and their 928 children, he was able to work out the properties of the bivariate normal distribution, and then have them verified by a mathematician. The origin of the term "regression" is also explained in Galton's book. His discussion runs to many pages, and makes interesting reading.

Galton worked with data for both sexes, and multiplied all female heights by 1.08 to make them comparable with male heights. The details are somewhat complicated, and in order to simplify the example sufficiently to permit its inclusion here, his findings are restated in terms of male heights only.

Let X_1 denote the height of the father, and X_2 the adult height of his son. Galton found that, individually, X_1 and X_2 were approximately normally distributed with roughly the same mean and variance:

$$\mu_1 = \mu_2 = 68; \quad \sigma_1^2 = \sigma_2^2 = 3.$$

He also found that if he considered only fathers of a given height x_1, their sons' heights were normally distributed, and their average height m_2 was a linear function of x_1. An increase of three inches in the father's height x_1 produced roughly an increase of one inch in the average heights of the sons. This implies that X_1 and X_2 have a bivariate normal distribution (Theorem 7.5.2); since $\beta = \frac{1}{3}$, we have $\rho = \beta\sigma_1/\sigma_2 = \frac{1}{3}$.

If the father's height is x_1, the mean height of his sons is

$$m_2 = \mu_2 + \rho\frac{\sigma_2}{\sigma_1}(x_1 - \mu_1) = 68 + \frac{1}{3}(x_1 - 68).$$

by (7.5.1). If the father is taller than average $(x_1 > 68)$, his sons will tend also to be taller than average $(m_2 > 68)$, but not so tall as their father. Similarly, a short father's sons will tend to be taller than he is, but shorter than average. In general, sons will be more mediocre in height than their fathers; their heights fall back, or regress, towards the mean of the population.

One might think that this regression of extreme heights towards the mean would thin the tails of the distribution, thus decreasing the overall variability in the sons' heights. This does not happen because the large number of fathers who are of near average height will produce some very tall and some very short sons, thus filling in the tails of the height distribution and compensating for the loss due to regression. Indeed, if it were not for regression towards the mean, the tallest fathers would produce some sons who were even taller than they, and the maximum height in the population would increase with each generation.

One can also consider the regression of the father's height X_1 on the son's height X_2. If we consider sons of a given height x_2, their fathers' heights will be normally distributed with mean

$$m_1 = \mu_1 + \rho \frac{\sigma_1}{\sigma_2}(x_2 - \mu_2) = 68 + \frac{1}{3}(x_2 - 68).$$

Fathers of tall sons will tend to be taller than average, but not so tall as their sons. This is as it should be, since many of the tall sons will have come from the very large number of fathers of near average height.

Effects of Paternal and Maternal Ages in Mongolism

Mongolism is a severe congenital disability associated with mental retardation as well as physical abnormalities, caused by representation of a portion of chromosome 21 three times instead of twice. The incidence of mongolism is greater among births to older parents. In fact, it is the age of the mother which is important, and not the age of the father. This was established by L.S. Penrose (The Relative Effects of Paternal and Maternal Age in Mongolism, Journal of Genetics 27 (1933), 219-224), and the following is a summary of his arguments.

The mother's age X_1 and the father's age X_2 were recorded at a large number of normal human births. The joint distribution of X_1 and X_2 was found to be approximately bivariate normal with parameters

$$\mu_1 = 31.25,\ \mu_2 = 33.83,\ \sigma_1 = 6.769,\ \sigma_2 = 7.711,\ \rho = 0.829.$$

Similarly, the ages were recorded at a large number of births of mongolian children. The variances and correlation coefficient were nearly the same, but the means were much larger:

$$\mu_1' = 37.25, \quad \mu_2' = 39.38.$$

Both parents tend to be older at mongolian births.

Because X_1 and X_2 are positively correlated ($\rho = 0.829$), if one parent is older, we would expect the other parent also to be older. If the mother's age is x_1, then the expected age of the father is

$$m_2 = \mu_2 + \rho\frac{\sigma_2}{\sigma_1}(x_1 - \mu_1).$$

In particular, if the mother's age is 37.25, the expected age of the father is

$$m_2 = 33.83 + 0.9444(37.25 - 31.25) = 39.50.$$

The average age of fathers at mongolian births was 39.38, and this is close to what one would expect. The increased paternal age at mongolian births can thus be accounted for by the tendency for older women to have older husbands.

If the father's age is x_2, the expected maternal age is

$$m_1 = \mu_1 + \rho\frac{\sigma_1}{\sigma_2}(x_2 - \mu_2).$$

In particular, for $x_2 = 39.38$ we have

$$m_1 = 31.25 + 0.7277(39.38 - 33.83) = 35.29.$$

The average age of mothers at mongolian births was 37.25, which is considerably larger than would be expected given the age of the fathers. Hence there appears to be a connection between the age of the mother and the incidence of mongolism.

The parameter values given above are estimates computed from data on parental ages at 727 births, of which 154 were mongolian. The above analysis can be refined to take into account the likely errors in these estimates. The difference between the observed paternal age 39.38 and the expected age 39.50 can be accounted for by possible estimation errors. However, the difference between observed and expected maternal age is real; that is, it is too large to have arisen merely as a result of errors in the parameter estimates. For further details, see Penrose's paper.

CHAPTER 8. GENERATING FUNCTIONS*

Suppose that the function $A(u)$ has a Taylor's series expansion about $u = 0$,

$$A(u) = a_0 + a_1u + a_2u^2 + \ldots + a_iu^i + \ldots ,$$

and that this series converges in some open interval containing the origin. Then $A(u)$ is called the <u>generating function</u> of the sequence $a_0, a_1, a_2, \ldots$. Generating functions have important applications in many branches of mathematics.

In Section 1, we give some preliminary definitions and results, and illustrate the use of generating functions in the solution of difference equations. In Sections 2 and 3 we define probability, moment, and cumulant generating functions, and consider some of their properties and uses. Section 4 gives some further applications, including a proof of the Central Limit Theorem in a special case. Finally, we discuss generating functions for bivariate distributions in Section 5.

8.1 Preliminary Results

Let $a_0, a_1, a_2, \ldots$ be a sequence of real numbers. Using these as coefficients, we may form a polynomial or power series in some variable u:

$$A(u) = a_0 + a_1u + a_2u^2 + \ldots + a_iu^i + \ldots \quad . \tag{8.1.1}$$

If (8.1.1) converges in some open interval containing $u = 0$, then $A(u)$ is called the <u>generating function</u> of the sequence $\{a_i\}$. The coefficient of u^i in the power series expansion of $A(u)$ is a_i, and hence we can generate the sequence $\{a_i\}$ by expanding $A(u)$ about $u = 0$ and examining the coefficients. No significance is attached to the variable u itself; it serves as merely the place-holder for the coefficients in (8.1.1).

The <u>exponential generating function</u> of the sequence $\{a_i\}$ is the generating function of the sequence $\{a_i/i!\}$:

---

* This chapter may be omitted at first reading.

$$B(u) = a_0 + a_1 u + a_2 \frac{u^2}{2!} + \dots + a_i \frac{u^i}{i!} + \dots, \tag{8.1.2}$$

provided that the series converges in some open interval containing $u = 0$. We can generate the series $\{a_i/i!\}$, and hence the series $\{a_i\}$, by examining the coefficients in the power series expansion of $B(u)$ about $u = 0$. It happens that, when the a_i's are probabilities, the ordinary generating function (8.1.1) is generally more convenient to work with. However, if the a_i's are moments, the exponential generating function has nicer properties.

Examples. The (ordinary) generating function of the infinite sequence $1,1,1,\dots$ is

$$A(u) = 1 + u + u^2 + \dots + u^i + \dots = \frac{1}{1-u},$$

and this geometric series converges for $|u| < 1$. The exponential generating function of this sequence is

$$B(u) = 1 + u + \frac{u^2}{2!} + \dots + \frac{u^i}{i!} + \dots = e^u,$$

and this series converges for all real u.

The generating function of the sequence $\{2^i\}$ is

$$A(u) = 1 + 2u + 2^2u^2 + \dots + 2^iu^i + \dots = \frac{1}{1-2u},$$

and the series converges for $|u| < \frac{1}{2}$. The exponential generating function of this sequence is

$$B(u) = 1 + 2u + \frac{2^2u^2}{2!} + \dots + \frac{2^iu^i}{i!} + \dots = e^{2u}$$

which converges for all real u.

The generating function of the sequence of binomial coefficients $\binom{n}{0},\binom{n}{1},\binom{n}{2},\dots$ is

$$A(u) = \binom{n}{0} + \binom{n}{1}u + \binom{n}{2}u^2 + \dots + \binom{n}{i}u^i + \dots = (1+u)^n$$

by the Binomial Theorem (2.1.9). This series is convergent for all u if n is a positive integer, and for $|u| < 1$ if n is negative or fractional. The exponential generating function of $\{\binom{n}{i}\}$ is

$$B(u) = \binom{n}{0} + \binom{n}{1}u + \binom{n}{2}\frac{u^2}{2!} + \dots + \binom{n}{i}\frac{u^i}{i!} + \dots,$$

for which there is no simple closed form expression. □

Several results from calculus are useful in the discussion of generating functions.

I. Convergence. If (8.1.1) converges when $u = u_0$ for some $u_0 \neq 0$, then (8.1.1) converges whenever $|u| < |u_0|$. The largest number R such that (8.1.1) converges whenever $|u| < R$ is called the radius of convergence of the series. If the series converges for all real values of u, we write $R = \infty$.

II. Differentiation. If (8.1.1) converges for $|u| < R$, then at each point in this interval, $A(u)$ has derivatives of all orders. These derivatives are represented by the series obtained when (8.1.1) is differentiated term-by-term, and have the same radius of convergence R. For instance,

$$\frac{d}{du} A(u) = a_1 + 2a_2 u + 3a_3 u^2 + \ldots + (i+1)a_{i+1}u^i + \ldots .$$

The first derivative of $A(u)$ is the generating function of the sequence $a_1, 2a_2, 3a_3, \ldots$. Repeated differentiation of (8.1.1) gives

$$a_i = \frac{1}{i!} \frac{d^i}{du^i} A(u) \bigg|_{u=0} .$$

III. Uniqueness. If two power series converge to the same sum

$$\sum a_i u^i = \sum b_i u^i$$

at every point in some interval $|u| < r$, then $a_i = b_i$ for $i = 0,1,2,\ldots$. Therefore, if two sequences $\{a_i\},\{b_i\}$ have the same generating function, they are identical.

IV. Multiplication. Suppose that $\sum a_i u^i$ and $\sum b_i u^i$ are convergent to sums $A(u)$ and $B(u)$ whenever $|u| < r$. Define

$$c_i = a_0 b_i + a_1 b_{i-1} + a_2 b_{i-2} + \ldots + a_i b_0; \quad i = 0,1,2,\ldots .$$

Then $\sum c_i u^i$ is convergent to $A(u)B(u)$ whenever $|u| < r$. The sequence $\{c_i\}$ is called the convolution of sequences $\{a_i\}$ and $\{b_i\}$. The generating function of $\{c_i\}$ is equal to the product of the generating functions of $\{a_i\}$ and $\{b_i\}$.

Difference Equations

One important use of generating functions is in the solution of difference equations and difference-differential equations. A complete discussion of this topic is beyond the scope of this book. However, we shall illustrate the technique by using generating functions to solve the difference-differential equation (4.4.1).

In Section 4.4, we defined $f_t(x)$ to be the probability that there will be x changes during a time interval of length t in a Poisson process, and showed that

$$\frac{d}{dt} f_t(x) = \lambda f_t(x-1) - \lambda f_t(x). \tag{8.1.3}$$

We defined $f_t(x)$ to be zero for all t whenever $x < 0$. Since there must be zero changes in zero time, we also have

$$f_0(0) = 1; \quad f_0(x) = 0 \quad \text{for} \quad x > 0. \tag{8.1.4}$$

We shall obtain the solution of (8.1.3) under the boundary conditions (8.1.4).

The generating function of the sequence $f_t(0), f_t(1), f_t(2), \ldots$ is

$$A(u) = \sum f_t(x) u^x.$$

We multiply all terms in (8.1.3) by u^x and add over all x to obtain

$$\frac{d}{dt} \sum f_t(x) u^x = \lambda u \sum f_t(x-1) u^{x-1} - \lambda \sum f_t(x) u^x.$$

This a differential equation in $A(u)$:

$$\frac{dA(u)}{dt} = \lambda u A(u) - \lambda A(u) = \lambda(u-1)A(u).$$

We thus have

$$\frac{d}{dt} \log A(u) = \frac{1}{A(u)} \cdot \frac{dA(u)}{dt} = \lambda(u-1).$$

The general solution of this differential equation is

$$A(u) = ce^{\lambda t(u-1)}$$

where c is a constant with respect to t.

For $t = 0$, we have

$$A(u) = c = \sum f_0(x) u^x = 1$$

by the boundary condition (8.1.4). Therefore,

$$A(u) = e^{\lambda t(u-1)} = e^{-\lambda t}e^{\lambda tu}.$$

We now expand $A(u)$ as a power series in u:

$$A(u) = e^{-\lambda t}[1 + \lambda tu + \frac{(\lambda tu)^2}{2!} + \ldots + \frac{(\lambda tu)^x}{x!} + \ldots].$$

Since $A(u)$ is the generating function of the sequence $f_t(0), f_t(1), f_t(2), \ldots,$ we may obtain $f_t(x)$ as the coefficient of u^x in this expansion. It follows that

$$f_t(x) = \frac{(\lambda t)^x e^{-\lambda t}}{x!} \quad \text{for} \quad x = 0,1,2,\ldots$$

as stated in Section 4.4.

Problems for Section 8.1

†1. For what sequence is $u^3 + 5u^5 + 12u^6$ the generating function? the exponential generating function?

2. For what sequence is $(1-u^2)^{-1}$ the generating function? the exponential generating function?

3. Find the convolution of the sequence $1,1,1,\ldots$ with itself, and its convolution with an arbitrary sequence $\{a_i\}$.

†4. Find the generating function and the exponential generating function for the sequence $1,2,3,4,\ldots$.

5. Consider two sequences of binomial coefficients, $\{\binom{a}{i}\}$ and $\{\binom{b}{i}\}$. Find the generating function of their convolution, and hence prove the hypergeometric identity (2.1.10).

†6. Find the generating function for the sequence of Fibonacci numbers $\{f_n\}$ defined by $f_0 = 0$, $f_1 = 1$, $f_n = f_{n-1} + f_{n-2}$ for $n \ge 2$. Expand the generating function in a power series, and verify that it gives the correct values for $f_0, f_1, \ldots, f_6$.

8.2 Probability Generating Functions

Let X be a discrete variate which takes only non-negative integer values, and define

$$a_x = P(X = x) = f(x); \quad x = 0,1,2,\ldots \quad .$$

The generating function of the sequence $\{a_x\}$ is called the probability generating function (p.g.f.) of X, and will be denoted by G_X. Note that

$$G_X(u) = \sum a_x u^x = \sum u^x f(x). \tag{8.2.1}$$

Upon comparing this with (5.1.3), we see that

$$G_X(u) = E(u^X). \tag{8.2.2}$$

The series (8.2.1) is convergent when $u = 1$; for then

$$\sum u^x f(x) = \sum f(x) = 1.$$

Hence, by result I of the preceding section, (8.2.1) is convergent whenever $|u| < 1$.

Because of the uniqueness result III, the probability generating function of X uniquely determines its distribution. If X and Y are non-negative integer valued variates and have the same p.g.f., then they have the same probability function. We may obtain $f(x)$ as the coefficient of u^x in the Taylor's series expansion of $G_X(u)$ about $u = 0$.

Table 8.2.1 lists the probability generating functions of four common discrete distributions. The p.g.f. of the hypergeometric distribution is more complicated, and has not been included. The p.g.f. of the negative binomial distribution is derived in the following example. The other entries in the table may be obtained similarly.

Table 8.2.1
Probability Generating Functions

Distribution of X	$f(x)$	Range	$G_X(u)$
Uniform on $1,2,\ldots,N$	$1/N$	$1 \le x \le N$	$u(1-u^N)/N(1-u)$
Binomial	$\binom{n}{x}p^x(1-p)^{n-x}$	$0 \le x \le n$	$(up+1-p)^n$
Poisson mean μ	$\mu^x e^{-\mu}/x!$	$0 \le x < \infty$	$e^{\mu(u-1)}$
Negative binomial	$\binom{-r}{x}p^r(p-1)^x$	$0 \le x < \infty$	$p^r(up-u+1)^{-r}$

Example 8.2.1. Derive the p.g.f. of the negative binomial distribution (4.2.4).

Solution. By (8.2.1) and (4.2.4), the p.g.f. is

$$G_X(u) = \sum u^x f(x) = \sum u^x \binom{-r}{x} p^r (p-1)^x$$

$$= p^r \sum \binom{-r}{x} [u(p-1)]^x \qquad (8.2.3)$$

where all sums extend from $x = 0$ to ∞. By the Binomial Theorem (2.1.9), this series converges for $|u(p-1)| < 1$ to give

$$G_X(u) = p^r [1 + u(p-1)]^{-r}$$

as given in the table.

Factorial Moments

We know that (8.2.1) converges for $|u| < 1$. Suppose now that (8.2.1) converges whenever $|u| < R$ where $R > 1$. Then, by result II of the preceding section, $G_X(u)$ has derivatives of all orders at $u = 1$. The ith derivative of $G_X(u)$ with respect to u is

$$G_X^{(i)}(u) = \frac{d^i}{du^i} \sum u^x f(x) = \sum f(x) \frac{d^i}{du^i} u^x$$

$$= \sum x(x-1)\dots(x-i+1)u^{x-i} f(x)$$

$$= \sum x^{(i)} u^{x-i} f(x).$$

Putting $u = 1$ gives

$$G_X^{(i)}(1) = \sum x^{(i)} f(x) = E\{X^{(i)}\} = g_i, \qquad (8.2.4)$$

which, by (5.3.1), is the ith factorial moment of X. All of the factorial moments may be obtained by repeatedly differentiating the p.g.f. and evaluating the derivatives at $u = 1$.

Consider the Taylor's series expansion of the function $G_X(1+u)$ about $u = 0$. The coefficient of u^i in this expansion is

$$\frac{1}{i!} \frac{d^i}{du^i} G_X(1+u)\Big|_{u=0} = \frac{1}{i!} G_X^{(i)}(1) = \frac{g_i}{i!} .$$

Hence $G_X(1+u)$ is the generating function of the sequence $\{\frac{g_i}{i!}\}$; it is the exponential generating function of the sequence $\{g_i\}$.

Example 8.2.2. Find the factorial moments and variance of the negative binomial distribution.

Solution. From Example 8.2.1, the series (8.2.3) is convergent for

$|u(p-1)| < 1$; that is, for $|u| < \frac{1}{1-p}$. Since $0 < p < 1$, the radius of convergence is greater than 1. The p.g.f. is

$$G_X(u) = p^r(up - u + 1)^{-r}.$$

Replacing u by $1+u$ gives

$$G_X(1+u) = p^r(up + p - u)^{-r} = p^r[p + u(p-1)]^{-r}$$
$$= [1 + u\,\frac{p-1}{p}]^{-r}.$$

Hence, by the Binomial Theorem (2.1.9),

$$G_X(1+u) = \sum \binom{-r}{i} [u\,\frac{p-1}{p}]^i = \sum (-r)^{(i)} (\frac{p-1}{p})^i \frac{u^i}{i!}.$$

The coefficient of $u^i/i!$ is the ith factorial moment:

$$g_i = (-r)^{(i)} (\frac{p-1}{p})^i = r(r+1)\ldots(r+i-1)(\frac{1-p}{p})^i.$$

In particular, we have

$$\mu = g_1 = r(\frac{1-p}{p}); \quad g_2 = r(r+1)(\frac{1-p}{p})^2.$$

Hence, by (5.3.2), the variance is

$$\sigma^2 = g_2 + \mu - \mu^2 = r(1-p)/p^2.$$

Sums of Independent Variates

Let X and Y be independent variates which take only non-negative integer values. Define

$$a_i = P(X=i); \quad b_i = P(Y=i); \quad c_i = P(X+Y=i)$$

for $i = 0,1,2,\ldots$. Then we have

$$c_i = P(X=0,Y=i) + P(X=1,Y=i-1) + P(X=2,Y=i-2) + \ldots + P(X=i,Y=0)$$
$$= a_0 b_i + a_1 b_{i-1} + a_2 b_{i-2} + \ldots + a_i b_0.$$

Hence the sequence $\{c_i\}$ is the convolution of the sequences $\{a_i\}$ and $\{b_i\}$. By result IV of Section 8.1, the generating function of $\{c_i\}$ is equal to the product of the generating functions of $\{a_i\}$ and $\{b_i\}$; that is,

$$G_{X+Y}(u) = G_X(u)G_Y(u). \tag{8.2.5}$$

Hence if X and Y are independent variates, the p.g.f. of their sum

$X+Y$ is equal to the product of their p.g.f.'s. This result may be extended by induction to any finite sum of independent variates.

Frequently, one wishes to find the distribution of a sum of n independent variates $X_1, X_2, \ldots, X_n$ with a common distribution. If their common probability function is f, then each of them has p.g.f.

$$G(u) = \sum u^x f(x).$$

The p.g.f. of their sum $S_n \equiv X_1 + X_2 + \ldots + X_n$ is then

$$G_{S_n}(u) = G_{X_1}(u) G_{X_2}(u) \ldots G_{X_n}(u) = [G(u)]^n. \tag{8.2.6}$$

This often provides the easiest method for deriving the distribution of S_n.

Example 8.2.3. Suppose that X and Y are independent binomial variates with parameters (n,p) and (m,p), respectively. Find the distribution of $X + Y$.

Solution. Using (8.2.5) and the binomial p.g.f. from Table 8.2.1, we obtain

$$\begin{aligned} G_{X+Y}(u) &= G_X(u) G_Y(u) \\ &= (up + 1 - p)^n (up + 1 - p)^m \\ &= (up + 1 - p)^{n+m} \\ &= \sum \binom{n+m}{i} (up)^i (1-p)^{n+m-i} \end{aligned}$$

by the Binomial Theorem. The coefficient of u^i is

$$P(X + Y = i) = \binom{n+m}{i} p^i (1-p)^{n+m-i}.$$

Hence $X + Y$ has a binomial distribution with parameters $(n+m, p)$. This is as expected; for if X is the number of heads in n tosses of a coin and Y is the number of heads in m different tosses, then $X + Y$ is the total number of heads in all $n + m$ tosses.

Example 8.2.4. Find the distribution of a sum of two independent Poisson variates.

Solution. Suppose that X and Y are independent Poisson variates with means μ and ν, respectively. Using (8.2.5) and Table 8.2.1, we obtain

$$G_{X+Y}(u) = G_X(u)G_Y(u)$$
$$= e^{\mu(u-1)} \cdot e^{\nu(u-1)} = e^{-(\mu+\nu)} \cdot e^{u(\mu+\nu)}$$
$$= e^{-(\mu+\nu)} \sum \frac{[u(\mu+\nu)]^i}{i!}.$$

The coefficient of u^i is

$$P(X + Y = i) = e^{-(\mu+\nu)}(\mu+\nu)^i/i!,$$

and hence $X + Y$ has a Poisson distribution with mean $\mu + \nu$. For another proof of this result, see Example 4.5.5.

Example 8.2.5. Find the distribution of the total score obtained when n balanced dice are rolled.

Solution. Let X_i denote the score on the ith die $(i = 1,2,\ldots,n)$. Then X_i has a uniform distribution on $1,2,\ldots,6$, and from Table 8.2.1, its p.g.f. is

$$G(u) = \frac{u(1-u^6)}{6(1-u)}.$$

Since the X_i's are independent, (8.2.6) implies that their sum $S \equiv X_1 + X_2 + \ldots + X_n$ has p.g.f.

$$G_S(u) = [G(u)]^n = \frac{u^n(1-u^6)^n}{6^n(1-u)^n}.$$

To determine the probability function of S, we expand $G_S(u)$ as a power series in u and examine the coefficients.

For $n = 2$, we obtain

$$G_S(u) = \frac{u^2(1-u^6)^2}{36(1-u)^2}$$

$$= \frac{1}{36}(u^2 + 2u^3 + 3u^4 + 4u^5 + 5u^6 + 6u^7 + 5u^8 + 4u^9 + 3u^{10} + 2u^{11} + u^{12}).$$

The coefficients of $u^2, u^3, u^4, \ldots, u^{12}$ are the probabilities of obtaining total scores $2,3,4,\ldots,12$ with two dice.

For general n, we apply the Binomial Theorem twice to obtain

$$G_S(u) = (\tfrac{u}{6})^n(1-u^6)^n(1-u)^{-n}$$
$$= (\tfrac{u}{6})^n \sum \binom{n}{i}(-1)^i u^{6i} \sum \binom{-n}{j}(-1)^j u^j.$$

We now rearrange the sum and substitute $j = k - 6i$ to obtain

$$G_S(u) = (\tfrac{u}{6})^n \sum\sum \binom{n}{i}\binom{-n}{j}(-1)^{i+j}u^{6i+j}$$

$$= (\tfrac{u}{6})^n \sum_k u^k \sum_i \binom{n}{i}\binom{-n}{k-6i}(-1)^{i+k}.$$

The coefficient of u^{n+k} in this expression is

$$P(S = n + k) = 6^{-n}\sum_i \binom{n}{i}\binom{-n}{k-6i}(-1)^{i+k}.$$

This is actually a finite sum; the ith term is zero whenever $i > n$ or $6i > k$.

Problems for Section 8.2

1. Derive the probability generating functions of the discrete uniform, binomial, and Poisson distributions (Table 8.2.1).
2. Obtain the factorial moments and variances of the binomial and Poisson distributions from their p.g.f.'s.
3. Let X have a discrete uniform distribution on values $1,2,\ldots,N$. Show that the factorial moments of X are

$$g_0 = 1, \quad g_N = (N-1)!, \quad g_i = 0 \quad \text{for} \quad i > N;$$

$$g_i = \frac{(N+1)^{(i+1)}}{N(i+1)} \quad \text{for} \quad i = 1,2,\ldots,N-1.$$

   Hence obtain the variance of X.

†4. (a) Find the p.g.f. for the geometric distribution (4.2.1).
   (b) Let $X_1, X_2, \ldots, X_r$ be independent variates having geometric distributions with the same probability parameter p. Show that their sum has a negative binomial distribution, and interpret this result in terms of waiting times in Bernoulli trials.

5. Suppose that X and Y are independent, and have negative binomial distributions with parameters (r_1, p) and (r_2, p), respectively. Show that $X + Y$ has a negative binomial distribution with parameters $(r_1 + r_2, p)$. Interpret this result in terms of waiting times in Bernoulli trials.
6. Consider a sequence of n Bernoulli trials. Define $X \equiv \sum Y_i$, where $Y_i = 1$ if the trial results in a success, and $Y_i = 0$ otherwise $(i = 1,2,\ldots,n)$. Find the p.g.f. of Y_i, and hence obtain

the p.g.f. and probability function of X.

7. Let X_n be the number of successes in n Bernoulli trials, and f_n be the probability function of X_n. Show that

$$f_{n+1}(x) = pf_n(x-1) + (1-p)f_n(x).$$

Multiply through by u^x and sum over x to show that

$$G_{n+1}(u) = (pu + 1 - p)G_n(u),$$

where G_n is the p.g.f. of X_n. Hence obtain expressions for the p.g.f. and probability function of X_n.

†8. Let X be a non-negative integer-valued variate, and define

$$a_x = P(X = x); \quad b_x = P(X > x) \quad \text{for} \quad x = 0,1,2,\ldots .$$

Let A and B denote the generating functions of the sequences $\{a_x\}$ and $\{b_x\}$, respectively. Show that

(i) $B(u) = [1 - A(u)]/(1-u)$;
(ii) $E(X) = B(1)$;
(iii) $\mathrm{var}(X) = 2B'(1) + B(1) - [B(1)]^2$.

Use these results to obtain the mean and variance of the geometric distribution (4.2.1).

9. Let G_n be the p.g.f. of the hypergeometric distribution (2.3.1). Show that

$$(1+uy)^a(1+y)^b = \sum_{n=0}^{\infty} \binom{a+b}{n} y^n G_n(u),$$

and hence obtain

$$G_n(u) = \frac{1}{(a+b)^{(n)}} \frac{d}{dy^n}(1+uy)^a(1+y)^b \bigg|_{y=0} .$$

*10. In Example 8.2.5, show that

$$P(S \le n + s) = 6^{-n}\sum_i \binom{n}{i}\binom{-n-1}{s-6i}(-1)^{i+s}.$$

Hint: Use the results proved in Problems 2.1.5(d) and 2.1.6.

*11. The Montmort distribution has probability function

$$f(x) = \frac{1}{x!}\left[1 - \frac{1}{1!} + \frac{1}{2!} - + \ldots \pm \frac{1}{(n-x)!}\right] \quad \text{for} \quad x = 0,1,\ldots,n.$$

Show that the probability generating function is

$$G(u) = \sum_{i=0}^{n} \frac{(u-1)^i}{i!}.$$

Hence find the mean, variance, and factorial moments, and consider the limit approached as $n \to \infty$.

Problems 12-18 involve conditional expectation (Section 5.6).

†12. Let Y and $X_1, X_2, \ldots$ be independent non-negative integer-valued variates, and consider

$$Z \equiv X_1 + X_2 + \ldots + X_Y;$$

that is, Z is the sum of a random number of variates X_i. Suppose that each of the X_i's has p.g.f. A, and Y has p.g.f. B. Show that Z has p.g.f.

$$G(u) = B\{A(u)\}.$$

13. (Continuation) Find the mean and variance of Z from its p.g.f., and also using (5.6.4) and (5.6.6).

14. In Problem 5.6.3, find the probability generating function for the total number of eggs in a field.

†15. At time 0 there is a single cell. With probability 1/2 this cell may divide, so that at time 1 there may be one cell (probability 1/2) or two cells (probability 1/2). Now each cell may or may not divide with equal probabilities, different cells being independent. Thus at time 2 there may be 1,2,3, or 4 cells. Show that the probabilities are 1/4, 3/8, 1/4, and 1/8, respectively. Find the probability distribution for the number of cells at time 3.

16. (Continuation) Let Y_n be the number of cells at time n, and let $G_n(t)$ be the probability generating function of Y_n. Show that

$$Y_{n+1} \equiv X_1 + X_2 + \ldots + X_{Y_n}$$

where $X_1, X_2, \ldots$ are independent variates with distribution

$$P(X_i = 1) = P(X_i = 2) = \frac{1}{2}.$$

Hence show that

$$G_{n+1}(t) = G_n\left(\frac{t(t+1)}{2}\right).$$

Find $G_3(t)$, and show that it yields the same distribution of Y_3 as that obtained in problem 15.

†17. (Continuation) Find the mean and variance of Y_n.

18. Consider the situation in problem 15, but now suppose that at each stage a cell either dies (probability 1/3), divides (probability

1/3) or remains unchanged. Show that

$$G_{n+1}(t) = G_n\left(\frac{1+t+t^2}{3}\right)$$

and hence find the mean and variance of Y_n. Verify your results by direct computation in the case $n = 2$.

8.3 Moment and Cumulant Generating Functions

The moment generating function (m.g.f.) of a variate X is defined by

$$M_X(u) = E(e^{uX}) = \begin{cases} \sum e^{ux} f(x) & \text{for } X \text{ discrete;} \\ \int_{-\infty}^{\infty} e^{ux} f(x)dx & \text{for } X \text{ continuous} \end{cases} \qquad (8.3.1)$$

provided that the sum or integral converges in some interval $|u| < r$ where r is positive. The ith derivative of $M_X(u)$ with respect to u is

$$M_X^{(i)}(u) = E\{\frac{d^i}{du^i} e^{uX}\} = E\{X^i e^{uX}\}.$$

Putting $u = 0$ gives

$$M_X^{(i)}(0) = E\{X^i\} = m_i$$

which, by (5.2.1), is the ith moment of X. The Taylor's series expansion of $M_X(u)$ about $u = 0$ is

$$M_X(u) = 1 + m_1 u + m_2 \frac{u^2}{2!} + \ldots + m_i \frac{u^i}{i!} + \ldots . \qquad (8.3.2)$$

Hence $M_X(u)$ is the generating function of the sequence $\{\frac{m_i}{i!}\}$; it is the exponential generating function of the sequence of moments $\{m_i\}$.

When X takes only non-negative integer values, its moment and probability generating functions are related as follows:

$$M_X(u) = G_X(e^u); \quad G_X(u) = M_X(\log u). \qquad (8.3.3)$$

Since G_X uniquely determines the distribution of X, the same is true of M_X. Also if X and Y are independent variates, then (8.2.5) gives

$$M_{X+Y}(u) = M_X(u)M_Y(u). \qquad (8.3.4)$$

If $X_1, X_2, \ldots, X_n$ are independent variates with the same m.g.f. $M(u)$, then the m.g.f. of their sum $S_n \equiv X_1 + X_2 + \ldots + X_n$ is

$$M_{S_n}(u) = M_{X_1}(u)M_{X_2}(u) \ldots M_{X_n}(u) = [M(u)]^n. \tag{8.3.5}$$

Probability generating functions are useful only for non-negative integer valued variates. Moment generating functions have most of the desirable properties of p.g.f.'s, with the added advantage that they maintain their usefulness for more general discrete and continuous variates. Results (8.3.4) and (8.3.5) for sums of independent variates continue to hold, and under quite general conditions, the moment generating function uniquely determines the distribution. However, it is difficult to obtain the probability function or p.d.f. from the moment generating function unless X is non-negative integer valued.

The probability generating functions of four discrete distributions were given in Table 8.2.1. To obtain the m.g.f.'s of these distributions, one need merely replace u by e^u. For instance, the m.g.f. of the binomial distribution with parameters (n,p) is

$$M_X(u) = G_X(e^u) = (pe^u + 1 - p)^n.$$

The moment generating functions of several continuous distributions are given in Table 8.3.1.

Table 8.3.1
Moment Generating Functions

Distribution of X	$f(x)$	Range	$M_X(u)$
Uniform on $(0,1)$	1	$0 < x < 1$	$\frac{1}{u}(e^u - 1)$
Exponential, mean θ	$\frac{1}{\theta}e^{-x/\theta}$	$0 < x < \infty$	$(1 - \theta u)^{-1}$
Gamma, parameters λ, p	$\frac{1}{\Gamma(p)}\lambda^p x^{p-1}e^{-\lambda x}$	$0 < x < \infty$	$(1 - \frac{u}{\lambda})^{-p}$
χ^2 with ν d.f.	$k_\nu x^{\frac{\nu}{2}-1} e^{-\frac{x}{2}}$	$0 < x < \infty$	$(1 - 2u)^{-\nu/2}$
Normal $N(\mu,\sigma^2)$	$\frac{1}{\sqrt{2\pi}\sigma}\exp\{-\frac{1}{2}(\frac{x-\mu}{\sigma})^2\}$	$-\infty < x < \infty$	$\exp(u\mu + \frac{u^2\sigma^2}{2})$

Example 8.3.1. Derive the m.g.f. of the standardized normal distribution, and find an expression for the ith moment.

Solution. Let $Z \sim N(0,1)$. By (8.3.1) and (6.6.2), the m.g.f. of Z is

$$M_Z(u) = E(e^{uZ}) = \frac{1}{\sqrt{2\pi}} \int_{-\infty}^{\infty} e^{uz}\, e^{-z^2/2}\, dz$$

$$= \frac{1}{\sqrt{2\pi}} \int_{-\infty}^{\infty} \exp\{-\tfrac{1}{2}(z^2 - 2uz)\}dz.$$

Now since

$$z^2 - 2uz = (z-u)^2 - u^2,$$

we have

$$M_Z(u) = \frac{1}{\sqrt{2\pi}} \int_{-\infty}^{\infty} \exp\{\tfrac{1}{2}u^2 - \tfrac{1}{2}(z-u)^2\}dz$$

$$= [\frac{1}{\sqrt{2\pi}} \int_{-\infty}^{\infty} \exp\{-\tfrac{1}{2}(z-u)^2\}dz]e^{u^2/2}.$$

The expression in square brackets is the total probability in a normal distribution $N(u,1)$, and therefore equals 1. Hence

$$M_Z(u) = e^{u^2/2} = 1 + \frac{u^2}{2} + \frac{1}{2!}(\frac{u^2}{2})^2 + \ldots + \frac{1}{i!}(\frac{u^2}{2})^i + \ldots \qquad (8.3.6)$$

Since only even powers of u appear, all odd moments of Z are zero. Equating coefficients of u^{2i} in this expression and (8.3.2) gives

$$\frac{m_{2i}}{(2i)!} = \frac{1}{i!2^i},$$

and therefore

$$m_{2i} = \frac{(2i)!}{i!2^i} = (2i-1)(2i-3)\ldots(3)(1)$$

as shown previously in Section 6.6.

Example 8.3.2. Derive the m.g.f. of the normal distribution $N(\mu,\sigma^2)$.

Solution. If $X \sim N(\mu,\sigma^2)$, we may write $X \equiv \mu + \sigma Z$ where Z is the standard form of X and has a standardized normal distribution. By (8.3.1), the m.g.f. of X is

$$M_X(u) = E(e^{uX}) = E(e^{u\mu} \cdot e^{u\sigma Z}).$$

Since $e^{u\mu}$ is a constant, we have

$$M_X(u) = e^{u\mu}E(e^{u\sigma Z}) = e^{u\mu}M_Z(u\sigma).$$

Now replacing u by $u\sigma$ in (8.3.6) gives

$$M_X(u) = e^{u\mu}e^{(u\sigma)^2/2} = \exp\{u\mu + \frac{1}{2}u^2\sigma^2\},$$

which is the result given in Table 8.3.1. □

The cumulant generating function is, by definition, the natural logarithm of the moment generating function:

$$K_X(u) = \log M_X(u). \tag{8.3.7}$$

The ith derivative of $K_X(u)$ at $u=0$ is called the ith cumulant of X, and will be denoted by κ_i.

$$\kappa_i = K_X^{(i)}(0) = \frac{d^i}{du^i} K_X(u)\bigg|_{u=0}.$$

Since $M_X(0)=1$, we have $\kappa_0=0$. The Taylor's series expansion of $K_X(u)$ about $u=0$ is

$$K_X(u) = \kappa_1 u + \kappa_2 \frac{u^2}{2!} + \ldots + \kappa_i \frac{u^i}{i!} + \ldots \quad . \tag{8.3.8}$$

Thus $K_X(u)$ is the exponential generating function of the sequence of cumulants $\{\kappa_i\}$.

Repeated differentiation of (8.3.7) with respect to u gives

$$\frac{dK}{du} = \frac{1}{M}\frac{dM}{du}; \qquad \frac{d^2K}{du^2} = -\frac{1}{M^2}\left(\frac{dM}{du}\right)^2 + \frac{1}{M}\frac{d^2M}{du^2}$$

$$\frac{d^3K}{du^3} = \frac{2}{M^3}\left(\frac{dM}{du}\right)^3 - \frac{3}{M^2}\frac{dM}{du}\frac{d^2M}{du^2} + \frac{1}{M}\frac{d^3M}{du^3}.$$

where $K = K_X(u)$ and $M = M_X(u)$. Setting $u=0$ gives

$$\kappa_1 = m_1 = \mu;$$

$$\kappa_2 = -m_1^2 + m_2 = E(X^2) - \mu^2 = \sigma^2$$

$$\kappa_3 = 2m_1^3 - 3m_1m_2 + m_3.$$

In general, the ith cumulant κ_i can be expressed as a function of the first i moments. Conversely, the ith moment m_i can be expressed as a function of the first i cumulants.

If X and Y are independent variates, we may take logarithms in (8.3.4) to obtain

$$K_{X+Y}(u) = K_X(u) + K_Y(u). \qquad (8.3.9)$$

The coefficients of u^i must be the same on both sides of this equation. Hence the ith cumulant of the sum X + Y is equal to the sum of the ith cumulants. This additivity is the main advantage which cumulants have over moments. In general, only first moments are additive.

Example 8.3.3. Find the cumulants of the Poisson distribution.

Solution. From (8.3.3) and Table 8.2.1, the moment generating function of the Poisson distribution with mean μ is

$$M_X(u) = G_X(e^u) = e^{\mu(e^u-1)}.$$

Hence, by (8.3.7), the cumulant generating function is

$$\begin{aligned} K_X(u) &= \log M_X(u) = \mu(e^u - 1) \\ &= \mu\left(u + \frac{u^2}{2!} + \ldots + \frac{u^i}{i!} + \ldots\right). \end{aligned}$$

Upon comparing this series with (8.3.8), we see that all of the cumulants are equal to μ.

Example 8.3.4. Find the cumulants of the normal distribution $N(\mu,\sigma^2)$.

Solution. If $X \sim N(\mu,\sigma^2)$, the cumulant generating function of X is

$$K_X(u) = \log M_X(u) = u\mu + \frac{u^2\sigma^2}{2}$$

from Table 8.3.1. The first two cumulants are $\kappa_1 = \mu$ and $\kappa_2 = \sigma^2$; all remaining cumulants are zero.

Laplace and Fourier Transforms

The Laplace transform of a function f is the function L defined by

$$L(u) = \int_0^\infty e^{-ux} f(x)dx$$

where u is a complex variable. Upon comparing this with (8.3.1), we see that the moment generating function of a non-negative variate is essentially the Laplace transform of its probability density function. The inversion formula for Laplace transforms and tables of Laplace transforms can sometimes be used to determine the p.d.f. which corresponds to a given moment generating function.

The moment and cumulant generating functions do not exist unless all of the moments of the distribution are finite. Hence there are many distributions that do not have moment generating functions - for instance, Student's distribution, the F distribution, and the distributions in Example 5.2.2. For this reason, the characteristic function is frequently used in place of the m.g.f.

The <u>characteristic function</u> of X is the function defined by

$$C_X(u) = E(e^{iuX}) = \begin{cases} \sum e^{iux} f(x) & \text{for } X \text{ discrete} \\ \int_{-\infty}^{\infty} e^{iux} f(x)dx & \text{for } X \text{ continuous} \end{cases}$$

where $i = \sqrt{-1}$. This is the Fourier transform of f. Since

$$e^{iux} = \text{cox } ux + i \sin ux$$

where $|\cos ux| \le 1$ and $|\sin ux| \le 1$, the characteristic function always exists. It uniquely determines the distribution, and its properties are similar to those of the moment generating function.

<u>Problems for Section 8.3</u>

†1. Let X be a discrete variate which takes the values 0,1, and 2 with probabilities 0.3, 0.1, and 0.6, respectively. Find the moment generating function of X. From it obtain the moment generating function of Y, where $Y \equiv 2X + 3$.

2. Let X be a continuous variate with p.d.f.

$$f(x) = 2x \quad \text{for} \quad 0 < x < 1.$$

Find the moment generating function of X, and use it to evaluate the mean and variance.

3. Derive the moment generating function for the continuous uniform,

exponential, gamma, and χ^2 distributions (Table 8.3.1).

4. Show by integration that, if $X \sim N(0,1)$, the m.g.f. of X^2 is $(1-2u)^{-1}$.

5. The coefficients of skewness and kurtosis were defined in Section 5.2. Show that

$$\gamma_1 = \kappa_3/\sigma^3; \qquad \gamma_2 = \kappa_4/\sigma^4.$$

†*6. Show that the cumulants of the binomial distribution with parameters (n,p) satisfy the following recursive formula:

$$\kappa_{r+1} = p(1-p)\frac{\partial}{\partial p}\kappa_r.$$

8.4 Applications

In this section we give some applications of moment and cumulant generating functions. We shall make use of the following two properties of moment and cumulant generating functions, which we state without proof.

I. Uniqueness. When (8.3.1) converges in an interval containing the origin, the moment (cumulant) generating function uniquely determines the probability function or p.d.f. of X. If two variates are shown to have the same moment (cumulant) g.f., they must have the same distribution.

II. Limiting distributions. If the moment (cumulant) g.f. of a variate X_n approaches the moment (cumulant) g.f. of a variate X as $n \to \infty$, then the probability function or p.d.f. of X_n approaches that of X as $n \to \infty$. One can investigate the limiting distribution of a sequence of variates $\{X_n\}$ by considering the limit approached by their m.g.f.'s.

Example 8.4.1. Let $X_1, X_2, \ldots, X_n$ be independent variates, each having an exponential distribution with mean θ, and let $T \equiv X_1 + X_2 + \ldots + X_n$. Show that $2T/\theta$ has a χ^2 distribution with $2n$ degrees of freedom.

Solution. From Table 8.3.1, the m.g.f. of X_i is

$$M(u) = (1-\theta u)^{-1}.$$

Hence, by (8.3.5), the m.g.f. of T is

$$M_T(u) = [M(u)]^n = (1 - \theta u)^{-n}.$$

Now define $Y \equiv 2T/\theta$. By (8.3.1), the m.g.f. of Y is

$$M_Y(u) = E(e^{uY}) = E(e^{2uT/\theta}) = M_T(\frac{2u}{\theta})$$
$$= (1 - 2u)^{-n}.$$

But, from Table 8.3.1, this is the m.g.f. of a χ^2 distribution with $\nu = 2n$ degrees of freedom. Hence by the uniqueness property, Y is distributed as $\chi^2_{(2n)}$.

Example 8.4.2. Show that a sum of two independent χ^2 variates has a χ^2 distribution.

Solution. Let X and Y be independent variates, with $X \sim \chi^2_{(n)}$ and $Y \sim \chi^2_{(m)}$. Their m.g.f.'s are

$$M_X(u) = (1 - 2u)^{-n/2}; \quad M_Y(u) = (1 - 2u)^{-m/2}$$

from Table 8.3.1. Hence, by (8.3.4), the m.g.f. of their sum is

$$M_{X+Y}(u) = M_X(u)M_Y(u) = (1 - 2u)^{-(n+m)/2}.$$

This is the m.g.f. of a χ^2 distribution with $n + m$ degrees of freedom. Hence, by the uniqueness property, $X + Y \sim \chi^2_{(n+m)}$. See Example 7.2.2 for a direct proof of this result.

Example 8.4.3. Show that a linear combination of independent normal variates has a normal distribution.

Solution. Let $X \sim N(\mu_1, \sigma_1^2)$ and $Y \sim N(\mu_2, \sigma_2^2)$, independently of X. We shall show that the m.g.f. of the linear combination $aX + bY$ is of the normal type. First, since aX and bY are independent, (8.3.4) gives

$$M_{aX+bY}(u) = M_{aX}(u)M_{bY}(u).$$

Now by (8.3.1) and Table 8.3.1, we have

$$M_{aX}(u) = E(e^{uaX}) = M_X(au)$$
$$= \exp[au\mu_1 + \frac{1}{2}(au)^2\sigma_1^2]$$

and similarly

$$M_{aY}(u) = \exp[bu\mu_2 + \frac{1}{2}(bu)^2\sigma_2^{\,2}].$$

It follows that

$$M_{aX+bY}(u) = \exp[u(a\mu_1 + b\mu_2) + \frac{1}{2}u^2(a^2\sigma_1^{\,2} + b^2\sigma_2^{\,2})].$$

This is the m.g.f. of a normal distribution with mean $\mu = a\mu_1 + b\mu_2$ and variance $\sigma^2 = a^2\sigma_1^{\,2} + b^2\sigma_2^{\,2}$. The desired result now follows from the uniqueness property of m.g.f.'s. For a direct proof, see Section 7.3.

Example 8.4.4. Use generating functions to establish the Poisson approximation to the binomial distribution.

Solution. Suppose that X has a binomial distribution with parameters n and p. By (8.3.3) and Table 8.2.1, the m.g.f. of X is

$$M_X(u) = G_X(e^u) = (pe^u + 1 - p)^n.$$

Upon substituting $p = \mu/n$, we obtain

$$M_X(u) = [1 + \frac{\mu}{n}(e^u - 1)]^n.$$

Now we keep μ and u fixed, and let $n \to \infty$. Since

$$\lim_{n\to\infty} (1 + \frac{a}{n})^n = e^a,$$

we have

$$\lim_{n\to\infty} M_X(u) = e^{\mu(e^u - 1)}.$$

But, by (8.3.3) and Table 8.2.1, the m.g.f. of a Poisson distribution with mean μ is

$$M_X(u) = G_X(e^u) = e^{\mu(e^u - 1)}.$$

Hence, by the second property of m.g.f.'s, the probability function of the binomial distribution approaches that of the Poisson distribution as $n \to \infty$. See Section 4.3 for a direct proof of this result.

Central Limit Theorem

As a final application, we shall prove the Central Limit Theorem (Section 6.7) in a special case.

Suppose that $X_1, X_2, \ldots, X_n$ are independent and identically distributed variates whose moment and cumulant generating functions exist. Define $S_n \equiv X_1 + X_2 + \ldots + X_n$, and let S_n^* be the standard form of S_n:

$$S_n^* \equiv \frac{S_n - n\mu}{\sqrt{n}\,\sigma}.$$

We shall show that, whatever the distribution of the X_i's, the cumulant g.f. of S_n^* approaches that of the standardized normal distribution as $n \to \infty$.

By (8.3.1), the moment generating function of S_n^* is

$$M_{S_n^*}(u) = E(e^{u(S_n - n\mu)/\sqrt{n}\sigma}) = E(e^{uS_n/\sqrt{n}\sigma} e^{-u\mu\sqrt{n}/\sigma}).$$

Since $e^{-u\mu\sqrt{n}/\sigma}$ is constant, we have

$$M_{S_n^*}(u) = e^{-u\mu\sqrt{n}/\sigma} E(e^{uS_n/\sqrt{n}\sigma}) = e^{-u\mu\sqrt{n}/\sigma} M_{S_n}(\frac{u}{\sqrt{n}\sigma}).$$

Now we take logarithms to obtain

$$K_{S_n^*}(u) = -\frac{u\mu\sqrt{n}}{\sigma} + K_{S_n}(\frac{u}{\sqrt{n}\sigma}).$$

But $X_1, X_2, \ldots, X_n$ are independent and have the same moment g.f. $M(u)$. Hence, by (8.3.5),

$$M_{S_n}(u) = [M(u)]^n$$

and taking logarithms gives

$$K_{S_n}(u) = nK(u)$$

where $K(u)$ is the cumulant g.f. of X_i. It follows that

$$K_{S_n^*}(u) = -\frac{u\mu\sqrt{n}}{\sigma} + nK(\frac{u}{\sqrt{n}\sigma}). \tag{8.4.1}$$

Now by (8.3.8) we have

$$K(u) = \kappa_1 u + \kappa_2 \frac{u^2}{2!} + \kappa_2 \frac{u^3}{3!} + \ldots$$

$$= u\mu + \frac{u^2\sigma^2}{2} + \sum_{j=3}^{\infty} \kappa_j \frac{u^j}{j!},$$

where $\kappa_1, \kappa_2, \kappa_3, \ldots$ are the cumulants of X_i. Hence (8.4.1) gives

$$K_{S_n^*}(u) = -\frac{u\mu\sqrt{n}}{\sigma} + n\left(\frac{u}{\sqrt{n}\sigma}\right)\mu + n\left(\frac{u}{\sqrt{n}\sigma}\right)^2 \frac{\sigma^2}{2} + n\sum_{j=3}^{\infty} \kappa_j \left(\frac{u}{\sqrt{n}\sigma}\right)^j \frac{1}{j!}$$

$$= \frac{u^2}{2} + \frac{1}{\sqrt{n}\sigma^3}\left[\sum_{j=3}^{\infty} \kappa_j \frac{u^j}{j!} (\sqrt{n}\sigma)^{3-j}\right]. \qquad (8.4.2)$$

For n sufficiently large, the terms of the series in square brackets are less in absolute value than the terms of the series

$$\sum_{j=3}^{\infty} \kappa_j \frac{u^j}{j!} .$$

The latter series is convergent by the assumption that $K(u)$ exists, and hence the series in (8.4.2) is convergent. It follows that

$$\lim_{n\to\infty} K_{S_n^*}(u) = \frac{u^2}{2} .$$

But, from Example 8.3.4, this is the cumulant generating function of $N(0,1)$. Hence, by the second property of moment and cumulant g.f.'s, the distribution of S_n^* approaches $N(0,1)$ as $n \to \infty$. This proves the Central Limit Theorem in the special case when the X_i's are identically distributed and their m.g.f. exists.

The coefficient of u^3 in (8.3.2) is

$$\frac{\kappa_3}{3!\sigma^3\sqrt{n}} = \frac{\gamma_1}{6\sqrt{n}}$$

where γ_1 is the coefficient of skewness of X_i (see Problem 8.3.5). If γ_1 is large in magnitude, then the distribution of X_i is heavily skewed, and normality is approached very slowly. However, if γ_1 is near zero, the order of the first neglected term in (8.4.2) is $\frac{1}{n}$ rather than $\frac{1}{\sqrt{n}}$, and the approach to normality is much more rapid. In particular, if the distribution of X_i is symmetrical about its mean, then $\gamma_1 = 0$ and the distribution of S_n will be close to normal for fairly small values of n. For instance, in Example 6.7.1 we found that the distribution of a sum of independent uniform variates was nearly normal for n as small as 3.

Problems for Section 8.4

†1. Let X have a χ^2 distribution with n degrees of freedom. Find the cumulant generating function of the standard form of X, and find its limit as $n \to \infty$. What can be concluded about the distribution of X when n is large?

8.5 Bivariate Generating Functions

Suppose that X and Y are discrete variates which take only non-negative integer values, and let f be their joint probability function. Their joint probability generating function is a function of two variables,

$$G_{X,Y}(u,v) = E(u^X v^Y) = \sum\sum u^x v^y f(x,y). \qquad (8.5.1)$$

The probabilities $f(x,y)$ may be generated by expanding G as a power series in u and v, and examining the coefficients.

The marginal probability generating function of one variable X is obtained by setting the other generating variable v equal to 1; for

$$G_{X,Y}(u,1) = E(u^X) = G_X(u) \qquad (8.5.2)$$

by (8.2.2). More generally, if a and b are any constants, then

$$G_{aX+bY}(u) = E(u^{aX+bY}) = E(u^{aX}u^{bY}) = G_{X,Y}(u^a,u^b). \qquad (8.5.3)$$

If X and Y are independent, then $f(x,y) = f_1(x)f_2(y)$ for all x and y, so that

$$G_{X,Y}(u,v) = \sum\sum u^x v^y f_1(x)f_2(y) = \sum u^x f_1(x)\sum v^y f_2(y).$$

Hence, by (8.2.1),

$$G_{X,Y}(u,v) = G_X(u)G_Y(v). \qquad (8.5.4)$$

Conversely, suppose that the joint p.g.f. factors:

$$G_{X,Y}(u,v) = H_1(u)H_2(v).$$

Then we may expand H_1 and H_2 as power series and equate coefficients of $u^x v^y$ in (8.5.1) to show that the joint p.f. factors,

$$f(x,y) = h_1(x)h_2(y),$$

for all x,y. Hence X and Y are independent. The factorization of the joint p.g.f. is therefore a necessary and sufficient condition for X and Y to be independent.

As in the univariate case, one can obtain factorial moments from the probability generating function. Differentiating (8.5.1) with respect to u and v gives

$$G_{X,Y}^{(i,j)}(u,v) = \frac{\partial^{i+j}}{\partial u^i \partial v^j} G_{X,Y}(u,v) = E\{X^{(i)}u^{X-i}Y^{(j)}v^{Y-j}\}$$

and putting $u = v = 1$ gives

$$G_{X,Y}^{(i,j)}(1,1) = E\{X^{(i)}Y^{(j)}\}. \qquad (8.5.5)$$

If we expand $G_{X,Y}(1+u,1+v)$ as a power series about $u = v = 0$, the coefficient of $u^i v^j$ will be $\frac{1}{i!j!} E\{X^{(i)}Y^{(j)}\}$.

Example 8.5.1. Multinomial Distribution.

Suppose that $X_1, X_2, \ldots, X_k$ have a multinomial distribution with parameters n and $p_1, p_2, \ldots, p_k$, where $\sum p_i = 1$ and $\sum X_i \equiv n$. The joint probability function of $X_1, X_2, \ldots, X_k$ is given by (4.6.2). Their joint p.g.f. is a function of k variables,

$$G(u_1, u_2, \ldots, u_k) = E(u_1^{X_1} u_2^{X_2} \ldots u_k^{X_k})$$

$$= \sum\sum\cdots\sum \binom{n}{x_1 x_2 \ldots x_k} (p_1u_1)^{x_1}(p_2u_2)^{x_2}\ldots(p_ku_k)^{x_k}$$

$$= (p_1u_1 + p_2u_2 + \ldots + p_ku_k)^n$$

by the Multinomial Theorem (2.1.12).

The marginal p.g.f. of X_1 is obtained by setting $u_2 = u_3 = \ldots = u_k = 1$:

$$G_{X_1}(u) = G(u,1,1,\ldots,1) = (p_1u + p_2 + p_3 + \ldots + p_k)^n.$$

Since $\sum p_i = 1$, we have

$$G_{X_1}(u) = (p_1u + 1 - p_1)^n.$$

Upon comparing this with Table 8.2.1, we conclude that X_1 has a binomial distribution with parameters n and p_1.

For $r < k$, the p.g.f. of $Y \equiv X_1 + X_2 + \ldots + X_r$ is

$$G_Y(u) = E(u^{X_1+X_2+\ldots+X_r}) = G(u,u,\ldots,u,1,\ldots,1)$$

$$= (p_1u + p_2u + \ldots + p_ru + p_{r+1} + \ldots + p_k)^n$$

$$= (pu + 1 - p)^n$$

where $p = p_1 + p_2 + \ldots + p_r$. Hence Y has a binomial distribution with parameters n and p. Furthermore, the joint p.g.f. of $Y, X_{r+1}, \ldots, X_k$ is

$$E(u^{X_1+X_2+\ldots+X_r} u_{r+1}^{X_{r+1}} \ldots u_k^{X_k}) = G(u,u,\ldots,u,u_{r+1},\ldots,u_k)$$

$$= (pu + p_{r+1}u_{r+1} + \ldots + p_ku_k)^n.$$

Thus $Y, X_{r+1}, \ldots, X_k$ jointly have a multinomial distribution with parameters n and $p, p_{r+1}, \ldots, p_k$. In the notation of Section 4.6, we are merely combining the first r classes into a single class $A = A_1 \cup A_2 \cup \ldots \cup A_r$ with probability $p = p_1 + p_2 + \ldots + p_r$, and Y represents the total frequency for the new class.

The first and second partial derivatives of G are

$$\frac{\partial G}{\partial u_i} = np_i(p_1u_1 + p_2u_2 + \ldots + p_ku_k)^{n-1};$$

$$\frac{\partial^2 G}{\partial u_i^2} = n(n-1)p_i^2(p_1u_1 + p_2u_2 + \ldots + p_ku_k)^{n-2};$$

$$\frac{\partial^2 G}{\partial u_i \partial u_j} = n(n-1)p_ip_j(p_1u_1 + p_2u_2 + \ldots + p_ku_k)^{n-2}.$$

Upon setting $u_1 = u_2 = \ldots = u_k = 1$ and using (8.5.5), we obtain

$$E(X_i) = np_i; \quad E\{X_i(X_i - 1)\} = n(n-1)p_i^2;$$

$$E(X_iX_j) = n(n-1)p_ip_j.$$

Now (5.2.3) and (5.4.3) give

$$\text{var}(X_i) = np_i(1 - p_i); \quad \text{cov}(X_i, X_j) = -np_ip_j.$$

Moment Generating Functions

The joint moment generating function of variates X and Y is defined by

$$M_{X,Y}(u,v) = E\{e^{uX+vY}\}. \tag{8.5.6}$$

If X and Y take only non-negative integer values, we have

$$M_{X,Y}(u,v) = G_{X,Y}(e^u, e^v),$$

and from (8.5.2) and (8.5.3) we obtain

$$M_X(u) = M_{X,Y}(u,0); \tag{8.5.7}$$

$$M_{aX+bY}(u) = M_{X,Y}(au,bu). \tag{8.5.8}$$

Furthermore, the factorization of the joint m.g.f. into a function of u times a function of v is a necessary and sufficient condition for X and Y to be independent variates. These results also hold for more general discrete and continuous variates.

As in the univariate case, we can obtain moments by differentiating the m.g.f. or expanding it in a power series about $u=v=0$. Differentiating (8.5.6) with respect to u and v gives

$$M_{X,Y}^{(i,j)}(u,v) = \frac{\partial^{i+j}}{\partial u^i \partial v^j} M_{X,Y}(u,v) = E(X^i Y^j e^{uX+vY}),$$

and putting $u=v=0$ gives

$$M_{X,Y}^{(i,j)}(0,0) = E(X^i Y^j) = m_{ij}.$$

The number m_{ij} is called a moment of order $i+j$, and will be the coefficient of $u^i v^j/i!j!$ in the series expansion of the joint m.g.f. about $u=v=0$.

There are two moments of order one:

$$m_{10} = E(X); \quad m_{01} = E(Y).$$

There are three moments of order two:

$$m_{20} = E(X^2); \quad m_{11} = E(XY); \quad m_{02} = E(Y^2).$$

In general, in the bivariate case, there are $r+1$ moments of order r.

Cumulant generating functions and characteristic functions

can also be defined for bivariate and multivariate distributions, but we shall not give details here.

Example 8.5.2. The Bivariate Normal Distribution

Let X and Y be continuous variates with means μ_1 and μ_2, non-zero variances σ_1^2 and σ_2^2, and correlation coefficient ρ. By the definition in Section 7.4, X and Y have a bivariate normal distribution if and only if every linear combination $aX+bY$ has a normal distribution. We shall deduce the joint m.g.f. of X and Y from this definition, and hence show that X and Y are independent if and only if $\rho = 0$.

Suppose that X and Y have a bivariate normal distribution. Then, for all constants a and b, $aX + bY$ has a normal distribution. From Table 8.3.1, the m.g.f. of $aX + bY$ is

$$M_{aX+bY}(u) = \exp(u\mu + \tfrac{1}{2}u^2\sigma^2)$$

where μ and σ^2 are the mean and variance of $aX + bY$:

$$\mu = a\mu_1 + b\mu_2; \quad \sigma^2 = a^2\sigma_1^2 + b^2\sigma_2^2 + 2ab\rho\sigma_1\sigma_2.$$

Now (8.5.8) gives

$$M_{X,Y}(au,bu) = M_{aX+bY}(u) = \exp(u\mu + \tfrac{1}{2}u^2\sigma^2)$$

$$= \exp\{u(a\mu_1 + b\mu_2) + \tfrac{1}{2}u^2(a^2\sigma_1^2 + b^2\sigma_2^2 + 2ab\rho\sigma_1\sigma_2)\}$$

$$= \exp\{(au)\mu_1 + (bu)\mu_2 + \tfrac{1}{2}(au)^2\sigma_1^2 + \tfrac{1}{2}(bu)^2\sigma_2^2 + (au)(bu)\rho\sigma_1\sigma_2\}.$$

Since this holds for all constants a and b, it follows that

$$M_{X,Y}(u_1,u_2) = \exp\{u_1\mu_1 + u_2\mu_2 + \tfrac{1}{2}u_1^2\sigma_1^2 + \tfrac{1}{2}u_2^2\sigma_2^2 + u_1u_2\rho\sigma_1\sigma_2\} \tag{8.5.9}$$

for all u_1 and u_2. Thus the definition uniquely determines the joint m.g.f. of X and Y.

Conversely, if we begin with the joint m.g.f. (8.5.9), we can use (8.5.8) to show that every linear combination of X and Y is normally distributed. Therefore the m.g.f. (8.5.9) uniquely determines the bivariate normal distribution.

The factorization of the joint m.g.f. is a necessary and sufficient condition for X and Y to be independent variates. Now

(8.5.9) factors into a function of u_1 times a function of u_2 if and only if the cross-product term $u_1u_2\rho\sigma_1\sigma_2$ is zero for all u_1 and u_2. Since $\sigma_1 > 0$ and $\sigma_2 > 0$, the joint m.g.f. factors if and only if $\rho = 0$. Hence, if X and Y have a bivariate normal distribution, a necessary and sufficient condition for their independence is $\rho = 0$.

Problems for Section 8.5

†1. The joint probability generating function of X and Y is

$$G(u,v) = \exp\{a(u-1) + b(v-1) + c(u-1)(v-1)\},$$

where $a > 0$, $b > 0$, and c are constants.

(a) Determine the marginal distributions of X and Y.

(b) Find the correlation coefficient of X and Y, and show that $c^2 \le ab$.

(c) Find the joint probability function of X and Y in the special case $a = b = c = 1$.

†2. A game consists in moving a peg n times. Each move is one unit to the north, south, east, or west, the direction being selected at random on each move. Find the expected value of D^2, where D is the distance of the peg from its starting point after n moves.

APPENDIX A. ANSWERS TO SELECTED PROBLEMS

1.3.2 $8/27,\ 1/27,\ 6/27;\quad (n-1)^3/n^3,\ (n-2)^3/n^3,\ n^{(3)}/n^3;\quad (n-1)^r/n^r,\ (n-2)^r/n^r,\ n^{(r)}/n^r.$

1.3.6 Possible outcomes: 2 sequences of four games, 8 sequences of five games, 20 sequences of six games, 40 sequences of seven games. Sequence of n games has probability 2^{-n}.
P(seven games needed) $= 40 \times 2^{-7}$.

2.1.1 0, 1, 120, 0.384, 120, 5040, 35, -84, 0

2.1.2 120, 34650, 2522520, 2.4609375

2.1.8 When like powers are collected, there are 4 terms of the form a^5, 12 terms a^4b, 12 terms a^3b^2, 12 terms a^3bc, 12 terms a^2b^2c, and 4 terms a^2bcd, for a total of 56. Coefficient $\binom{5}{1\,0\,2\,2} = 30$.

2.2.1 3/7, 5/21, 1/42, 1/7

2.2.3 $\binom{10}{2}/\binom{100}{2};\quad 1-\binom{90}{2}/\binom{100}{2}$

2.2.7 $1-p$ where $p = P(\text{no pair}) = \binom{n}{2r}2^{2r}/\binom{2n}{2r}$.

2.2.9 $\binom{5}{2}5^{(2)}5^{(3)}/10^{(5)};\quad \binom{4}{2}5^{(2)}5^{(3)}/10^{(5)};\quad 1/10;\quad [\binom{5}{3}-\binom{4}{3}]/\binom{10}{3}.$

2.2.13 8/47; 4/47

2.3.3 $\binom{20-d}{5}/\binom{20}{5} + \binom{d}{1}\binom{20-d}{4}\binom{16-d}{5}/\binom{20}{5}\binom{15}{5}$

2.3.6 0.0020, 0.0470, 0.0564, 0.0051

2.3.8 $[\binom{2}{2}\binom{3}{1}\binom{4}{0} + \binom{2}{1}\binom{3}{2}\binom{4}{0} + \binom{2}{1}\binom{3}{1}\binom{4}{1}]/\binom{9}{3} = 0.3929$

2.3.10 $\binom{13}{6\,4\,2\,1}\binom{39}{7\,9\,11\,12}/\binom{52}{13\,13\,13\,13}$

2.4.2 0.010, 0.720, 0.810, 125/900, $(5\times5\times5 + 4\times5\times5)/900$

2.4.6 $3^{10}/4^{10}$; P(2 or more cars in every lot) = P(occupancy numbers 2,2,2,4 or 2,2,3,3 in any order) $= \frac{4!}{3!1!}\,\frac{10!}{2!2!2!4!}\,4^{-10} + \frac{4!}{2!2!}\,\frac{10!}{2!2!3!3!}\,4^{-10}$.

2.5.3 $f(x) = P(x \text{ under } 18) = \binom{300}{x}\binom{7700}{40-x}/\binom{8000}{40} \approx \binom{40}{x}\left(\frac{3}{80}\right)^x\left(\frac{77}{80}\right)^{40-x};$
$f(4) \approx 0.0457;\quad f(0)+f(1) \approx 0.5546$

3.1.3 $P(DS) = 0.3;\quad P(DS\bar{S}_W) = 0.1$

3.1.6 $[3\binom{4}{2}\binom{48}{4} - 3\binom{4}{2}^2\binom{44}{2} + \binom{4}{2}^3]/\binom{52}{6} = 0.1670$

3.2.3 $2^{-x};\quad 1-2^{-x}$

3.2.6 $1-p^4(1+4q+10q^2) - q^4(1+4p+10p^2);\quad p^4(1+4q+10q^2) + 20p^5q^3/(1-2pq)$

3.2.9 $(1-p_1)/(1-p_1p_2)$

3.3.3 0, 7/8, 7/8; 1/8, 43/64, 31/64

3.3.5 $\frac{4}{9} : \frac{5}{9}$ or 4:5

3.4.1 5/13

3.4.4 $\binom{9}{2}/\binom{18}{4}$; $\binom{10}{3}/[\binom{20}{6} - 2^6\binom{10}{6}]$

3.4.7 $.4\times.7\times.9/(.4\times.7\times.9 + .6\times.3\times.9 + .6\times.7\times.1) = 0.553$; 0.117

3.4.9 $\binom{i}{x}p^x(1-p)^{i-x}$; n_i/N; $\sum\binom{i}{x}p^x(1-p)^{i-x}\cdot n_i/N$; $(y+1)n_{y+1}/\sum in_i$

3.5.1 98/1097

3.5.4 17/18

3.5.7 $\frac{9}{10}\cdot\frac{7}{8}\cdot\frac{5}{25}/(\frac{9}{10}\cdot\frac{7}{8}\cdot\frac{5}{25} + \frac{1}{10}\cdot\frac{1}{8}\cdot\frac{20}{25}) = \frac{63}{67}$

3.5.11 $1 - (p_1+p_2)/2$; $[(1-p_1)^n + (1-p_2)^n]/2$; $(1-p_1)^n/[(1-p_1)^n + (1-p_2)^n]$

3.6.1 $[16\binom{39}{13\ 13\ 13} - 72\binom{26}{13} + 96 - 24]/\binom{52}{13\ 13\ 13\ 13} = 0.252\times10^{-10}$

3.6.4 $S_1 - S_2 + S_3 -+\ldots\pm S_{N-1}$ where $S_r = \binom{N-r}{r}\frac{(N-r)!}{N!}$

3.6.7 $\sum_{i=0}^{10}(-1)^i\binom{10}{i}(1-\frac{i}{10})^{20} = 0.2147$

3.6.10 251/1024

3R1 0.262, 0.078

3R3 $(1-.069)^{10} = 0.489$; 6.314%; 0.953

3R5 $\binom{4}{2}p^2(1-p)^2$; $\binom{3}{1}p(1-p)^2$; $\binom{2}{1}p(1-p)$; $\binom{4}{2}p^2(1-p)^2/[1-p^4-(1-p)^4]$

3R7 P(3 in 1st week, 0 in 2nd week)+P(2,1,0)+P(1,2,0)+P(1,1,1,0) = 0.0751

3R9 n-1 players must be eliminated, one per game; $(n-1)/\binom{n}{2}$

3R11 1-a, (1-a-b)/(1-a), (1-a-b-c)/(1-a-b); b(1-a-b-c)/(1-a-b)

4.1.1 $f(x) = \frac{9}{24}, \frac{8}{24}, \frac{6}{24}, 0, \frac{1}{24}$ for $x = 0,1,2,3,4$.

4.1.4 $f(x) = F(x)-F(x-1) = 2^{-x}$ for $x = 1,2,\ldots$; $f(5) = \frac{1}{32}$; $1-F(4) = \frac{1}{16}$

4.1.7 $f(d^2) = \frac{20}{64}, \frac{30}{64}, \frac{12}{64}, \frac{2}{64}$ for $d^2 = 0,4,16,36$.

4.1.10 $k = 2/n(n+1)$; $F(x) = x(x+1)/n(n+1)$ for $x = 1,2,\ldots,n$

4.2.2 $f(x) = \frac{1}{b+g-x+1}\binom{b}{b-1}\binom{g}{x-b}/\binom{b+g}{x-1}$ for $x = b,b+1,\ldots,b+g$.

4.2.6 $[\binom{5}{19-x}4^{19-x}/\binom{20}{19-x}]\cdot\frac{3(19-x)}{x+1}$ for $x = 14,15,\ldots,18$

4.2.7 $P(X > r) = n^{(r)}/n^r$; $f(x) = P(X>r-1) - P(X > r) = (r-1)n^{(r-1)}/n^r$ for $r = 2,3,\ldots,n+1$.

4.2.9 $f(x) = \sum_{i=1}^{n}(-1)^{i-1}\binom{n-1}{i-1}(1-\frac{i}{n})^{x-1}$ for $x = n,n+1,\ldots$.

4.3.2 $1 - (.99)^n - .01n(.99)^{n-1} \ge 0.95$ for $n \ge 473$; 0.8024

4.4.1 0.4408, 19.81

4.4.4 $1 - e^{-\mu} - \mu e^{-\mu} \ge 0.95$ for $\mu \ge 4.744$

4.5.1 6/37; $f(2,2) + f(3,2) + f(3,3) = 15/37$

4.5.4 0.45, 0.20, 2/3; $f(0,3)f(1.3) + f(1,3)f(0,3) = 0.005$

4.5.7 $f(x,y)=\binom{5}{x}(.1)^x(.9)^{5-x}\binom{5}{y}(.3)^y(.7)^{5-y}$;
$P(X>Y)=f(1,0)+f(2,0)+\ldots+f(5,4) = 0.1009$

4.5.10 binomial (m,p) distribution

4.5.13 $(x/N)^n$, $(N-y)^n/N^n$, $(x-y)^n/N^n$; $N^nf_1(x) = x^n - (x-1)^n$;
$N^nf_2(y)=(N+1-y)^n-(N-y)^n$; $N^nf(x,y)=(x-y+1)^n-2(x-y)^n$ for $y = x$;
$N^nf(x,y)=(x-y+1)^n-2(x-y)^n+(x-y-1)^n$ for $y < x$.

4.6.1 $\binom{10}{4\ 4\ 2}(.5)^4(.3)^4(.2)^2 = 0.0638$

4.6.3 $(1-\theta^2)^8$; $\binom{8}{2\,4\,2}[\theta^2]^2[2\theta(1-\theta)]^4[(1-\theta)^2]^2 = 6720\theta^8(1-\theta)^8$. Set derivative w.r.t. θ equal to zero, giving $\theta = 0.5$.

4R1 $1 - 1.25e^{-.25}$; $e^{-.25}$

4R3 $\binom{10}{3}(.33)^3(.67)^7$; $\binom{9}{6}(.33)^3(.67)^7$; $e^{-\mu} = 0.67$ so $\mu = 0.4005$ and $1 - e^{-\mu} - \mu e^{-\mu} = 0.0617$.

4R5 $1 - 1.45e^{-.45}$; $1.9e^{-.9}$; $\frac{9}{19}$; $\binom{6}{1\,2\,3}p_0 p_1^2 p_2^3$ where $p_0 = e^{-.9}$, $p_1 = .9e^{-.9}$, $p_2 = 1 - p_0 - p_1$; $(\log 2)/\lambda = 462.1$ feet.

4R8 0.3174, 0.1868, 0.1538

4R11 $\sum[\binom{n}{x}2^{-n}]^2 = 2^{-2n}\binom{2n}{n}$ by (2.1.10)

4R14 0.3431. Define A_i = event "no offspring with ith seed type" and use (3.6.1).

5.1.1 By charging \$1.00 per car. Second scheme pays \$0.867 and third pays \$0.9725 per car on average.

5.1.5 \$3.60, \$2.50

5.1.10 $f(x) = (1-\theta)\theta^r$ for $x = \lambda^r P$ $(r = 0,1,\ldots,n-2)$; $f(x) = \theta^{n-1}$ for $x = \lambda^{n-1}P$; $E(X) = \sum xf(x) = \frac{P(1-\theta)}{1-\lambda\theta} + P(\lambda\theta)^{n-1}\frac{\theta(1-\lambda)}{1-\lambda\theta} = E_n$, say. As n increases, E_n decreases to E_∞. In order that $E_n > kP$ for all n, we must have $E_\infty \geq kP$.

5.2.1 $\mu = 2.4$; $\sigma^2 = 1.68, 1.061$. More probability in tails for sampling with replacement; see Example 2.5.1.

5.2.3 $\mu = \sigma^2 = 2$

5.2.5 $\mu = \frac{1}{2}$ for all N; $\mathrm{var}(X) = (N^2 - 1)/12N^2 \to 1/12$ as $N \to \infty$.

5.3.2 Define $X \equiv$ # failures before 100th success. X has a negative binomial distn with $r = 100$, $p = P(\leq 2 \text{ errors}) = 5e^{-2}$. $E(X) = r(1-p)/p = 47.8$. Expected total number of pages is 147.8.

5.3.5 $n^2 - 4n(n-1)p(1-p)$

5.4.1 10, 22/3

5.4.4 16, 10.8; $7 \leq X \leq 25$

5.4.7 $\mu\sum c_i$, $\sum c_i^2\sigma_i^2$; $c_i = \sigma_i^{-2}/\sum\sigma_i^{-2}$; $c_i = 1/n$ and $Y \equiv \sum X_i/n \equiv \bar{X}$.

5.4.10 $(n-1)\alpha + n\mu + \beta\mu n(n-1)/2$; $n\sigma^2[1 + \beta(n-1) + \beta^2(n-1)(2n-1)/6]$.

5.5.2 $X \equiv \sum X_i$ where $X_i = 1$ if ith home visited and $X_i = 0$ otherwise. $E(X) = \sum p_i$, $\mathrm{var}(X) = \sum p_i(1-p_i) = \sum p_i - \sum p_i^2$. For $\sum p_i$ fixed, $\sum p_i^2$ is minimized when p_i's are all equal.

5.5.6 $X_{ijk} = 1$ if triangle (i,j,k) is complete, 0 otherwise. $E(X_{ijk}X_{rst}) = p^3$ (3 common vertices), p^5 (2 common vertices), or p^6. Mean $\binom{n}{3}p^3$; variance $\binom{n}{3}p^3[1+3(n-3)p^2-(3n-8)p^3]$.

5.6.1 np and $np(1-p)$ where $p = (p_1+p_2)/2$. All clouds might be seeded, or none might be seeded, making a comparison impossible.

5.6.3 $r\mu(1-p)/p$; $r\mu(\mu+p)(1-p)/p^2$.

5R1 $1 - (5/6)^4 = 0.518$; $1 - (35/36)^{24} = 0.491$

5R4 $p = 1\cdot[1\cdot p + \frac{1}{k}(1-p)] - m\cdot[0\cdot p + \frac{k-1}{k}(1-p)]$ gives $m = \frac{1}{k-1}$.

5R7 $f(x) = k\binom{n}{x}p^x(1-p)^{n-x}$ for $x = 1,2,\ldots,n$, where $k = 1/[1-(1-p)^n]$ $E(X) = npk$; $E\{X^{(2)}\} = n^{(2)}p^2k$; $\mathrm{var}(X) = npk[np(1-k) + 1 - p]$.

6.1.1 $k = 1$, $F(x) = 3x^2 - 2x^3$ for $0 < x < 1$; 0.5440; 0.0288, 0.028796; $m = \mu = 0.5$, $\sigma^2 = 0.05$

6.1.4 $f(x) = 2xe^{-x^2}$ for $0 < x < \infty$; $\mu = \Gamma(3/2) = \sqrt{\pi}/2$; $var(X) = 1 - \frac{\pi}{4}$.

6.1.7 Area $A \equiv \sin\frac{X}{2}\cos\frac{X}{2} \equiv \frac{1}{2}\sin X$; p.d.f. $\frac{2k}{\sqrt{1-4a^2}}[\text{Sin}^{-1}2a][\pi - \text{Sin}^{-1}2a]$

for $0 < a < \frac{1}{2}$; $E(A) = 12/\pi^3$.

6.1.10 $k = 1$; $f(x) = nx^{n-1}(1+x^n)^{-2}$ for $x > 0$; $Q_\alpha = (\frac{\alpha}{1-\alpha})^{1/n}$.

6.1.14 $k = 1/2e$; $F(x) = 1-(x^2+4x+5)e^{-x-1}/2$ for $x > -1$; $m = 1.674$

6.2.1 $P(X > 1) = e^{-.5} = P(X > 2 | X > 1)$

6.2.5 $f(x) = nx^{n-1}$ for $0 < x < 1$

6.4.1 $m_r = \theta^r\Gamma(1+\frac{r}{\beta})$; $\mu = \theta\Gamma(1+\frac{1}{\beta})$; $\sigma^2 = \theta^2\Gamma(1+\frac{2}{\beta}) - \mu^2$;

median $m = \theta(\log 2)^{1/\beta}$; $\mu > m$ for $\beta < 3.44$.

6.4.4 $xS(x) = x\int_x^\infty f(t)dt \le \int_x^\infty tf(t)dt \to 0$ as $x \to \infty$ because

$\mu = \int_0^\infty tf(t)dt < \infty$. $r(x) = \int_x^\infty (t-x)\frac{f(t)}{S(x)}\,dt$; integrate by parts.

For (b), note that $\frac{1}{r(x)} = S(x)/\int_x^\infty S(t)dt = -\frac{d}{dx}\log\int_x^\infty S(t)dt$.

6.6.1 $F(2.5) - F(-2.5) = 0.9876$ from Table B2.

6.6.4 $.05 = F(\frac{31.5-\mu}{\sigma})$, $.85 = F(\frac{32.3-\mu}{\sigma})$; hence $\frac{31.5-\mu}{\sigma} = -1.645$,

$\frac{32.3-\mu}{\sigma} = 1.036$ from Table B1. Solving gives $\sigma = 0.2984$, $\mu = 31.991$. $P(V > 32.2) = 0.2420 = p$, say; $\binom{10}{3}p^3(1-p)^7$.

6.6.7 $S_{10} \equiv X_1 + X_2 + \ldots + X_{10}$ where X_i's are independent $N(165,100)$. Hence $S_{10} \sim N(1650,1000)$; $P(S_{10} > 1750) = 1 - F(\frac{100}{\sqrt{1000}}) = 0.0008$.

6.6.10 $S_n \equiv X_1 + X_2 + \ldots + X_n \sim N(2n, \frac{n}{4})$; $P(S_n > 7) = 1 - F\{(7-2n)/\sqrt{\frac{n}{4}}\}$.

$g(x) = P(S_x > 7) - P(S_{x-1} > 7)$.

6.6.13 The hazard first increases, then decreases; point of maximum hazard occurs earlier for β small. Specimens first experience positive ageing (deterioration), then negative ageing (improvement).

6.7.1 0.412, 0.240, 0.0127

6.7.4 0.093

6.7.7 $X_i \equiv$ # slices on burger, $Y_i \equiv$ # slices from pickle;

$X \equiv X_1 + X_2 + \ldots + X_{500} \approx N(650,355)$; $Y \equiv Y_1 + Y_2 + \ldots + Y_{100} \approx N(700,40)$

$Y - X \approx N(50,395)$; $P(Y - X \ge 0) \approx 1-F(-50/\sqrt{395}) = 0.9941$.

6.8.1 Approximate 0.171, 0.248, 0.0547; exact 0.1719, 0.2461, 0.0537.

6.8.4 $X \approx N(50,25)$; $P\{|X - 50| \ge 7\} \approx 0.194$. If coin were balanced, such a large deviation would occur in about 20% of similar experiments, so the result does not give strong evidence of bias. See Chapter 12.

6.8.7 0.758, 0.980; $X \approx N(.05n, .0475n)$, so that

$.95 = P(.049n \le X \le .051n) \approx 2F(\frac{.001n}{\sqrt{.0475n}}) - 1$.

Table B1 gives $F(-1.960) = 0.975$, so $n = 182476$.

6.9.1 Table B4 gives $x_1 = 0.8312$, $x_2 = 12.83$.

6.9.5 Put $Y \equiv 2X/\theta$ and use (6.1.11). In (b), put $Y_i \equiv 2X_i/\theta$, so Y_i's are independent $\chi^2_{(2)}$, and $T \equiv \sum Y_i \sim \chi^2_{(2n)}$ by (6.9.3). In (c), $n = 15$, $\theta = 100$, so $T/50 \sim \chi^2_{(30)}$; $P(T > 2000) = P(\chi^2_{(3)} > 40)$ ≈ 0.10 from Table B4.

6.9.8 0.250, 3.00

6.9.9 $x_1 = 1.697$, $x_2 = -2.750$ from Table B3. Values for N(0,1) are 1.645 and -2.576 from Table B1.

6.9.13 Survivor function $S(t) = P(T > t) = P(\chi^2_{(2p)} > 2\lambda t)$ by Problem 12(b). For $p < 1$, $h(x)$ decreases as x increases (improvement); for $p = 1$, $h(x)$ is constant (no ageing); for $p > 1$, $h(x)$ increases with x (deterioration). Changing λ alters the scales on the two axes, but does not alter the shape of the hazard function.

6R1 0.3012; $(0.3012)^2$; $\sum[p(1-p)^x]^2 = \frac{p}{2-p}$ where $p = 1-e^{-6/300}$.

6R4 $k = 1$, $f(x) = 2x(1+x)^{-3}$ for $x > 0$; $k = (1-e^{-\theta})^{-1}$, $f(x) = k\theta e^{-\theta \sin x}\cos x$ for $0 < x < \pi/2$.

6R7 $S_{20} \approx N(0,30)$; $P\{|S_{20}| \le 5\} \approx 2F(5.5/\sqrt{30}) - 1 = 0.685$.

6R10 $X \equiv$ # defectives out of n; $X \sim$ binomial $(n,.015)$; $0.8 \le P(n-X \ge 100) = f(0)+f(1)+\ldots+f(n-100)$ where $f(x) = \binom{n}{x}(.015)^x(.985)^{n-x}$. $n \ge 102$ by trial and error.

6R12 0.0668, 0.1587; 0.3768; $\bar{D} \sim N(3.4,.04)$ so $P(\bar{D} > 4) = 0.0013$; $\binom{10}{2\;7\;1}(.1587)^2(.7745)^7(.0668) = 0.1012$.

GR15 $P(|X-41| \ge 17) \approx 0.0003$ where $X \sim$ binomial $(82,\frac{1}{2})$; $(.99)^{82} = 0.4386$; $(.99)^{820} = 0.0003$.

7.1.1 $k = 1/2$, $f(x,y) = x + y$ for $0 < x < 1$ and $0 < y < 1$; $f_1(x) = (2x+1)/2$ and $f_1(x|y) = 2(x+y)/(2y+1)$ for $0 < x < 1$; 1/8, 3/8, 1/3.

7.1.4 3/4, 1/2, 2/3, 1/4.

7.1.7 $k = 12/7$, $f_1(x) = (12x^2 + 6x)/7$, $f_2(y) = (4 + 6y)/7$; $E(X) = 5/7$, $\text{var}(X) = 23/490$, $E(Y) = 4/7$, $\text{var}(Y) = 23/294$, $\text{cov}(X,Y) = -1/294$, $\rho(X,Y) = -0.0561$.

7.1.11 $f_1(x|t) = (\lambda t)^x e^{-\lambda t}/x!$ for $x = 0,1,2,\ldots$; $f(x,t) = f_1(x|t)f_2(t)$ where $f_2(t) = \frac{1}{\theta} e^{-t/\theta}$; $f_1(x) = \int_0^\infty f(x,t)dt = p(1-p)^x$ where $p = (1 + \lambda\theta)^{-1}$.

7.2.2 $A \equiv XY/2$ where $f(x,y) = 1$ for $0 < x < 1$, $0 < y < 1$; p.d.f. of A is $-2\log 2a$ for $0 < a < 1/2$.

7.2.6 Define $U \equiv X_1/X_2$ to complete 1-1 transformation.

$$g(u,r) = f(x_1,x_2)\cdot\left|\frac{\partial(x_1,x_2)}{\partial(u,r)}\right| = \frac{ur^2}{1+u^2} e^{-r^2/2} \cdot \frac{r}{1+u^2} \quad \text{for } u > 0, \; r > 0.$$

$$g_2(r) = \int_0^\infty g(u,r)du = \frac{1}{2} r^3 e^{-r^2/2} \quad \text{for } r > 0.$$

7.2.9 $P\{X_{(1)} > x\} = \prod_{i=1}^{n} [1 - F_i(x)] = \prod_{i=1}^{n} e^{-\lambda_i x} = e^{-x\sum\lambda_i}$. Hence c.d.f. of $X_{(1)}$ is $1 - e^{-x\sum\lambda_i}$ for $x > 0$. For $n = 5$, $\lambda_i = \frac{1}{5000}$ we have $P(X_{(1)} > 2000) = e^{-2}$.

7.2.11 $G(y,z) = [F(z)]^n - [F(z) - F(y)]^n$;
$g(y,z) = n(n-1)[F(z) - F(y)]^{n-2} f(y)f(z)$ for $z \ge y$.

7.2.14 $f(n) = 0.00668, 0.59038, 0.40294$ for $n = 0,1,2$.

8.1.1 0,0,0,1,0,5,12,0,0,...; 0,0,0,6,0,600,8640,0,0,... .

8.1.4 $(1-u)^{-2}$; $(1+u)e^u$.

8.1.6 $F(u) = u + uF(u) + u^2F(u)$ gives $F(u) = u(1 - u - u^2)^{-1}$

8.2.4 $G(u) = p\sum[u(1-p)]^x = p[up - u + 1]^{-1}$;
$G_{S_r}(u) = [G(u)]^r$ which is p.g.f. of negative binomial distn.

8.2.8 $a_x = b_{x-1} - b_x$ where $b_{-1} = 1$; multiply by u^x and sum to get $A(u) = -1 + uB(u) - B(u)$. For (ii) and (iii), find $A'(u)$ and $A''(u)$, then set $u = 1$.

8.2.12 Given that $Y = y$, u^Z has expected value $[A(u)]^y$. P.g.f. of Z is $E(u^Z) = \sum[A(u)]^y b_y$ where $b_y = P(Y = y)$. This sum gives $B(u) = \sum u^y b_y$ but with u replaced by $A(u)$.

8.2.15 $128f(n) = 16,28,28,25,16,10,4,1$ for $n = 1,2,3,\ldots,8$.

8.2.17 $\mu_n = G_n'(1) = (\frac{3}{2})^n$; $\sigma_n^2 = G_n''(1) + \mu_n - \mu_n^2 = \frac{1}{3}[(\frac{3}{2})^{2n} - (\frac{3}{2})^n]$.

8.3.1 $.3 + .1e^u + .6e^{2u}$; $.3e^{3u} + .1e^{5u} + .6e^{7u}$.

8.3.6 $K(u) = \log(pe^u + 1 - p) = n\log(1-p) + n\log(1 + ae^u)$ where $a = \frac{p}{1-p}$;

$$\log(1 + ae^u) = \sum_{i=1}^{\infty} \frac{(-1)^{i-1}(ae^u)^i}{i} = \sum_{i=1}^{\infty} \frac{(-1)^{i-1}a^i}{i} \sum_{r=0}^{\infty} \frac{(ui)^r}{r!}$$

$$= \sum_{r=0}^{\infty} \frac{u^r}{r!} \sum_{i=1}^{\infty} (-1)^{i-1}a^i i^{r-1}.$$

Hence $\kappa_r = n\sum_{i=1}^{\infty}(-1)^{i-1}a^i i^{r-1}$. Differentiate w.r.t. p to get result.

8.4.1 Standard form of X has cumulant generating function

$$u\sqrt{\frac{n}{2}} - \frac{n}{2}\log(1 - \frac{2u}{\sqrt{2n}}) = \frac{u^2}{2} + \sqrt{\frac{2}{n}} \sum_{r=3}^{\infty} \frac{u^r}{r}(\sqrt{\frac{2}{n}})^{r-3} \to \frac{u^2}{2}$$

as $n \to \infty$. The distribution of X is approximately normal for n large.

8.5.1 $X \sim$ Poisson mean a, $Y \sim$ Poisson mean b; $\rho(X,Y) = c/\sqrt{ab}$, and $\rho^2 \le 1$ gives $c^2 \le ab$; $f(x,y) = e^{-1}/x!$ for $y = x = 0,1,2,\ldots$ and $f(x,y) = 0$ otherwise, so that X has a Poisson distribution with mean 1 and Y is identically equal to X.

8.5.2 $E(D^2) = n$.

APPENDIX B. TABLES

TABLES B1, B2

Standardized Normal Distribution

$$F(x) = P\{N(0,1) \le x\} = \int_{-\infty}^{x} \frac{1}{\sqrt{2\pi}} e^{-u^2/2} du$$

Table B1 gives the value x whose cumulative probability $F(x)$ is the sum of the corresponding row and column headings. Example: the value x such that $F(x) = .64$ is 0.358 (from row .6 and column .04 of Table B1).

Table B2 gives the cumulative probability $F(x)$, where x is the sum of the corresponding row and column headings. Example: the cumulative probability at value 0.36 is $F(.36) = .6406$ (from row .3 and column .06 of Table B2).

TABLE B1

Percentiles of the Standardized Normal Distribution

F	.00	.01	.02	.03	.04	.05	.06	.07	.08	.09
.5	.000	.025	.050	.075	.100	.126	.151	.176	.202	.228
.6	.253	.279	.305	.332	.358	.385	.412	.440	.468	.496
.7	.524	.553	.583	.613	.643	.674	.706	.739	.772	.806
.8	.842	.878	.915	.954	.994	1.036	1.080	1.126	1.175	1.227
.9	1.282	1.341	1.405	1.476	1.555	1.645	1.751	1.881	2.054	2.326

x	1.960	2.576	3.090	3.291	3.891	4.417	4.892
F	.975	.995	.999	.9995	.99995	.999995	.9999995
2(1-F)	.05	.01	.002	.001	.0001	.00001	.000001

Source: R.A. Fisher and F. Yates, *Statistical Tables for Biological, Agricultural and Medical Research*, Table I; published by Longman Group Ltd., London (previously published by Oliver and Boyd, Edinburgh); reprinted by permission of the authors and publishers.

TABLE B2

Standardized Normal Cumulative Distribution Function

x	.00	.01	.02	.03	.04	.05	.06	.07	.08	.09
.0	.5000	.5040	.5080	.5120	.5160	.5199	.5239	.5279	.5319	.5359
.1	.5398	.5438	.5478	.5517	.5557	.5596	.5636	.5675	.5714	.5753
.2	.5793	.5832	.5871	.5910	.5948	.5987	.6026	.6064	.6103	.6141
.3	.6179	.6217	.6255	.6293	.6331	.6368	.6406	.6443	.6480	.6517
.4	.6554	.6591	.6628	.6664	.6700	.6736	.6772	.6808	.6844	.6879
.5	.6915	.6950	.6985	.7019	.7054	.7088	.7123	.7157	.7190	.7224
.6	.7257	.7291	.7324	.7357	.7389	.7422	.7454	.7486	.7517	.7549
.7	.7580	.7611	.7642	.7673	.7703	.7734	.7764	.7794	.7823	.7852
.8	.7881	.7910	.7939	.7967	.7995	.8023	.8051	.8078	.8106	.8133
.9	.8159	.8186	.8212	.8238	.8264	.8289	.8315	.8340	.8365	.8389
1.0	.8413	.8438	.8461	.8485	.8508	.8531	.8554	.8577	.8599	.8661
1.1	.8643	.8665	.8686	.8708	.8729	.8749	.8770	.8790	.8810	.8830
1.2	.8849	.8869	.8888	.8907	.8925	.8944	.8962	.8980	.8997	.90147
1.3	.90320	.90490	.90658	.90824	.90988	.91149	.91309	.91466	.91621	.91774
1.4	.91924	.92073	.92220	.92364	.92507	.92647	.92785	.92922	.93056	.93189
1.5	.93319	.93448	.93574	.93669	.93822	.93943	.94062	.94179	.94295	.94408
1.6	.94520	.94630	.94738	.94845	.94950	.95053	.95154	.95254	.95352	.95449
1.7	.95543	.95637	.95728	.95818	.95907	.95994	.96080	.96164	.96246	.96327
1.8	.96407	.96485	.96562	.96638	.96712	.96784	.96856	.96926	.96995	.97062
1.9	.97128	.97193	.97257	.97320	.97381	.97441	.97500	.97558	.97615	.97670
2.0	.97725	.97778	.97831	.97882	.97932	.97982	.98030	.98077	.98124	.98169
2.1	.98214	.98257	.98300	.98341	.98382	.98422	.98461	.98500	.98537	.98574
2.2	.98610	.98645	.98679	.98713	.98745	.98778	.98809	.98840	.98870	.98899
2.3	.98928	.98956	.98983	.990097	.990358	.990613	.990863	.991106	.991344	.991576
2.4	.991802	.992024	.992240	.992451	.992656	.992857	.993053	.993244	.993431	.993613
2.5	.993790	.993963	.994132	.994297	.994457	.994614	.994766	.994915	.995060	.995201
2.6	.995339	.995473	.995604	.995731	.995855	.995975	.996093	.996207	.996319	.996427
2.7	.996533	.996636	.996736	.996833	.996928	.997020	.997110	.997197	.997282	.997365
2.8	.997445	.997523	.997599	.997673	.997744	.997814	.997882	.997948	.998012	.998074
2.9	.998134	.998193	.998250	.998305	.998359	.998411	.998462	.998511	.998559	.998605
3.0	.998650	.998694	.998736	.998777	.998817	.998856	.998893	.998930	.998965	.998999

Source: A. Hald, *Statistical Tables and Formulas* (1952), Table II; reprinted by permission of John Wiley & Sons, Inc.

TABLE B3

Percentiles of Student's (t) Distribution

$$F(x) = P(t_{(\nu)} \leq x) = \int_{-\infty}^{x} (1 + \frac{u^2}{\nu})^{-\frac{\nu+1}{2}} du \cdot \frac{\Gamma(\frac{\nu+1}{2})}{\sqrt{\pi\nu}\ \Gamma(\frac{\nu}{2})}$$

The body of the table gives the values x corresponding to selected values of the cumulative probability (F) and degrees of freedom (ν).

ν \ F	.60	.70	.80	.90	.95	.975	.99	.995	.9995
1	.325	.727	1.376	3.078	6.314	12.706	31.821	63.657	636.619
2	.289	.617	1.061	1.886	2.920	4.303	6.965	9.925	31.598
3	.277	.584	.978	1.638	2.353	3.182	4.541	5.841	12.924
4	.271	.569	.941	1.533	2.132	2.776	3.747	4.604	8.610
5	.267	.559	.920	1.476	2.015	2.571	3.365	4.032	6.869
6	.265	.553	.906	1.440	1.943	2.447	3.143	3.707	5.959
7	.263	.549	.896	1.415	1.895	2.365	2.998	3.499	5.408
8	.262	.546	.889	1.397	1.860	2.306	2.896	3.355	5.041
9	.261	.543	.883	1.383	1.833	2.262	2.821	3.250	4.781
10	.260	.542	.879	1.372	1.812	2.228	2.764	3.169	4.587
11	.260	.540	.876	1.363	1.796	2.201	2.718	3.106	4.437
12	.259	.539	.873	1.356	1.782	2.179	2.681	3.055	4.318
13	.259	.538	.870	1.350	1.771	2.160	2.650	3.012	4.221
14	.258	.537	.868	1.345	1.761	2.145	2.624	2.977	4.140
15	.258	.536	.866	1.341	1.753	2.131	2.602	2.947	4.073
16	.258	.535	.865	1.337	1.746	2.120	2.583	2.921	4.015
17	.257	.534	.863	1.333	1.740	2.110	2.567	2.898	3.965
18	.257	.534	.862	1.330	1.734	2.101	2.552	2.878	3.922
19	.257	.533	.861	1.328	1.729	2.093	2.539	2.861	3.883
20	.257	.533	.860	1.325	1.725	2.086	2.528	2.845	3.850
21	.257	.532	.859	1.323	1.721	2.080	2.518	2.831	3.819
22	.256	.532	.858	1.321	1.717	2.074	2.508	2.819	3.792
23	.256	.532	.858	1.319	1.714	2.069	2.500	2.807	3.767
24	.256	.531	.857	1.318	1.711	2.064	2.492	2.797	3.745
25	.256	.531	.856	1.316	1.708	2.060	2.485	2.787	3.725
26	.256	.531	.856	1.315	1.706	2.056	2.479	2.779	3.707
27	.256	.531	.855	1.314	1.703	2.052	2.473	2.771	3.690
28	.256	.530	.855	1.313	1.701	2.048	2.467	2.763	3.674
29	.256	.530	.854	1.311	1.699	2.045	2.462	2.756	3.659
30	.256	.530	.854	1.310	1.697	2.042	2.457	2.750	3.646
40	.255	.529	.851	1.303	1.684	2.021	2.423	2.704	3.551
60	.254	.527	.848	1.296	1.671	2.000	2.390	2.660	3.460
120	.254	.526	.845	1.289	1.658	1.980	2.358	2.617	3.373
∞	.253	.524	.842	1.282	1.645	1.960	2.326	2.576	3.291

Source: R.A. Fisher and F. Yates, *Statistical Tables for Biological, Agricultural and Medical Research*, Table III; published by Longman Group Ltd., London (previously published by Oliver and Boyd, Edinburgh); reprinted by permission of the authors and publishers.

TABLE B4

Percentiles of the Chi-Square (χ^2) Distribution

$$F(x) = P(\chi^2_{(\nu)} \le x) = \int_0^x u^{\frac{\nu}{2}-1} e^{-\frac{u}{2}} du/2^{\frac{\nu}{2}} \Gamma(\tfrac{\nu}{2})$$

The body of the table gives the values x corresponding to selected values of the cumulative probability (F) and degrees of freedom (ν).

ν \ F	.005	.01	.025	.05	.10	.25	.5
1	$.0^4 3927$	$.0^3 1571$	$.0^3 9821$	$.0^2 3932$	.01579	.1015	.4549
2	.01003	.02010	.05064	.1026	.2107	.5754	1.386
3	.07172	.1148	.2158	.3518	.5844	1.213	2.366
4	.2070	.2971	.4844	.7107	1.064	1.923	3.357
5	.4117	.5543	.8312	1.145	1.610	2.675	4.351
6	.6757	.8721	1.237	1.635	2.204	3.455	5.348
7	.9893	1.239	1.690	2.167	2.833	4.255	6.346
8	1.344	1.646	2.180	2.733	3.490	5.071	7.344
9	1.735	2.088	2.700	3.325	4.168	5.899	8.343
10	2.156	2.558	3.247	3.940	4.865	6.737	9.342
11	2.603	3.053	3.816	4.575	5.578	7.584	10.34
12	3.074	3.571	4.404	5.226	6.304	8.438	11.34
13	3.565	4.107	5.009	5.892	7.042	9.299	12.34
14	4.075	4.660	5.629	6.571	7.790	10.17	13.34
15	4.601	5.229	6.262	7.261	8.547	11.04	14.34
16	5.142	5.812	6.908	7.962	9.312	11.91	15.34
17	5.697	6.408	7.564	8.672	10.09	12.79	16.34
18	6.265	7.015	8.231	9.390	10.86	13.68	17.34
19	6.844	7.633	8.907	10.12	11.65	14.56	18.34
20	7.434	8.260	9.591	10.85	12.44	15.45	19.34
21	8.034	8.897	10.28	11.59	13.24	16.34	20.34
22	8.643	9.542	10.98	12.34	14.04	17.24	21.34
23	9.260	10.20	11.69	13.09	14.85	18.14	22.34
24	9.886	10.86	12.40	13.85	15.66	19.04	23.34
25	10.52	11.52	13.12	14.61	16.47	19.94	24.34
26	11.16	12.20	13.84	15.38	17.29	20.84	25.34
27	11.81	12.88	14.57	16.15	18.11	21.75	26.34
28	12.46	13.56	15.31	16.93	18.94	22.66	27.34
29	13.12	14.26	16.05	17.71	19.77	23.57	28.34
30	13.79	14.95	16.79	18.49	20.60	24.48	29.34

TABLE B4 (continued)

Percentiles of the Chi-Square (χ^2) Distribution

ν \ F	.75	.9	.95	.975	.99	.995	.999
1	1.323	2.706	3.841	5.024	6.635	7.879	10.83
2	2.773	4.605	5.991	7.378	9.210	10.60	13.82
3	4.108	6.251	7.815	9.348	11.34	12.84	16.27
4	5.385	7.779	9.488	11.14	13.28	14.86	18.47
5	6.626	9.236	11.07	12.83	15.09	16.75	20.52
6	7.841	10.64	12.59	14.45	16.81	18.55	22.46
7	9.037	12.02	14.07	16.01	18.48	20.28	24.32
8	10.22	13.36	15.51	17.53	20.09	21.96	26.13
9	11.39	14.68	16.92	19.02	21.67	23.59	27.88
10	12.55	15.99	18.31	20.48	23.21	25.19	29.59
11	13.70	17.28	19.68	21.92	24.72	26.76	31.26
12	14.85	18.55	21.03	23.34	26.22	28.30	32.91
13	15.98	19.81	22.36	24.74	27.69	29.82	34.53
14	17.12	21.06	23.68	26.12	29.14	31.32	36.12
15	18.25	22.31	25.00	27.49	30.58	32.80	37.70
16	19.37	23.54	26.30	28.85	32.00	34.27	39.25
17	20.49	24.77	27.59	30.19	33.41	35.72	40.79
18	21.60	25.99	28.87	31.53	34.81	37.16	42.31
19	22.72	27.20	30.14	32.85	36.19	38.58	43.82
20	23.83	28.41	31.41	34.17	37.57	40.00	45.32
21	24.93	29.62	32.67	35.48	38.93	41.40	46.80
22	26.04	30.81	33.92	36.78	40.29	42.80	48.27
23	27.14	32.01	35.17	38.08	41.64	44.18	49.73
24	28.24	33.20	36.42	39.36	42.98	45.56	51.18
25	29.34	34.38	37.65	40.65	44.31	46.93	52.62
26	30.43	35.56	38.89	41.92	45.64	48.29	54.05
27	31.53	36.74	40.11	43.19	46.96	49.64	55.48
28	32.62	37.92	41.34	44.46	48.28	50.99	56.89
29	33.71	39.09	42.56	45.72	49.59	52.34	58.30
30	34.80	40.26	43.77	46.98	50.89	53.67	59.70

For $\nu > 30$, $\sqrt{\frac{9\nu}{2}}\,\{[\frac{1}{\nu}\,\chi^2_{(\nu)}]^{1/3} - 1 + \frac{2}{9\nu}\}$ is approximately N(0,1).

Source: E.S. Pearson and H.O. Hartley (editors), *Biometrika Tables for Statisticians, vol. I,* Table 8; Cambridge University Press (3rd edition, 1966); reprinted by permission of the Biometrika Trustees.

TABLE B5

Percentiles of the Variance Ratio (F) Distribution

n numerator and m denominator degrees of freedom

$$F(x) = P\{F_{n,m} \le x\} = \int_0^x (\frac{n}{m}u)^{\frac{n}{2}-1} (1+\frac{n}{m}u)^{-\frac{n+m}{2}} \frac{n}{m}\, du \cdot \Gamma(\frac{n+m}{2})/\Gamma(\frac{n}{2})\Gamma(\frac{m}{2})$$

90th Percentiles (F = .9)

m \ n	1	2	3	4	5	6	8	12	24	∞
1	39.86	49.50	53.59	55.83	57.24	58.20	59.44	60.70	62.00	63.33
2	8.53	9.00	9.16	9.24	9.29	9.33	9.37	9.41	9.45	9.49
3	5.54	5.46	5.39	5.34	5.31	5.28	5.25	5.22	5.18	5.13
4	4.54	4.32	4.19	4.11	4.05	4.01	3.95	3.90	3.83	3.76
5	4.06	3.78	3.62	3.52	3.45	3.40	3.34	3.27	3.19	3.10
6	3.78	3.46	3.29	3.18	3.11	3.05	2.98	2.90	2.82	2.72
7	3.59	3.26	3.07	2.96	2.88	2.83	2.75	2.67	2.58	2.47
8	3.46	3.11	2.92	2.81	2.73	2.67	2.59	2.50	2.40	2.29
9	3.36	3.01	2.81	2.69	2.61	2.55	2.47	2.38	2.28	2.16
10	3.28	2.92	2.73	2.61	2.52	2.46	2.38	2.28	2.18	2.06
12	3.18	2.81	2.61	2.48	2.39	2.33	2.24	2.15	2.04	1.90
15	3.07	2.70	2.49	2.36	2.27	2.21	2.12	2.02	1.90	1.76
20	2.97	2.59	2.38	2.25	2.16	2.09	2.00	1.89	1.77	1.61
25	2.92	2.53	2.32	2.18	2.09	2.02	1.93	1.82	1.69	1.52
30	2.88	2.49	2.28	2.14	2.05	1.98	1.88	1.77	1.64	1.46
40	2.84	2.44	2.23	2.09	2.00	1.93	1.83	1.71	1.57	1.38
60	2.79	2.39	2.18	2.04	1.95	1.87	1.77	1.66	1.51	1.29
120	2.75	2.35	2.13	1.99	1.90	1.82	1.72	1.60	1.45	1.19
∞	2.71	2.30	2.08	1.94	1.85	1.77	1.67	1.55	1.38	1.00

95th Percentiles (F = .95)

m \ n	1	2	3	4	5	6	8	12	24	∞
1	161	200	216	225	230	234	239	244	249	254
2	18.5	19.0	19.2	19.2	19.3	19.3	19.4	19.4	19.5	19.5
3	10.1	9.55	9.28	9.12	9.01	8.94	8.84	8.74	8.64	8.53
4	7.71	6.94	6.59	6.39	6.26	6.16	6.04	5.91	5.77	5.63
5	6.61	5.79	5.41	5.19	5.05	4.95	4.82	4.68	4.53	4.36
6	5.99	5.14	4.76	4.53	4.39	4.28	4.15	4.00	3.84	3.67
7	5.59	4.74	4.35	4.12	3.97	3.87	3.73	3.57	3.41	3.23
8	5.32	4.46	4.07	3.84	3.69	3.58	3.44	3.28	3.12	2.93
9	5.12	4.26	3.86	3.63	3.48	3.37	3.23	3.07	2.90	2.71
10	4.96	4.10	3.71	3.48	3.33	3.22	3.07	2.91	2.74	2.54
12	4.75	3.88	3.49	3.26	3.11	3.00	2.85	2.69	2.50	2.30
15	4.54	3.68	3.29	3.06	2.90	2.79	2.64	2.48	2.29	2.07
20	4.35	3.49	3.10	2.87	2.71	2.60	2.45	2.28	2.08	1.84
25	4.24	3.38	2.99	2.76	2.60	2.49	2.34	2.16	1.96	1.71
30	4.17	3.32	2.92	2.69	2.53	2.42	2.27	2.09	1.89	1.62
40	4.08	3.23	2.84	2.61	2.45	2.34	2.18	2.00	1.79	1.51
60	4.00	3.15	2.76	2.52	2.37	2.25	2.10	1.92	1.70	1.39
120	3.92	3.07	2.68	2.45	2.29	2.17	2.02	1.83	1.61	1.25
∞	3.84	2.99	2.60	2.37	2.21	2.10	1.94	1.75	1.52	1.00

TABLE B5 (continued)

Percentiles of the Variance Ratio (F) Distribution

n numerator and m denominator degrees of freedom

99th Percentiles (F = .99)

m \ n	1	2	3	4	5	6	8	12	24	∞
1	4052	4999	5403	5625	5764	5859	5982	6106	6234	6366
2	98.5	99.0	99.2	99.2	99.3	99.3	99.4	99.4	99.5	99.5
3	34.1	30.8	29.5	28.7	28.2	27.9	27.5	27.1	26.6	26.1
4	21.2	18.0	16.7	16.0	15.5	15.2	14.8	14.4	13.9	13.5
5	16.3	13.3	12.1	11.4	11.0	10.7	10.3	9.89	9.47	9.02
6	13.74	10.92	9.78	9.15	8.75	8.47	8.10	7.72	7.31	6.88
7	12.25	9.55	8.45	7.85	7.46	7.19	6.84	6.47	6.07	5.65
8	11.26	8.65	7.59	7.01	6.63	6.37	6.03	5.67	5.28	4.86
9	10.56	8.02	6.99	6.42	6.06	5.80	5.47	5.11	4.73	4.31
10	10.04	7.56	6.55	5.99	5.64	5.39	5.06	4.71	4.33	3.91
12	9.33	6.93	5.95	5.41	5.06	4.82	4.50	4.16	3.78	3.36
15	8.68	6.36	5.42	4.89	4.56	4.32	4.00	3.67	3.29	2.87
20	8.10	5.85	4.94	4.43	4.10	3.87	3.56	3.23	2.86	2.42
25	7.77	5.57	4.68	4.18	3.86	3.63	3.32	2.99	2.62	2.17
30	7.56	5.39	4.51	4.02	3.70	3.47	3.17	2.84	2.47	2.01
40	7.31	5.18	4.31	3.83	3.51	3.29	2.99	2.66	2.29	1.80
60	7.08	4.98	4.13	3.65	3.34	3.12	2.82	2.50	2.12	1.60
120	6.85	4.79	3.95	3.48	3.17	2.96	2.66	2.34	1.95	1.38
∞	6.64	4.60	3.78	3.32	3.02	2.80	2.51	2.18	1.79	1.00

99.9th Percentiles (F = .999)

m \ n	1	2	3	4	5	6	8	12	24	∞
1*	405*	500*	540*	563*	576*	586*	598*	611*	623*	637*
2	998	999	999	999	999	999	999	999	999	999
3	167	149	141	137	135	133	131	128	126	124
4	74.1	61.3	56.2	53.4	51.7	50.5	49.0	47.4	45.8	44.1
5	47.2	37.1	33.2	31.1	29.8	28.8	27.6	26.4	25.1	23.8
6	35.5	27.0	23.7	21.9	20.8	20.0	19.0	18.0	16.9	15.8
7	29.2	21.7	18.8	17.2	16.2	15.5	14.6	13.7	12.7	11.7
8	25.4	18.5	15.8	14.4	13.5	12.9	12.0	11.2	10.3	9.34
9	22.9	16.4	13.9	12.6	11.7	11.1	10.4	9.57	8.72	7.81
10	21.0	14.9	12.6	11.3	10.5	9.92	9.20	8.45	7.64	6.76
12	18.6	13.0	10.8	9.63	8.89	8.38	7.71	7.00	6.25	5.42
15	16.6	11.3	9.34	8.25	7.57	7.09	6.47	5.81	5.10	4.31
20	14.8	9.95	8.10	7.10	6.46	6.02	5.44	4.82	4.15	3.38
25	13.9	9.22	7.45	6.49	5.88	5.46	4.91	4.31	3.66	2.89
30	13.3	8.77	7.05	6.12	5.53	5.12	4.58	4.00	3.36	2.59
40	12.6	8.25	6.60	5.70	5.13	4.73	4.21	3.64	3.01	2.23
60	12.0	7.76	6.17	5.31	4.76	4.37	3.87	3.31	2.69	1.90
120	11.4	7.32	5.79	4.95	4.42	4.04	3.55	3.02	2.40	1.54
∞	10.8	6.91	5.42	4.62	4.10	3.74	3.27	2.74	2.13	1.00

* For m = 1, the 99.9th percentiles are 1000 times the tabulated values

Source: R.A. Fisher and F. Yates, *Statistical Tables for Biological, Agricultural and Medical Research*, Table V; published by Longman Group Ltd., London (previously published by Oliver and Boyd, Edinburgh); reprinted by permission of the authors and publishers.

INDEX TO VOLUME I